SEMICONDUCTING
LEAD CHALCOGENIDES

MONOGRAPHS IN SEMICONDUCTOR PHYSICS

1969

Volume 1
HEAVILY DOPED SEMICONDUCTORS
by Viktor I. Fistul'

Volume 2
LIQUID SEMICONDUCTORS
by V. M. Glazov, S. N. Chizhevskaya, and N. N. Glagoleva

Volume 3
SEMICONDUCTING II-VI, IV-VI, and V-VI COMPOUNDS
by N. Kh. Abrikosov, V. F. Bankina, L. V. Poretskaya,
L. E. Shelimova, and E. V. Skudnova

Volume 4
SWITCHING IN SEMICONDUCTOR DIODES
by Yu. R. Nosov

1970

Volume 5
SEMICONDUCTING LEAD CHALCOGENIDES
by Yu. I. Ravich, B. A. Efimova, and I. A. Smirnov

In preparation
ORGANIC SEMICONDUCTORS AND BIOPOLYMERS
by L. I. Boguslavskii and A. V. Vannikov

SEMICONDUCTING LEAD CHALCOGENIDES

Yu. I. Ravich, B. A. Efimova, and I. A. Smirnov
Institute of Semiconductors
Academy of Sciences of the USSR, Leningrad

Edited by L. S. Stil'bans
Institute of Semiconductors
Academy of Sciences of the USSR, Leningrad

Translated from Russian by Albin Tybulewicz
Editor, *Soviet Physics—Semiconductors*

PLENUM PRESS • NEW YORK–LONDON • 1970

Yurii Isaakovich Ravich was born in 1937 in the city of Pskov. In 1959 he was graduated from the Physics Department of the Leningrad State University. He then joined the Institute of Semiconductors of the Academy of Sciences of the USSR, where he investigated the theory of photoelectric phenomena in semiconductors. At present, Ravich is a senior member of the scientific staff of the Institute. He is working on the band structure and carrier scattering mechanisms in semiconducting materials suitable for thermoelectric converters. Among his publications is a book on the "Photomagnetic Effect in Semiconductors and Its Applications."

Bella Anatol'evna Efimova was born in 1932 in Volgograd. After graduation from the Ural State University in 1954, she joined the Institute of Semiconductors. At present she is a senior member of the scientific staff of the Institute and is working on thermoelectric properties of semiconducting materials.

Igor' Aleksandrovich Smirnov was born in 1932 in Leningrad. In 1955 he was graduated from the M. I. Kalinin Leningrad Polytechnical Institute and promptly joined the Institute of Semiconductors, where he is now the head of the Laboratory for Thermal Phenomena. His major field of study is the thermal properties of solids. In 1964 Smirnov was awarded the A. F. Ioffe Prize for Solid State Physics.

The original Russian text, published by Nauka Press in Moscow in 1968 as a part of a series on "Physics of Semiconductors and Semiconducting Devices," has been corrected by the authors for this edition. The English translation is published under an agreement with Mezhdunarodnaya Kniga, the Soviet book export agency.

Равич Юрий Исаакович,
Ефимова Белла Анатольевна,
Смирнов Игорь Александрович

Методы исследования полупроводников
в применении к халькогенидам свинца
PbTe, PbSe и PbS

METODY ISSLEDOVANIYA POLUPROVODNIKOV
V PRIMENENII K KHAL'KOGENIDAM SVINTSA
PbTe, PbSe I PbS

Library of Congress Catalog Card Number 77-107542

SBN 306-30426-0

© 1970 Plenum Press, New York
A Division of Plenum Publishing Corporation
227 West 17th Street, New York, N. Y. 10011

United Kingdom edition published by Plenum Press, London
A Division of Plenum Publishing Company, Ltd.
Donington House, 30 Norfolk Street, London W.C.2, England

All rights reserved

No part of this publication may be reproduced in any form
without written permission from the publisher

Printed in the United States of America

PREFACE TO THE AMERICAN EDITION

Many new papers on lead chalcogenides have been published in the two years since the completion of the manuscript of the Russian edition of this book. A considerable amount of new information has become available on the mechanisms of carrier scattering. Higher values of the thermoelectric figure of merit have been reported and more accurate physical parameters of lead chalcogenides have been obtained. This new material required a revision or a fuller treatment of some of the topics.

In the preparation of the American edition we have not attempted to give references to all the new work but we have selected only those papers which alter substantially or give important new information on subjects dealt with in the Russian edition. We hope that the inclusion of the new material will make the book more useful and we are grateful to Plenum Publishing Corporation for the opportunity to make corrections and additions in the American edition.

Yu. I. Ravich
B. A. Efimova
I. A. Smirnov

EDITOR'S PREFACE

The last decade has seen radical changes in our understanding of the physical properties of semiconductors. It has been established that the energy spectrum of electrons is much more complex than had originally been predicted: in many cases, there are several energy bands with different parameters. It has been found that the effective carrier mass, which had been assumed to be constant for a given material, depends on the carrier energy, temperature, pressure, and even the nature and number of defects. Our understanding of the mechanism of the motion and scattering of carriers, recombination mechanisms, and interaction with electromagnetic radiation has also changed. New applications of semiconducting materials have been discovered and old ones have been extended; these include high-power devices, devices sensitive to infrared radiation, and lasers. The visible evidence of the progress is in the form of hundreds of publications, some of which report extremely refined and comprehensive investigations of semiconducting materials.

A scientist concerned with investigations or applications of semiconducting materials or devices cannot ignore these publications because of the possibility of repeating work already done or of committing serious errors. On the other hand, a beginner would require years to obtain a thorough understanding of the literature in his own narrow subject, and in many cases this process would be like the chase of a tortoise by Achilles in the paradox of Zeno of Elea (495-435 B.C.).* For this reason, we should always welcome the appearance of monographs, such as the present one, which summarize the results of decades of investigations of semi-

---

*See, for example, E. Kasner and J. Newman, Mathematics and the Imagination, Simon and Schuster, New York (1950), pp. 37-38, 57-58.

conducting materials that are scientically and technologically important. Such monographs help young scientists throughout the world to reach a certain level of understanding of their subject much more easily.

The importance of the present monograph is greatly enhanced by the fact that the available textbooks and monographs do not include the recent data on semiconducting materials, and do not describe the latest theoretical achievements and the new methods for analyzing the experimental data. This gap is filled by the theoretical briefs which preface each chapter of the present book. This treatment extends the monograph's usefulness to a very wide range of readers, including scientists, engineers, postgraduate and undergraduate students.

The monograph describes in detail the optical properties of semiconductors, phenomena in quantizing magnetic fields, the methods of calculating effective masses (the $\mathbf{k}\cdot\mathbf{p}$ perturbation theory), methods for calculating the parabolicity of energy bands (a simplified variant of the Kane theory), as well as group-theory methods for analyzing the influence of the symmetry of a crystal on the energy spectrum of electrons.

The emphasis is on the presentation of the results of investigations of the physical (optical, electrical, photoelectric, thermal, thermoelectric, and magnetic) properties of lead chalcogenides and on the structure of their energy bands, carrier scattering mechanisms, and phonon spectra, which can be deduced from the experimental data and theoretical calculations. The physicochemical properties of lead chalcogenides are described, as well as methods of synthesizing these materials and growing their single crystals. The applications of these materials are also briefly reviewed. Certain phenomena, which are interesting from the point of view of the physics of semiconductors as a whole, were observed most clearly — and for the first time — in lead chalcogenides. Such phenomena include, for example, the strong temperature dependence of the effective mass and the large relativistic corrections which are necessary in the energy band structure calculations. Lead chalcogenides were found to be convenient materials in the first observations of the de Haas−van Alphen effect, helicon waves, and quantum oscillations of the absorption of ultrasound.

The book is written by leading specialists in theoretical and experimental investigations of semiconductors, and, therefore, it is not simply a compilation of the available data but a critical and generalized interpretation of them.

B. A. Efimova contributed §§1.1-1.4 and 7.1; I. A. Smirnov was responsible for §§1.5 and 4.2; the remainder of the book was written by Yu. I. Ravich.

L. S. Stil'bans

CONTENTS

Notation	1
Introduction	5
CHAPTER I. PHYSICOCHEMICAL PROPERTIES	13
§1.1. Crystal Structure and the Nature of Chemical Binding	13
§1.2. Preparation Methods	15
A. Preparation of Polycrystalline Materials	16
B. Methods for Single-Crystal Growth	21
§1.3. Phase Diagrams	26
§1.4. Control of the Stoichiometry and the Doping Methods	31
§1.5. Diffusion	36
A. Diffusion in Undoped Single Crystals	39
B. Diffusion in Doped Single Crystals	40
C. Diffusion in Polycrystalline Samples	40
CHAPTER II. OPTICAL AND PHOTOELECTRIC PROPERTIES	43
§2.1. Absorption and Reflection of Light	43
A. Optical Parameters and Measurement Methods	43
B. Microscopic Processes of Interaction between Light and a Semiconductor	48
C. Fundamental Absorption Edge	51
D. Dispersion and Absorption at Long Wavelengths	56
E. Dispersion and Absorption at Short Wavelengths	63

§2.2 Photoelectric Effects. Carrier Recombination
and Recombination Radiation.   66
    A. Photoelectric Effects and the Determination
of the Recombination Parameters   66
    B. Recombination Mechanisms   72
    C. Spectral Characteristics of Photoelectric
Effects and of Recombination Radiation . . .   76
    D. Stimulated Recombination Radiation.   79
    E. Photoelectric Emission.   82

CHAPTER III. ELECTRICAL PROPERTIES   85
§3.1. Mobility   85
    A. Mobility for Various Scattering
Mechanisms   85
    B. Temperature Dependence of the Mobility
at Moderate and High Temperatures.
Scattering by Lattice Vibrations   90
    C. Mobility at Low Temperatures. Scattering
by Ionized Impurities, Vacancies, and
Other Lattice Defects   100
    D. Role of Scattering of Carriers on Each
Other.   106
    E. Mobility in Solid Solutions   108
    F. Mobility in Thin Films   110
§3.2. Hall Effect and Other Galvanomagnetic
Phenomena   110
    A. Hall Effect in the Extrinsic Conduction
Region   112
    B. Special Features of the Hall Effect in p-type
PbTe   118
    C. Hall Effect in Mixed and Intrinsic
Conduction Regions   122
    D. Resistance in a Magnetic Field and the
Planar Hall Effect   124
§3.3. Influence of Deformation on Electrical
Properties.   129
    A. Electrical Properties under Hydrostatic
Pressure   130
    B. Piezoresistance.   133

§3.4. Properties of p−n Junctions 139
    A. Current−Voltage Characteristic. Determination of Longitudinal Phonon Energy from Tunnel Effect Measurements . 140
    B. Barrier Capacitance. Determination of the Static Permittivity 145

CHAPTER IV. THERMOELECTRIC AND THERMAL PROPERTIES 149

§4.1. Thermoelectric Power and Thermomagnetic Effects 149
    A. Temperature and Carrier-Density Dependences 153
    B. Determination of the Effective Mass from Thermoelectric Power Measurements 159
    C. Investigations of the Band Edge Structure by Thermoelectric Power Measurements in Strong Magnetic Fields 162
    D. Determination of the Effective Mass from Nernst−Ettingshausen Effect Measurements 169
    E. Investigation of Scattering Mechanisms by Thermoelectric Methods. Electron−Electron Collisions 171

§4.2. Thermal Properties 175
    A. Electronic Thermal Conductivity 177
    B. Ambipolar Diffusion 184
    C. Influence of a Complex Valence Band on the Thermal Conductivity of p-type PbTe 189
    D. Photon Thermal Conductivity 193
    E. Lattice Thermal Conductivity and Specific Heat 194
    F. Thermal Expansion of the Lattice and its Relationship to the Thermal Conductivity . . 207
    G. Influence of Impurities on the Thermal Conductivity of Lead Chalcogenides at Temperatures $T \geq \theta$. Solid Solutions 209
    H. Phonon Spectrum of PbTe 216

CHAPTER V. MAGNETIC PROPERTIES	221
§5.1. Diamagnetism and Paramagnetism	223
A. Magnetic Susceptibility	223
B. de Haas – van Alphen Effect	225
C. Paramagnetism of Free Carriers. Knight Shift	226
§5.2. Shubnikov – de Haas Effect	227
A. Conditions for the Observation of Resistance Oscillations in Strong Magnetic Fields	227
B. Determination of the Anisotropy Coefficient and of the Number of Ellipsoids	231
C. Determination of Effective Mass and Band Nonparabolicity	233
D. Spin Splitting and the Determination of the Effective g Factors	236
§5.3. Tunnel Effect in a Magnetic Field	237
§5.4. Magneto-Optical Effects	240
A. Faraday Effect	241
B. Infrared Reflection in a Magnetic Field	244
C. Fundamental Absorption in a Magnetic Field	244
D. Recombination (Laser) Radiation in a Magnetic Field	249
§5.5. Low-Frequency Oscillations in a Magnetic Field	249
A. Cyclotron Resonance	250
B. Reflection of Microwaves in a Strong Magnetic Field	257
C. Helicon Waves	259
D. Oscillations of Ultrasound Absorption	261
CHAPTER VI. BAND STRUCTURE AND SCATTERING MECHANISMS (THEORY AND CONCLUSIONS FROM EXPERIMENTAL DATA)	263
§6.1. General Energy Band Structure	263
A. Interpretation of the Band Structure from Experimental Data	263
B. Methods for Calculating Band Structure. Relativistic Corrections and Spin – Orbit Interaction	267

   C. Application of Band Structure Calculation Methods to Lead Chalcogenides 280
   D. Comparison of Calculated Values of Energy Gaps with Experimental Data 289
 §6.2. Structure of the Energy Band Edges 292
   A. Effective Masses 292
   B. Energy Band Nonparabolicity. Kane and Cohen Models 299
   C. Values of the g Factors 306
 §6.3. Dependence of the Band Structure Parameters on Temperature, Deformation, and Composition. 310
   A. Influence of Pressure on the Forbidden Band Width and Effective Masses. Deformation Potentials 310
   B. Temperature Dependences of the Forbidden Band Width and Effective Masses 312
   C. Forbidden Band Width of Solid Solutions . . . 317
 §6.4. Carrier Scattering Mechanisms 319

CHAPTER VII. APPLICATIONS OF LEAD CHALCOGENIDES 323
 §7.1. Thermoelectric Properties 323
   A. Thermoelectric Figure of Merit of Lead Chalcogenides and their Alloys 323
   B. Thermoelectric Generators Based on Lead Telluride 330
 §7.2. Infrared Technology 334
   A. Photoresistors 334
   B. Photodiodes 341
   C. Lasers 343
 §7.3. Strain Gauges 345

APPENDIX A. Formulas for a Nonparabolic Band 347
APPENDIX B. Methods Used in Investigations of Lead Chalogenides 351
APPENDIX C. Main Properties of Lead Chalcogenides . . . 353

LITERATURE CITED 355

SUBJECT INDEX 373

NOTATION

a	lattice constant
A	numerical factor in Hall coefficient [cf. Eq. (3.33)]
$\mathscr{A}$	reciprocal of the period of quantum oscillations in a magnetic field [cf. Eq. (5.12)]
b	ratio of electron and hole mobilities
b_p	ratio of mobilities of light and heavy holes
b, c, and d	parameters governing magnetoresistance
c	velocity of light
c_{11}, c_{12}, and c_{44}	elastic moduli
C	capacitance of a p-n junction
C_P	specific heat at constant pressure
C_V	specific heat at constant volume
$\mathscr{D}$	diffusion coefficient of impurities or free carriers
$\mathscr{D}^*$	detectivity of a photodetector for white light
$\mathscr{D}^*_\lambda$	detectivity of a photodetector for monochromatic light of wavelength λ
e	absolute value of the electron charge
E	electric field intensity
$\mathscr{E}$	energy of free carriers, measured from the bottom of the conduction band
$\mathscr{E}_D$	activation energy of impurity diffusion
$\mathscr{E}_g$	forbidden band width: $\mathscr{E}^*_g = \mathscr{E}_g / k_0 T$
$\mathscr{E}_{gi}$	effective width of the interaction forbidden band (cf. §6.2B)
$\mathscr{E}_{1g}$	constant representing the change in $\mathscr{E}_g$ under isotropic deformation
f	Fermi distribution function

F_i	Fermi integrals [cf. Eq. (3.9)]
g	spectroscopic splitting factor, representing effect of magnetic fields
H	magnetic field intensity
$\mathcal{H}$	Hamiltonian
I	electric current
$\mathcal{J}_0$	total light flux (quanta·cm^{-2}·sec^{-1})
j	electric current density
k	crystal momentum of electrons or phonons
k_0	Boltzmann's constant
K	effective-mass anisotropy coefficient
$\mathcal{K}$	extinction coefficient
l	mean free path
l	oscillatory (Landau) quantum number for electrons in magnetic fields
L	Lorenz number
L_0	Lorenz number for elastic scattering
m_{11}, m_{12}, and m_{44}	elastoresistance constants
m_0	free-electron mass
m*	effective mass of carriers
m_c^*	cyclotron effective mass [cf. Eq. (5.3)]
m_χ	electric-susceptibility (conductivity) effective mass [cf. Eq. (2.22)]
m_d^*	density-of-states effective mass [cf. Eq. (4.9)]
m_F^*	Faraday effective mass
m_0^*	effective mass in a nonparabolic band near an extremum
n	electron density
n_i	intrinsic density of free carriers
n_d and n_a	donor and acceptor concentrations
N	number of equivalent extrema
$\mathcal{N}$	refractive index (real component)
$\mathcal{N}^*$	complex refractive index
p	hole density
$\hat{p}$	momentum operator
P	pressure
Q	Nernst-Ettingshausen coefficient
r	scattering parameter [cf. Eq. (3.3)]
r_{sc}	screening radius (for screening of charge by free carriers)

R	Hall coefficient
$\mathscr{R}$	reflection coefficient
$\mathscr{R}_r$	recombination velocity (number of recombination events per cm^3 per sec)
s	scattering cross section
t	temperature in °C
T	absolute temperature
$\mathscr{T}$	transmission coefficient of light
v	velocity of carriers
V	potential or difference of potentials
U	potential energy
$W_e = 1/\chi_e$	thermal resistivity due to carriers
W_{ee}	component of W_e due to electron-electron collisions
W_0	component of W_e due to elastic scattering mechanisms
$W_l = 1/\chi_l$	lattice thermal resistivity
Z	thermoelectric figure of merit [cf. Eq. (7.1)]
α	absorption coefficient of light
α_0	thermoelectric power
α_∞	thermoelectric power in strong magnetic fields
β	quantum yield (efficiency) of photoconductivity
γ	oscillation damping constant
γ_G	Grüneisen constant
δ	linear thermal expansion coefficient
$\Delta \equiv \partial \ln m^*/\partial \ln T$	
$\varepsilon^* = \varepsilon_1 + i\varepsilon_2$	complex permittivity
ε_0	static permittivity
ε_∞	high-frequency permittivity
ζ	Fermi energy: $\zeta^* = \zeta/k_0 T$ reduced Fermi level
η	density (in g/cm^3, etc.)
ϑ	angle
θ	Debye temperature for acoustical modes
$\theta_{l,t}$	Deby temperature for longitudinal and transverse optical modes
$\varkappa$	thermal conductivity
$\varkappa_a$	ambipolar diffusion component of thermal conductivity

$\varkappa_e$	free-carrier component of thermal conductivity
$\varkappa_l$	lattice component of thermal conductivity
λ	wavelength of electromagnetic radiation in vacuum
μ	carrier mobility
μ_B	Bohr magneton
ξ	compressibility
Ξ	deformation potential constant, occurring in the expression for carrier mobility [cf. Eq. (3.20)], which is a combination of the constants Ξ_d and Ξ_u, defined by Eq. (3.54)
π_{11}, π_{12}, and π_{44}	piezoresistance constants
Π	Peltier coefficient
ρ	density of states
σ	electrical conductivity
τ	relaxation time of carriers
τ_{ph}	relaxation time of phonons
τ_0	carrier lifetime
τ_n and τ_p	electron and hole lifetimes in the presence of traps
τ_r	carrier lieftime in the radiative recombination case
χ	electric susceptibility
ψ	wave function
ω	angular frequency of oscillations
ω_c	cyclotron frequency [cf. Eq. (5.2)]
ω_j	natural frequency of an oscillator
ω_l and ω_t	frequencies of longitudinal and transverse optical phonons
ω_p	plasma frequency [cf. Eq. (2.19)]

INTRODUCTION

The semiconducting properties of lead chalcogenides (compounds of lead with sulfur, selenium, and tellurium) have been known for about one hundred years. In 1865, Stefan observed the thermoelectric power of natural PbS crystals (galenite) and reported different signs of the thermoelectric power for different samples. (We now know that these different signs indicate n- and p-type conduction.) Rectification at a contact between a metal and a semiconductor (PbS) was discovered by F. Braun in 1874 [311], and used in the early stages of the development of point-contact crystal detectors in radio engineering. The photovoltaic effect at a metal - semiconductor contact was also observed and exploited in lead chalcogenides.

Thorough investigations of the physical properties of lead chalcogenides began relatively recently, since the Second World War. To a great extent, these investigations have been stimulated by the development of lead chalcogenide infrared detectors and thermoelectric devices. Lead sulfide photoresistors were developed and carefully investigated in Germany before the beginning of the Second World War, and during the war they were used in guidance systems. Since the war, lead sulfide photoresistors have been developed in the USA, USSR, and in other countries. Lead telluride and selenide photoresistors were produced later.

The introduction of lead chalcogenide photoresistors was a major advance in infrared technology. The main advantages of lead chalcogenide detectors, compared with those used earlier, is their high sensitivity and fast response. Moreover, lead chalcogenide photoresistors cover a wider range of wavelengths, require lower operating voltages, are simpler to use, and are cheaper.

These properties of lead chalcogenide photoresistors are responsible for the very wide use of these devices. Many automatic units used in industry and in military equipment are based on semiconducting photoresistors. Lead selenide photoresistors, which are sensitive at longer wavelengths and have a faster response than lead sulfide devices, are also used in airborne thermal (infrared) mapping systems.

Thermoelectric devices made of lead chalcogenides represent another important application of these compounds. The high thermoelectric power of PbS was noted by Seebeck back in 1822. In 1928, Ioffe was the first to suggest that the use of semiconducting materials in thermoelectric generators would be advantageous. In 1940, Maslakovets reported the construction of a thermoelement, with both arms made of PbS, whose efficiency was 3%.

At the end of the Second World War, work began in the USSR on the development of thermoelectric generators as dc sources for the supply of radio receivers in nonelectrified regions of the country, using the heat of a kerosene lamp. Thermoelectric generators using solar radiation, radioactive heat sources, heat of nuclear reactions, and gas burners are being investigated both in the USSR and elsewhere. The power output of such generators can reach tens and hundreds of kilowatts.

In the first models of semiconductor thermoelectric generators, the negative arm of a thermoelement was made of PbS and the positive arm of ZnSb. The first semiconductor cooling units consisted of thermoelements made of PbTe (negative arm) and ZnSb (positive arm). Later, lead chalcogenides were replaced in thermoelectric cooling units by Bi_2Te_3-type materials, which were found to be more efficient at low temperatures. However, lead chalcogenides, particularly PbTe, are still among the best materials used in the construction of thermoelectric generators, particularly those working at moderately high temperatures (600-1000°K).

Successes in the investigations and applications of lead chalcogenides were achieved after the development of methods for the preparation of relatively pure single-crystal samples of these materials. The first investigations of lead salts were carried out on natural crystals of galenite (PbS) as well as on synthetic compacted or fine-grained samples. W. D. Lawson was the first to grow synthetic single crystals of lead chalcogenides by the Bridgman (Stockbarger) method. At present, single crystals of the highest quality are obtained by growth from the vapor phase or by pulling from the melt (the Czochralski method).

The technique for the preparation of thin films of lead chalcogenides was developed in connection with their use in photoresistors. Usually, films used in photoresistors were polycrystalline and investigations of such films did not yield the information on the basic properties of lead chalcogenides. In the early sixties, single-crystal epitaxial films appeared and were exploited both in devices and in investigations of the physical properties of lead chalcogenides.

By the mid-sixties, a wealth of experimental material had been accumulated on the electrical, thermoelectric, optical, photoelectric, galvanomagnetic, thermal, and magnetic properties of lead chalcogenides. The parameters of these materials, which are important in the understanding of the fundamental properties and in applications of these semiconductors, have also been measured: these parameters include the forbidden band width, optical absorption coefficient, refractive index, permittivity, carrier density, carrier mobility, thermoelectric power, thermal conductivity, and piezoresistance. Those methods for investigating the properties of semiconductors which can be applied to lead chalcogenides are shown in a schematic form in Appendix B. Reviews of the physical and physicochemical properties of lead chalcogenides can be found in [1-5].

Investigations of the fundamental absorption edge, carried out by Scanlon and others, demonstrated that the forbidden band of lead chalcogenides is relatively narrow: 0.3-0.4 eV at room temperature. This forbidden band width governs the spectral range of applications of photoresistors made of these materials.

The forbidden band width increases with increasing temperature. This is why the spectral range of the photosensitivity of lead chalcogenides in the infrared region can be extended by cooling. The positive temperature coefficient of the forbidden band width of lead chalcogenides is also important in the use of lead chalcogenides in thermoelectric devices because the widening of the forbidden band with increasing temperature raises the temperature at which an appreciable amount of minority carriers appears in the material (this is undesirable, because the presence of these carriers reduces the thermoelectric figure of merit of a semiconductor).

Investigations of the magneto-optical absorption near the fundamental absorption edge have made it possible to obtain more accurate values of the forbidden band widths of lead chalcogenides at low temperatures.

Investigations of the fundamental absorption edge have also established that the conduction and valence band extrema are located at the same point in the k space and that direct optical transitions between the extrema are permitted by the selection rules. This is important, in particular, in the understanding of the recently discovered properties of lead chalcogenides, such as the small effective carrier masses, the approximate equality of the electron and hole effective masses, the strong temperature dependence of the effective masses, and the strong nonparabolicity of the conduction and valence bands.

Measurements of the magnetoresistance (in weak magnetic fields), carried out by Allgaier, have played an important role in investigations of the band structure of lead chalcogenides. These measurements established that the constant-energy surfaces near the extrema of the conduction and valence bands of PbTe are in the shape of prolate ellipsoids strongly elongated along the <111> directions. This conclusion has been confirmed by measurements of the piezoresistance, cyclotron resonance, tunnel effect in magnetic fields, and various quantum oscillations in magnetic fields (Shubnikov-de Haas and de Haas-van Alphen effects; oscillations of the absorption of ultrasound). Studies of these quantum oscillations have also demonstrated that PbSe and PbS have prolate constant-energy ellipsoids elongated along the <111> directions but that their elongation is much weaker; the number of ellipsoids in all three chalcogenides is four, i.e., the extrema are located at the boundary (edge) of the Brillouin zone along the <111>directions.

Electron and hole mobilities in lead salts at moderate and high temperatures were investigated by Putley and others, who found a temperature dependence approximately given by $T^{-5/2}$. This temperature dependence has been attributed to the scattering of carriers by the acoustical vibrations of the lattice, taking into account the temperature dependence of the effective mass. Later calculations and analysis of the experimental data have led to the conclusion that, in addition to the acoustical scattering, the polar scattering by the optical vibrations also plays an important role.

are being made to allow for the band nonparabolicity in the analysis of various experimental results.

Among the other investigations of lead chalcogenides we ought to mention studies of the recombination mechanisms, including radiative recombination, which have also yielded information on the carrier lifetime. At low temperatures, the dominant recombination mechanism in pure samples is in the form of interband (band-band) radiative transitions and this circumstance has made it possible to construct lasers of lead chalcogenides and their alloys with other substances. PbSe and PbTe-SnTe lasers emit the longest wavelengths known. Thus, the development of laser technology opens up new possibilities for exploiting lead salts.

Theoretical calculations of the band structure of lead chalcogenides were carried out in the mid-sixties. The most fruitful theoretical calculations were those based on the augmented-plane-wave or pseudopotential methods. The results of calculations of energy gaps between bands at various points in the Brillouin zone, effective masses, and g factors are in good agreement with the experimental data obtained, for example, from measurements of the ultraviolet reflection, photoelectric emission, and the Shubnikov-de Haas effect. Relativistic corrections, of which only the spin-orbit interaction has been allowed in earlier work, have been found to be important in the calculations of the band structures of lead chalcogenides.

It follows from this general discussion that lead chalcogenides are important both from the technical point of view and from the point of view of the physics of semiconductors as a whole, and that a monograph on the properties of lead chalcogenides may be of interest to a wide circle of physicists and engineers.

The authors are grateful to B. Ya. Moizhes, who read the manuscript and made many valuable suggestions. They are grateful also to N. S. Baryshev and F. M. Berkovskii for their valuable advice and to G. E. Steshchenko for his help in the preparation of the manuscript.

CHAPTER I

PHYSICOCHEMICAL PROPERTIES

Lead chalcogenides, i.e., compounds of lead with elements of group VI (sulfur, selenium, and tellurium), belong to a class of semiconducting compounds described by the general formula $A^{IV}B^{VI}$. The physicochemical properties of lead chalcogenides have much in common: they are isomorphous, have the same type of chemical binding, can be prepared by methods which are (in many respects) similar, have similar phase diagrams, and can be doped in a similar manner.

The similarity of the properties of lead chalcogenides is manifested also in a clear correlation between the variation of the degree of ionicity and such parameters as their lattice constants, melting points, and densities. This similarity allows us to consider the physicochemical properties of all three lead chalcogenides by referring to the properties of only the one or two most thoroughly investigated of these compounds without any great loss in generality. For the reader interested in the properties of all three compounds, reference is made throughout to the relevant literature.

1.1. Crystal Structure and the Nature of Chemical Binding

Lead chalcogenides have the cubic NaCl-type lattices of the m3m symmetry class. The unit cell is a face-centered cube and the coordination number for all the atoms is six. X-ray diffraction investigations show that the lattice constant varies monotonically from lead sulfide through lead selenide in steps of about 10%: its value is 5.94, 6.12, and 6.50 Å for PbS, PbSe, and PbTe, respectively. The values of the densities, calculated from the x-ray diffraction data, are in good agreement with the experimental results: they are 7.6, 8.3, and 8.2 g/cm^2 for PbS, PbSe, and PbTe, respectively.

Lead chalcogenide crystals are opaque and have a characteristic metallic luster. They are all highly brittle and are easily cleaved along the (100) planes. The cleavage tendency is stronger at lower temperatures and it disappears practically completely at high temperatures (above 700°C for PbS, above 350°C for PbSe, and above 300°C for PbTe [3]).

The nature of chemical binding in lead chalcogenides is mixed ionic-covalent, i.e., they are usually classified as polar semiconductors. In contrast to the purely covalent compounds, in which an electron pair is shared equally by both interacting atoms, in polar bonds the electrons reside for a longer time in the field of one or other of the two nuclei.

In compounds with the mixed type of binding, the properties are governed by the dominant (ionic or covalent) component. Until fairly recently, lead chalcogenides were regarded as typical semiconductors with a predominantly ionic type of binding [2, 3]. This conclusion was based on the following observations: the structure of lead chalcogenides is isomorphous with the structure of typical ionic compounds, the distances between the nearest neighbors are closer to the ionic than to the covalent radii, and the values of the high-frequency and static permittivities differ strongly (cf., §2.1D). However, later investigations of the mechanism of carrier scattering (cf. §6.4) have demonstrated conclusively that the covalent type of binding predominates in these crystals. According to the theory, crystals with a predominantly ionic type of binding should exhibit carrier scattering mainly by one of the optical types of phonon. Investigations of the carrier scattering in lead chalcogenides have shown that carriers are scattered both by the optical and acoustical phonons, and the role of the acoustical scattering is very important.

The predominance of the covalent component of the binding may be associated with the strong polarizability of lead chalcogenides. It is known that ions of sulfur, selenium, and tellurium are among the most easily deformed. Consequently, the polarizabilities of lead chalcogenides are considerably higher than those of all other compounds with the NaCl-type structure, i.e., the ionicity of the bonds in lead chalcogenides should be considerably weaker than in other isomorphous compounds [6].

No satisfactory quantitative estimates of the polarity of binding in lead chalcogenides are available. Callen's effective charges, representing the interaction of carriers with the polar optical vibrations, are approximately 0.2e (cf. §3.1B), and they decrease somewhat from PbS to PbTe. However, we must bear in mind that these effective charges do not fully reflect the true distribution of the electron density in the lattice. Approximate calculations of Pauling [7], based on the electronegativities of ions, also suggest a small contribution of the ionic component (20%), but these estimates are not very reliable.

Thus, accurate information on the ionic component of the chemical binding in lead chalcogenides is not yet available but an analysis of all the published data, including the results of investigations of the carrier scattering, shows clearly that the chemical binding in lead chalcogenides has both ionic and covalent components.

1.2. Preparation Methods

The semiconducting properties of lead chalcogenides were discovered and investigated first using fine-grained and fired samples as well as natural crystals of galenite (PbS). Natural crystals of PbSe (clausthalite) and PbTe (altaite) have hardly been investigated because they are quite rare and usually too small for convenient investigation.

An important stage in the investigation of the properties of lead chalcogenides has been the development of methods for the preparation of synthetic single crystals. The problem was first solved by Lawson [8, 9] who used the Bridgman-Stockbarger method to grow lead chalcogenides. Later, single crystals were prepared successfully using other methods (zone leveling, growth from the vapor phase, and the Czochralski method). However, the Bridgman-Stockbarger method, first suggested by Lawson, has been found to be technically the simplest and is still the most widely used in the preparation of single crystals of these compounds.

The possibilities of improving the techniques for the preparation of polycrystalline samples have also been investigated. Apart from the importance of such investigations — especially in practical

TABLE 1.1. Dissociation Energies of Vapor Molecules $\mathscr{E}_d$ and Saturation Vapor Pressure P of Lead Chalcogenides [14-18, 71-73]

Compound	$\mathscr{E}_d$, eV	P, mm Hg					
		$T=500°C$	$T=600°C$	$T=700°C$	$T=800°C$	$T=900°C$	$T=1000°C$
PbTe	2.2	$5.0 \cdot 10^{-5}$	$4.1 \cdot 10^{-3}$	$6.5 \cdot 10^{-2}$	—	—	—
PbSe	2.7	$7.0 \cdot 10^{-5}$	$2.5 \cdot 10^{-3}$	$5.6 \cdot 10^{-2}$	—	—	—
PbS	3.3	—	$1.7 \cdot 10^{-3}$	$2.8 \cdot 10^{-2}$	0.46	4.4	26.0

applications (photoresistors, thermoelements) — polycrystalline ingots constitute the basic raw material used in the growth of single crystals. In view of this, we shall consider first the methods for the preparation of polycrystalline materials.

A. Preparation of Polycrystalline Materials

Lead chalcogenides are synthesized from materials whose purity is not less than 99.99-99.90%. Soviet industry produces the following grades of materials of suitable purity: lead of grade SOO STU 99-114-63; metallic tellurium of grade T-A1, prepared in accordance with the Soviet State Standard (GOST) 9514-60; metallic selenium of rectifier grade, prepared in accordance with the Soviet State Standard (GOST) 6738-53; sulfur of grade OSCh VZ.

To prepare very pure materials, containing fewer than 10^{16} cm^{-3} foreign impurities, these elements are subjected to additional purification by distillation or zone melting. Special attention is paid to the removal of oxides, which is achieved by remelting in vacuum or in a stream of hydrogen. The necessary information on the methods for purification of these elements is given in [10].

The methods used in the synthesis of lead chalcogenides are governed by the following factors:

1) the high volatility of one of the components (sulfur, selenium, or tellurium);
2) the strong evaporation and partial dissociation of the vapor during melting (Table 1.1);

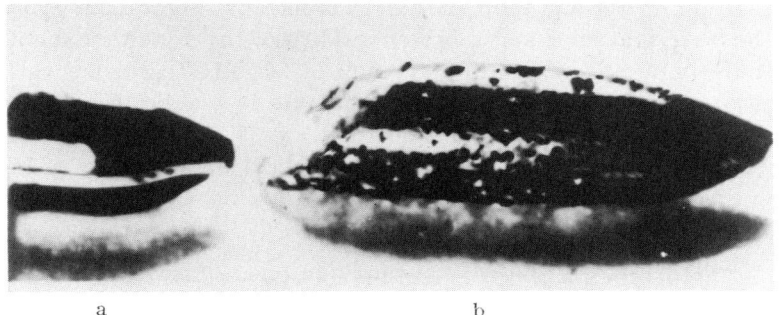

Fig. 1.1. Influence of the cleanness of an ampoule on the quality of PbTe ingots [10]: a) ingot prepared in a specially cleaned ampoule; b) ingot prepared in an ampoule of ordinary cleanness.

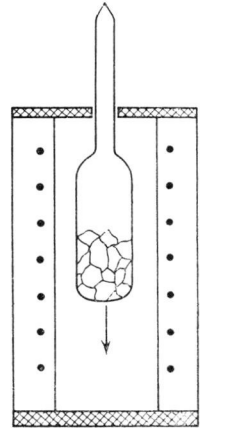

Fig. 1.2. An ampoule used in the rapid melting of lead chalcogenides.

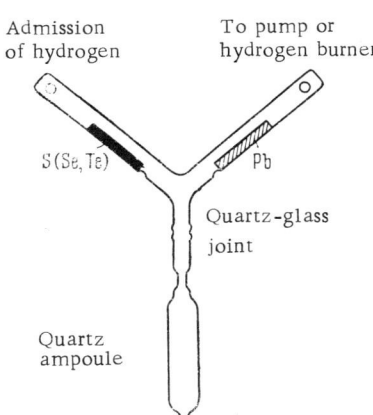

Fig. 1.3. Y-shaped container for avoiding oxidation.

3) the influence of oxygen on the properties of lead chalcogenides and on the melting process itself.

Because of these factors, lead chalcogenides are usually synthesized from the elements in evacuated and sealed quartz ampoules. Before use each ampoule is carefully cleaned in order to remove dust, grease, and other possible contaminants [10]. The influence of the cleanness of an ampoule on the quality of the ingots obtained can be seen quite clearly in Fig. 1.1.

An ampoule containing a suitable mixture is evacuated to 10^{-2}-10^{-3} mm Hg and then sealed. To obtain a higher purity of the synthesized material and a better quality of the resultant ingot, the ampoule is pumped down to a vacuum of 10^{-6}-10^{-7} mm Hg while being slightly heated. The sealed ampoule is placed in a furnace and heated to a temperature exceeding, by 30-50 deg C, the melting point of the synthesized compound (cf. §1.3). The use of higher temperatures produces ingots of poorer quality and increases the chance of explosion.

The synthesis of PbTe can be accelerated considerably by lowering the ampoule into a previously heated furnace. However, in the case of lead selenide and particularly lead sulfide (whose volatile components have high vapor pressures) rapid heating of the ampoule increases the chance of explosion. For this reason, the synthesis is accelerated as follows. An ampoule of special shape (Fig. 1.2) is lowered into a heated furnace so that its upper end remains initially quite cold and the excess amount of the volatile component condenses in that cold end. As the reaction proceeds, the ampoule is gradually lowered into the furnace and the whole of its volume is heated.

When the melting point of the compound is reached, the ampoule is kept at this point for 30-60 min and the melt is periodically stirred. To reduce liquation (segregation) and to obtain more uniform ingots, the ampoule should be cooled rapidly.

During all the stages of the synthesis of lead chalcogenides, it is very important to protect the material being synthesized from contact with oxygen. The presence of oxygen contaminates lead chalcogenides and produces stable p-type conduction [3]. Moreover, oxides react with quartz and may cause an explosion or the cracking of the ampoule.

Contact with oxygen is particularly difficult to avoid during the preparation of the original mixture before melting, especially because of the easy oxidation of lead. The formation of an oxide film on lead can be prevented if all the preparatory stages are carried out in a special box filled with an inert gas. Good results are also obtained when a mixture is prepared using a Y-shaped container (Fig. 1.3). Appropriate amounts of the elements are placed in the separate branches of the Y-shaped container. After a preliminary pumping, the elements are melted in a hydrogen atmo-

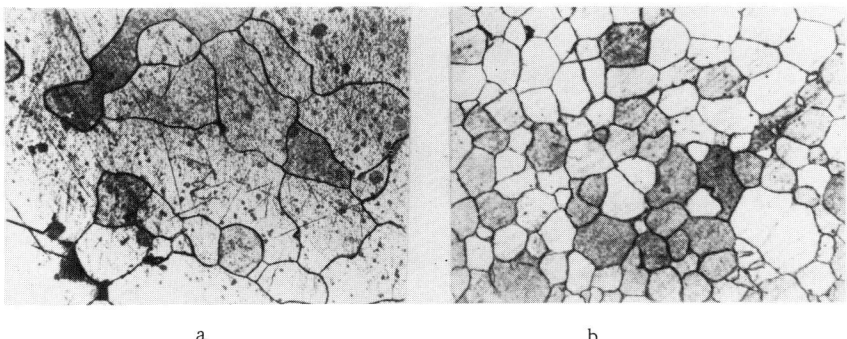

Fig. 1.4. Photomicrographs of PbTe sections (× 100) [13]: a) section of an ingot prepared by the standard melting method; b) section of an ingot prepared using ultrasonic stirring of the melt.

sphere so that they flow gradually into the lower part of a quartz ampoule. Then, the melting is carried out as usual [8].

To check the quality of an ingot so obtained, (its structure, the number of phases it contains, and its homogeneity) by metallographic analysis, one should use the following etchants [11, 12]:

1) 10 g KOH, 10 g glycol, 1 g H_2O_2 (for PbTe and PbSe);
2) 1 g HNO_3, 1 g H_2O (for PbTe, PbSe, and PbS);
3) 10 g HNO_3, 30 g HCl, 1 g acetic acid (for PbS).

Figure 1.4a shows a photomicrograph of a section of PbTe prepared by the usual melting method. Similar photomicrographs have been obtained also for sections of lead selenide and sulfide ingots. Polycrystalline ingots of these compounds usually have poor mechanical properties (they are porous and contain microcracks) and they are fairly inhomogeneous as regards composition. The quality of these ingots can be improved using ultrasonic or vibration stirring of the melt during solidification. The effect of ultrasound on the crystallization process is manifested by an increase of the number and rate of formation of growth nuclei. This gives rise to homogeneous fine-grained ingots with good mechanical properties.

Ultrasonic stirring of molten PbTe was used successfully by Weinstein [13]. The ultrasonic treatment of the melt is begun at a temperature of about 1100°C and continued until solidification. The best results are obtained using ultrasound of about 20 kc fre-

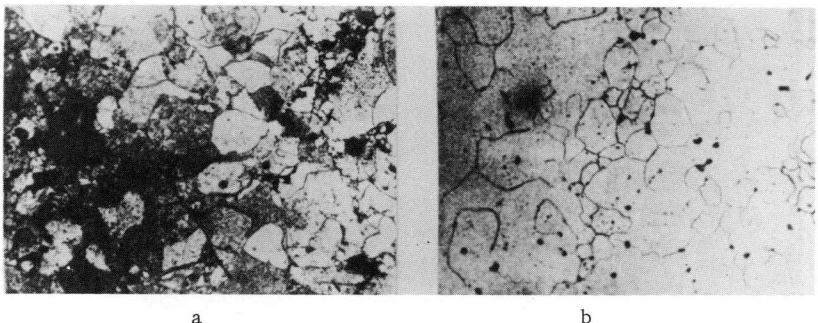

Fig. 1.5. Photomicrographs of hot-pressed PbTe samples (×100): a) un-annealed sample; b) sample annealed at T = 600°C.

quency and 10-40 W power. Figure 1.4b shows a photomicrograph of a section obtained for an ingot prepared from a melt subjected to ultrasonic stirring. We can see that the effect of the ultrasound has been to suppress almost completely the porosity and microcracks, as well as to reduce and equalize the grain dimensions (compare with Fig. 1.4a).

A considerable improvement in the strength and homogeneity of polycrystalline samples is obtained also when powder metallurgy methods are used or ingots are pulled under pressure. Samples of this type are widely employed in thermoelectric devices and are sometimes used successfully in investigations of physical properties.

The method of preparation of polycrystalline samples by the powder metallurgy technique is relatively simple. An ingot, prepared by the fusion of the components in vacuum, is ground in an agate or a porcelain mortar until the particle size is reduced to 0.1-0.3 mm. The resultant powder is poured into a mold, heated to 300-500°C, and the material is subjected for several minutes to a pressure of 5-7 metric tons/cm^2. Samples prepared in this way are then subjected to a homogenizing annealing in an inert-gas atmosphere. Figure 1.5 shows photomicrographs of sections of annealed and unannealed samples of lead telluride, from which the quality of the samples prepared by the powder metallurgy method can be judged.

When samples are prepared by the hot-pressing (powder metallurgy) method, there is always a danger of the formation of an oxide film on the surfaces of individual grains, particularly during the pouring of the powder into a hot mold. This can be avoided by

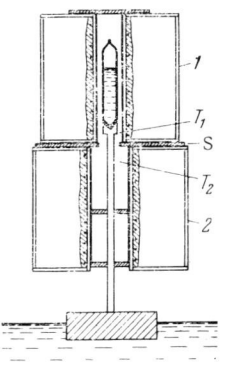

Fig. 1.6. Furnace for growing single crystals by the Bridgman-Stockbarger method.

a preliminary briquetting of a sample in a cold mold, followed by hot pressing or firing in a chemically reducing atmosphere. Good results are also obtained when all the operations are carried out in an atmosphere of a dry inert gas, such as argon [19]. However, this complicates considerably the preparation technique, and is therefore used relatively rarely.

In the extrusion method the temperatures and pressures are approximately the same as in the hot-pressing method. The extrusion is usually carried out in an inert-gas atmosphere or in a stream of dry inert gas.

B. Methods for Single-Crystal Growth

Lead chalcogenide single crystals are mainly prepared by the following methods:

1) the Bridgman-Stockbarger method;
2) the method of slow cooling of the melt;
3) the zone leveling method;
4) the Czochralski method;
5) the vapor phase method.

The most widely used Bridgman-Stockbarger method was first employed by Lawson [8] to grow lead telluride single crystals. Later, Lawson employed successfully the same method to grow lead selenide and sulfide single crystals [9].

In the Bridgman-Stockbarger method, the growth of a crystal takes place in a sealed ampoule during gradual solidification of the melt. This method is convenient because it makes it possible to avoid the loss of the material by evaporation and condensation. The quartz ampoule has a pointed end (to reduce the initial number of growth nuclei); it is filled with a polycrystalline material, evacuated, and sealed. To grow a single crystal, a double furnace, shown schematically in Fig. 1.6, is employed. One furnace (1) is kept at a temperature slightly higher than the melting point of the chalcogenide, while the other furnace (2) is kept below the melting

Fig. 1.7. Photomicrograph of a section of PbTe, showing the mosaic structure of a crystal (× 350) [30].

point. A screen S helps to establish a strong temperature gradient and to smooth out the crystallization front. According to Lawson [8], the temperature gradient should not be less than 25 deg/cm in order to grow a single crystal. The rate of lowering of the ampoule should be within the limits 0.1-10 cm/h.

Thermal conditions close to those which apply in the Bridgman - Stockbarger method are obtained also in the growth of single crystals by slow cooling of the melt, using a natural gradient in a furnace [20]. A substance is placed in a graphite crucible, which is enclosed in an evacuated sealed quartz ampoule. This system is then cooled, together with the furnace, at a rate of 20-30 deg/h. This method is very simple and does not require any device to generate mechanical motion, but the rate of growth and the temperature gradient are not easy to control, which makes it difficult to prepare high-quality single crystals.

These growth methods make it possible to prepare relatively large (20-40 mm long) crystals of all three lead chalcogenides. However, the structure of such crystals is not always perfect and the dislocation density is relatively high (of the order 10^6 cm^{-2}) [21, 23]. The mosaic structure is frequently observed, i.e., each large crystal is found to consist of a multitude of many slightly tilted (by 1-2°) crystallites [24, 25].

The mosaic structure of crystals can be seen quite clearly in metallographic investigations because the density of dislocations is much higher along the mosaic boundaries (cf. "dislocation net-

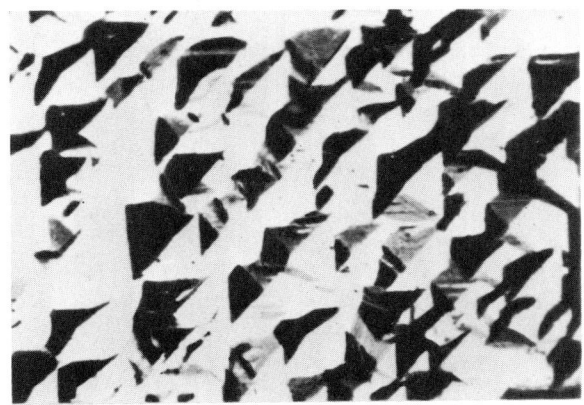

Fig. 1.8. Photomicrograph of a section of a PbTe single crystal prepared by Crocker (× 350) [30].

works" in Fig. 1.7). The dislocation patterns in lead chalcogenides are revealed by etching a freshly cleaved or well-polished surface. Dislocation etching can be carried out using the following mixture [26]: 1 g HCl, 3 g 10% solution of thiourea (the etching takes 1-10 min at 60°C). Surfaces are polished using a mixture consisting of 10 g HNO_3, 30 g HCl, 1 g acetic acid (the surface is treated for several minutes at 50°C and then washed in a 10% solution of acetic acid). The compositions of other etchants, which can be used to reveal dislocations, are given in [27-29].

After the etching, etch pits measuring 3-7 μ appear at the point of emergence of the dislocations [22]; these pits can be easily seen with a standard metallographic microscope.

Crocker [30] was able to prepare single crystals of PbTe practically free of dislocation networks (mosaic structure). He achieved this by repeated application of the Bridgman–Stockbarger method, removing the upper end of the ingot (where lead accumulates during the solidification of the melt, §1.4) after each run. Figure 1.8 shows a photomicrograph of a crystal prepared by Crocker.

The zone leveling method was used recently to grow single crystals of lead chalcogenides [31, 32]. To prevent the condensation of the volatile component on the walls of the ampoule, the whole ampoule is heated to a temperature slightly higher than the melting point of the volatile component. A molten zone is established

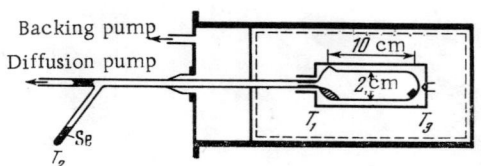

Fig. 1.9. Prior's apparatus [36] for growing PbSe single crystals from the vapor phase.

by a separate heater and its temperature should be much higher than the melting point of the compound.

The principal disadvantage of all the methods described so far is that the crystallization front is in close contact with the wall of a crucible or an ampoule. Tiny imperfections of or contamination on the crucible or ampoule walls may produce additional nuclei and disturb the single-crystal structure of a growing material [33]. Moreover, if the thermal compressions of the growing crystal and the crucible are appreciably different, large internal stresses may appear during the cooling of a crystal and these stresses may increase the density of dislocations and produce microcracks.

In view of these difficulties, the method of free pulling of PbTe crystals from the melt, using the Czochralski method, has attracted much attention [34, 35]. The evaporation of the substance from the surface of the liquid phase is prevented by a specially selected flux (B_2O_3) which covers the surface of the melt during pulling; a thin layer of graphite powder can also be used. This method can be employed to obtain large crystals of lead telluride and selenide.

In addition to the crystallization methods based on the solidification of the melt, one can successfully grow single crystals of lead chalcogenides from the vapor phase [36-38]. Crystals prepared by this method are usually purer than those grown from the melt and have fewer defects.

Figure 1.9 shows an arrangement used by Prior [36] to grow lead selenide single crystals. An initial ingot of PbSe is vaporized in the high-temperature part of a container, where the temperature is T_1. Crystallization centers are formed and single crystals grow in that part of the container where the temperature is lower (T_3). The apparatus must be extremely clean because if the

Fig. 1.10. External appearance of crystals grown by Prior (×7) [36]: a) well-formed sample; b) crystal of ordinary quality.

walls of the container are contaminated, the simultaneous formation of many growth centers will give rise to many crystals. The composition of a crystal grown from the vapor phase is governed by the selenium vapor pressure and the thermal conditions during growth. To establish a controlled selenium vapor pressure, a small amount of elemental selenium, kept at a temperature T_2, is placed in a side tube. The constancy of the composition of the growing crystal is maintained by the precise stabilization of the temperature difference $\Delta T = T_1 - T_3$ and by the control of the selenium vapor pressure which is governed by the temperature T_2. To grow a single crystal successfully, the temperature difference should be very small (ΔT should be of the order of 1 deg), which means that the temperature stabilization must be highly precise.

The quality of crystals grown from the vapor phase depends considerably on the selection of the crystallization temperature. At high crystallization temperatures, crystals grow quickly but they have a large number of defects, compared with the number of defects in crystals obtained from the melt. A very low crystallization temperature is inconvenient because the growth process is then very slow. Prior [36] grew PbSe at 775°C and obtained good-quality single crystals, whose volume was about 0.2 cm^3, in two or three days. The external appearance (habit) of the crystals grown by Prior is shown in Fig. 1.10.

Single crystals of lead chalcogenides have also been prepared by chemical methods. However, at present, these methods are used relatively rarely because they produce very small crystals and it is very difficult to control their composition. Information on the chemical methods for growing single crystals is given in [39].

The recently developed technique for growing single-crystal epitaxial films of lead chalcogenides [40, 41] is very interesting. The optical and electrical properties of such films are very similar to those of bulk single crystals and this makes it possible to use such epitaxial films in various experimental investigations and practical applications. The technique of preparation of single-crystal films is described in detail in, for example, [42].

1.3. Phase Diagrams

The semiconducting properties of synthesized crystals are governed, to a considerable degree, by their phase diagrams. For a given purity of the component elements, both the type of conduction and the carrier density usually depend on the method of preparation. This is because changes in conditions during the preparation produce, in accordance with the phase diagram of the system, changes in the proportions of the principal components.

In the case of lead chalcogenides, one of the components has a high vapor pressure (Te, Se, S) and the phase diagram is plotted using three coordinates, P, T, and x (the pressure, temperature, and composition, respectively), i.e., each phase of a given composition exists in a definite range of temperatures and vapor pressures of the volatile component. When P or x is constant, we obtain two-dimensional $T-x$ and $P-T$ diagrams.

The main results of investigations of the $T-x$ diagrams of the Pb-Te, Pb-Se, and Pb-S systems (at atmospheric pressure) by thermal and metallographic methods were reported by Hansen and Anderko [43]. The use of these methods has established the existence of the compounds PbTe, PbSe, and PbS and has shown that they are the only compounds in these systems. Moreover, investigations of these diagrams have yielded approximate estimates of the melting points T_{mp} (to within ±16 deg). However, in order to synthesize lead chalcogenides with specified properties, more accurate data on the phase diagrams and the homogeneity regions of the compounds, as well as accurate values of the melting points T_{mp}, are required. Therefore, later investigations have been concerned with details of the $P-T-x$ diagram, particularly near the regions of existence of the three chalcogenides.

Bloem and Kröger [44-47] investigated the properties of lead sulfide in equilibrium with the vapor phase and plotted the projections of the phase diagram onto the $T-x$ and $P-T$ planes in the temperature range 730-1130°C. Brebrick and Scanlon [48] obtained data for lower temperatures (730-290°C). Similar results for lead telluride and selenide were reported in [49-58].

In all the cited investigations, the $P-T-x$ diagrams were plotted employing the standard methods of phase analysis, as well as using measurements of the electrical properties by means of which it is possible to detect a departure from the stoichiometric composition amounting to 0.001 at.%. Analysis of the crystal − vapor phase equilibrium has been carried out using the law of mass action, which relates the activity in the vapor phase with the concentration of defects in solids. The interaction between a crystal and the vapor phase can be described by equations of the following type:*

$$\tfrac{1}{2}(S_2)_{gas} \rightleftarrows S_S + V^-_{Pb} + \text{hole},$$
$$(Pb)_{gas} \rightleftarrows Pb_{Pb} + V^+_S + \text{electron}.$$

The equilibrium constants of these reactions are, according to the law of mass action:

$$K_1 = N_{V^-_{Pb}}\, p/P^{1/2}_{S_2}, \quad K_2 = N_{V^+_S} n/P_{Pb}.$$

We can use this approach to consider all other processes which are possible in a crystal with an excess of one of the components or in the presence of a dopant. A detailed analysis of these processes and determination of the constants of the various reactions in the case of lead sulfide are given in Kröger's monograph [57].

Calculation methods and the technique of plotting phase diagrams from the results of electrical measurements are described in [50, 53]. The main assumptions used in such a procedure are listed below.

---

*To make these equations more definite, the subscripts refer to lead sulfide but these reactions apply equally well to all three lead chalcogenides.

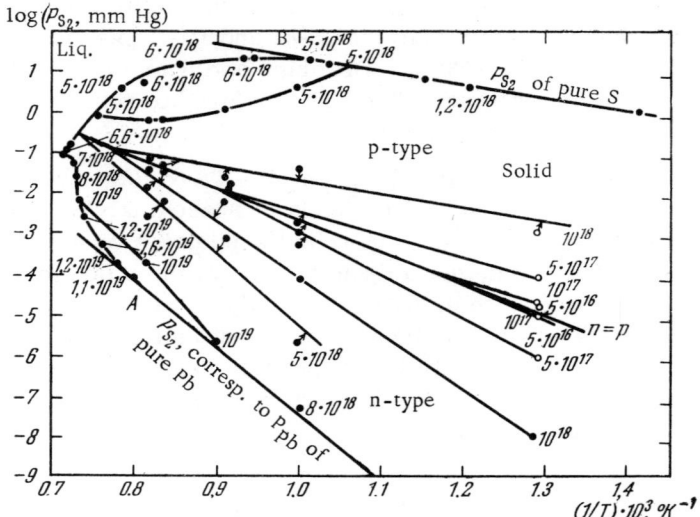

Fig. 1.11a. P – T – x diagram of PbS [45].

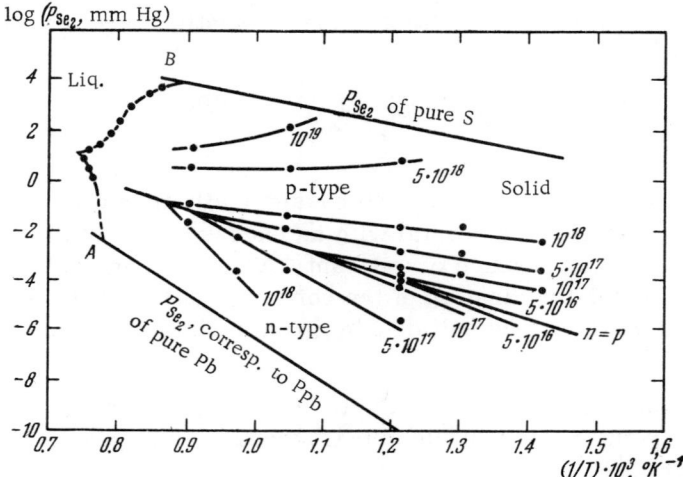

Fig. 1.11b. P – T – x diagram of PbSe [56].

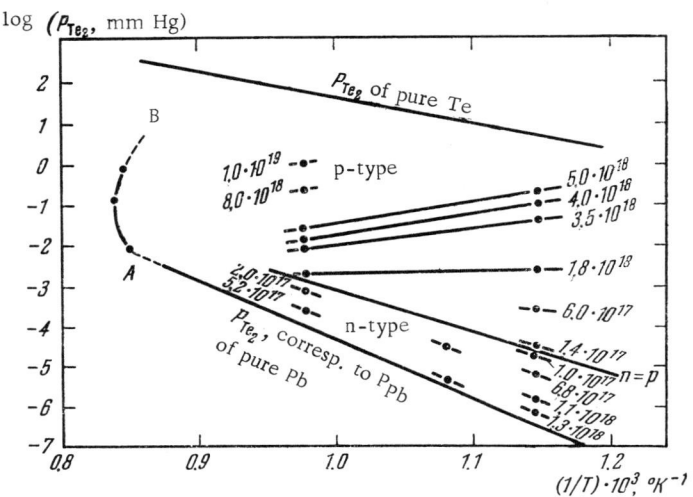

Fig. 1.11c. P − T − x diagram of PbTe [58].

1. It is postulated that the equilibrium composition of a crystal at a given temperature and pressure can be "frozen" by rapid cooling to room temperature.

2. The main types of defect produced at high temperatures are the anion and cation vacancies (Schottky defects), which form a corresponding number of donor and acceptor centers:

$$V_S \rightleftarrows V_S^+ + \text{electron}; \qquad V_{Pb} \rightleftarrows V_{Pb}^- + \text{hole}.$$

According to optical measurements [2], second ionization of vacancies does not occur. It must be stressed that the model used to describe defects is somewhat simplified. Recently, it was reported that excess lead atoms can occupy interstices in lead chalcogenides [54]. Allowance for defects of this type may alter somewhat the established limits of the stability of these compounds.

3. According to the results obtained from the Hall effect (cf. §3.2), all donors and acceptors are fully ionized at room temperature and the carrier density is equal to the difference between the concentrations of defects.

The P − T − x diagrams of lead chalcogenides are shown in Figs. 1.11a-1.11c. An analysis of these diagrams shows that lead chalcogenides are compounds which may exist not only when the ratio of the components is stoichiometric but also in a narrow range of compositions (up to 0.1 at. %) near the stoichiometric ratio. It is found that the composition with the highest melting point is not stoichiometric. In the Pb – S system, the highest melting point, equal to 1127°C, is observed in the presence of excess lead (49.969 at.% S) [45]. In contrast to the sulfide, the maximum melting points in the Pb – Se and Pb – Te systems (1078°C [56] and 924°C [53], respectively) are shifted in the direction of compositions with an excess of Se (50.005 at.%) or Te (50.002 at.%) [49, 53]. Phase diagrams of this type, i.e., with a deviation of the highest melting point from the stoichiometric composition, are typical of compounds with a tendency to form defects and are frequently found among chalcogenides (SnTe, GeTe, Sb_2Te_3, etc.). The shift of the highest melting point may be attributed to the difference between the free energies of formation of the anion and cation vacancies, which are practically always unequal [59]. This means that the shift of the highest melting point away from the stoichiometric composition should be regarded as the rule rather than the exception.

Investigations of the stability regions of lead chalcogenides have shown that these regions, measured relative to the stoichiometric composition, change from lead sulfide to selenide and telluride in accordance with the shift of the highest melting point. The stability region of lead sulfide is shifted in the direction of lead [45], while in the case of lead telluride it is shifted in the direction of tellurium [50]. Lead selenide occupies an intermediate position: its region of existence is spread approximately symmetrically on both sides of the stoichiometric composition [49]. In accordance with general thermodynamic principles, these changes in the positions of the stability region are in agreement with changes in the sign of the difference between the energies of formation of the anion and cation defects [51].

The limits of the stability regions of lead chalcogenides, which govern the solubilities of the components, depend strongly on temperature. According to Bloem and Kröger [45], the maximum deviation of the composition of solid lead sulfide from the stoichiometric ratio is observed at 1090°C and it lies on the lead side ($1.6 \cdot 10^{19}$ cm^{-3} excess lead atoms)* but at 840°C the maximum

---

* According to recent results of Strauss [520], the maximum solubility of Pb in PbS is somewhat higher: $\geq 2.4 \cdot 10^{19}$ cm^{-3} at $T \approx 900°C$ and $1.3 \cdot 10^{18}$ cm^{-3} at $T \approx 350°C$.

deviation lies on the sulfur side ($6.8 \cdot 10^{18}$ cm^{-3} excess sulfur atoms). Goldberg and Mitchell [49] reported the highest solubility of the components in PbSe ($1.4 \cdot 10^{19}$ cm^{-3} excess lead or selenium atoms) at a temperature of the order of 600°C. According to Brebrick and Allgaier [50], the maximum solubility of the components in lead telluride amounts to $3.3 \cdot 10^{18}$ cm^{-3} excess lead atoms and $7.6 \cdot 10^{18}$ cm^{-3} excess tellurium atoms, and it is observed near 775°C. At lower and higher temperatures, the solubility of the components in lead chalcogenides is much less.

Changes in the properties and compositions of a crystal interacting with the vapor phase can be represented by projections of the phase diagram onto the P − T plane (Figs. 1.11a–1.11c). The three-phase curve, AB, represents the region of compositions of solids which can be obtained from the melt when all three phases (solid, liquid, and vapor) are in equilibrium. Curves within the region bounded by the AB curves represent definite compositions of a solid, which are in equilibrium at a given value of temperature and vapor pressure of the volatile component. The $n=p$ line represents crystals of stoichiometric composition, and its intersection with the three-phase curve AB gives the temperature at which the liquid phase is in equilibrium with the stoichiometric solid (T = 1087°C for PbS [45], T = 1 044°C for PbSe [56], and T = 860°C for PbTe [50]).

It is evident from Figs. 1.11a–1.11c that, at high temperatures, all the curves representing various compositions approach the $n=p$ line, i.e., the composition of a crystal depends strongly on the partial vapor pressure of the volatile component. When the temperature is lowered, the difference between equilibrium pressures, corresponding to different compositions, increases and this makes it easier to prepare crystals with definite properties by annealing them at specified values of temperature and vapor pressure.

1.4. Control of the Stoichiometry and the Doping Methods

The special features of the phase diagrams of lead chalcogenides make it difficult to prepare uniform crystals of near-stoichiometric composition. Only at the point corresponding to the maximum melting temperature does the liquid phase retain its composition during solidification (this is known as the "invariant point") and homogeneous ingots of this composition can be prepared by any method of crystallization from the melt. At all other points of the phase diagram, including that corresponding to the stoichiometric composition, the difference between the composition of the melt and the composition of the solid results in a gradual change of the composition of an ingot during solidification. For example, the growth of lead sulfide from the stoichiometric melt produces initially a solid phase with an excess of lead and having n-type conduction. As the melt solidifies, the composition of the liquid phase gradually shifts in the direction of increasing concentration of sulfur. This results in a corresponding change in the composition of the solid phase and may give rise to p-type conduction. In the cases of lead selenide and telluride, whose maximum melting points are shifted in the direction of the chalcogen, variations in the composition are opposite in nature: the first to precipitate are p-type crystals containing an excess of selenium or tellurium; next, the composition of the crystals shifts in the direction of increasing lead content and the conductivity may change its sign from p- to n-type. The crystallization of molten lead chalcogenides yields stoichiometric crystals only when the melt has a definite excess of one of the components and this excess is kept constant throughout the crystallization process. Thus, for example, near-stoichiometric PbTe crystals are obtained from a melt containing about 0.4 at.% excess Pb [53]. However, it is difficult to use this technique to prepare stoichiometric crystals because the present methods of crystallization from the melt (cf. §1.2) do not allow us to maintain a constant composition of the liquid phase. Consequently, crystals prepared by the Bridgman – Stockbarger method, the slow cooling method, or any other similar method usually exhibit a considerable inhomogeneity of their properties as well as an appreciable deviation from stoichiometry, and have free-carrier densities of 10^{18}-10^{19} cm^{-3} even when the original materials are extremely pure [10, 63]. To increase the homogeneity of chalcogenide crystals and to obtain near-stoichiometric compositions, it is usual to heat the grown ingots at some definite temperature and vapor pressure of the volatile component. A controlled vapor pressure is produced in a two-temperature furnace, shown schematically in Fig. 1.12. The temperature T_1 is the temperature at which a crystal is annealed; the temperature T_2 establishes and maintains a given vapor pressure. Apart from changes in and

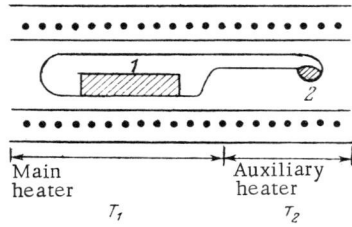

Fig. 1.12. Furnace for annealing single crystals at a given vapor pressure of the volatile component. 1) Sample; 2) elemental Te (Se, S).

homogenization of the properties of a crystal, such an annealing process relieves mechanical stresses which are frequently produced during solidification.

The establishment of a crystal – vapor phase equilibrium alters the composition of a crystal in accordance with the $P - T - x$ diagram of the system. Stoichiometric samples are obtained when the annealing temperature and the vapor pressure correspond to the $n = p$ line in the phase diagram (Figs. 1.11a-1.11c). The selection of one of these parameters (T or P) is, generally speaking, arbitrary, but we must bear in mind that at high temperatures it is more difficult to keep the vapor pressure stable during annealing. Moreover, the annealing of crystals at high temperatures produces a large number of defects in them. On the other hand, if the annealing temperature is too low, the time necessary to establish a crystal – vapor phase equilibrium becomes very long. At 1200°K, an equilibrium state for a lead sulfide film 1 mm thick is established in 0.5 h; at 1000°K, we need 3 h; and at 800°K, 20 h [3]. In the cases of lead selenide and telluride, the time necessary for the establishment of equilibrium at the same temperatures is considerably longer. Thus, the time for the establishment of equilibrium in a film of lead selenide 1 mm thick is 50-100 h at 800°K [56]. For these reasons, the annealing temperature is usually not lower than 800-900°K.

Single crystals of compositions very close to the stoichiometric can be prepared also by heat treatment of samples containing excess Pb in the form of a second (usually finely dispersed) phase. Miller et al. [521] reported that single crystals of n-type PbTe with $n = 4 \cdot 10^{16}$ cm^{-3} could be prepared by annealing of two-phase ingots at $T \approx 825°$C until these ingots became fully homogeneous. The number of defects in such crystals was reduced by an additional low-temperature annealing at $T < 400°$C.

Bearing in mind the physicochemical properties of lead chalcogenides, we may conclude that the most suitable method for the preparation of especially pure and defect-free crystals is the method of growth from the vapor phase. The lowering of the temperature at which the crystal is formed results in a corresponding reduction in the number of defects, and the absence of contact between the growing surface and the walls of an ampoule makes it possible to avoid mechanical stresses and reduces considerably the dislocation density. The control of the composition of a crystal by the stabilization of the growth temperature and of the vapor pressure of the volatile component makes it possible to prepare, in principle, single crystals of any dimensions which are homogeneous and have specified properties. The method of growth of crystals from the vapor phase makes it possible to approach very closely the stoichiometric composition and to reduce the carrier density in unannealed samples to 10^{16}-10^{17} cm^{-3}.

Depending on the growth and heat-treatment conditions, the composition of a crystal may vary within the limits of the solubility of the components. Since an excess of lead or chalcogenide atoms gives rise to a corresponding number of donor or acceptor centers, it follows that the carrier density in lead chalcogenides free of foreign impurities may reach values of 10^{18}-10^{19} cm^{-3}. If materials with a higher carrier density are required, various impurities are used as dopants. Much work on the influence of impurities on the properties of lead chalcogenides is reported in [60-67].

According to the results reported by Koval'chik and Maslakovets [61], halogens (Cl, Br, I) produce deep donors by means of which the carrier density can be increased up to $(1-2) \cdot 10^{20}$ cm^{-3}. Doping with halogens is usually carried out by introducing a halide ($PbCl_2$, etc.) and a suitable amount of Pb or an element such as Sn, Ge, Mg, Pt, Ni [61]. Halogen atoms, which have seven free electrons, replace hexavalent tellurium, i.e., the introduction of one halogen molecule gives rise to two free electrons.

Elements of group V (Bi, Sb) replace lead and act as donor impurities, producing carrier densities up to 10^{20} cm^{-3} [45, 67, 522, 523]. In agreement with the donor nature of Bi, lead sulfide crystals doped with this impurity exhibit a considerable increase in the equilibrium sulfur vapor pressure necessary to produce p-type conduction [45]. In contrast to halogens, the introduction of elements of group V produces a lower carrier density than that predicted by a calculation based on the assumption that one carrier is contributed by each impurity atom. This is attributed in [67] to the formation of neutral Bi_2Te_3-type complexes. Moreover, Bi-doped materials exhibit much lower electron mobilities than those found in halogen-doped compounds.

According to [68], Ta, Zn, Al, Ga, Mn, U, Nb, and Ti act as donors in lead telluride but details of the mechanism of the effect of these impurities have not been investigated. Interstitial Cu [65, 66], and Fe [64] ions can act as donors.

The principal acceptor impurities, capable of producing hole densities up to $1.5 \cdot 10^{20}$ cm^{-3}, are Na, Li, and Tl [64, 68]. These impurities are usually introduced simultaneously with some excess tellurium and they are very sensitive to the presence of oxygen. Acceptor levels are also produced by silver, which is usually introduced in the form of Ag_2Te. The acceptor nature of silver was demonstrated convincingly by Bloem and Kröger [45] in a physicochemical investigation of lead sulfide. According to Koval'chik and Maslakovets [61], the doping of lead chalcogenides with silver produces only one hole for every two molecules of Ag_2Te. This may be due to silver atoms, being located partly at the lead sites and partly at interstices.

Oxygen, sulfur, and selenium are analogs of Te and they behave like excess Te atoms [61]; these impurities can occupy sites in the tellurium sublattice. Their presence produces stable p-type conduction at hole densities up to $(3-4) \cdot 10^{18}$ cm^{-3}.

Investigations of the temperature dependence of the Hall coefficient (§3.2A) show that the carrier density in lead chalcogenides usually depends weakly on temperature in a wide range of temperatures and impurity concentrations. This is because the energy levels of most of the impurities merge with the edge of the nearest allowed band. Some investigators [46, 47, 62] have reported the existence, in lead chalcogenides, of impurity levels with an activation energy of 0.02-0.04 eV.

The zone leveling method can be used successfully in the preparation of doped crystals. Kaidanov [32] and others have used this method to prepare PbTe single crystals with electron densities up to $1 \cdot 10^{20}$ cm^{-3}. To increase the effective segregation coefficient of impurities, the rate of motion of the molten zone is increased (up to 15 cm/h). Ingots obtained in this way are single crystals or bicrystals and the length of that part of an ingot which is electrically uniform can reach 60-70 mm.

Doped single-crystal and polycrystalline samples, of fairly homogeneous composition, are also obtained when various synthesis or growth methods are combined with a suitable heat treatment [3, 69, 70]. The time necessary to reach homogeneity depends on the rate of diffusion of the dopant at the annealing temperature as well as on the crystal structure and dimensions of a sample. Fine-grained samples are affected most strongly and rapidly by heat treatment. In view of this, powder metallurgy (hot pressing) and extrusion methods are widely used in the preparation of uniformly doped materials, which are required for scientific and practical applications.

1.5. Diffusion

A diffusion flux of particles in a solid is produced by a concentration gradient of the particles, a temperature gradient, mechanical stresses, and other factors. It is necessary to know the diffusion coefficients of impurities for the correct selection of the heat-treatment conditions, in the preparation of diffused p−n junctions, in calculating the service life of thermoelectric devices, etc. Apart from the technical aspects, investigations of the diffusion of impurities gives information on the nature of the motion of particles in a solid (diffusion along interstices, vacancies, block boundaries, etc.).

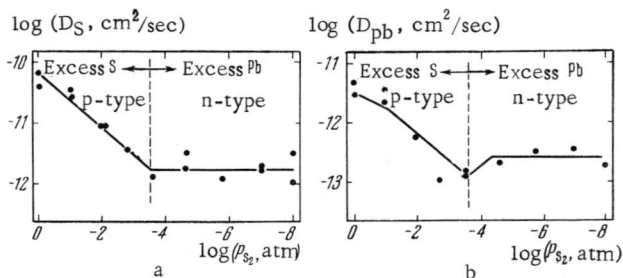

Fig. 1.13. Self-diffusion coefficients of S^{35} (a) and Pb^{210} (b) for PbS as a function of the S vapor pressure at T = 700°C.

TABLE 1.2. Self-Diffusion Coefficients of Pb [77] and S [78] in PbS Single Crystals under Controlled Sulfur Vapor Pressure

Composition	Diffusant	D_0, cm²/sec	$\mathscr{E}_D$, eV	T, °C*	Diffusion mechanism
Stoichiom. (n = p)	Pb	$8.6 \cdot 10^{-5}$	1.52	500-800	—
$\geq 10^{18}$ cm^{-3} excess Pb (n-type)		$2.6 \cdot 10^{-5}$	1.35		Pb^{2+} donors along interstices
10^{18} cm^{-3} excess S (p-type)		$5.5 \cdot 10^{-7}$	1.01		Pb along vacancies (V_{Pb}^-)
Stoichiom. (n = p)	S	$6.8 \cdot 10^{-5}$	1.38	500-750	—
10^{18} cm^{-3} excess Pb (n-type)		$1.69 \cdot 10^{-6}$	1.16		Neutral defect pairs ($V_{Pb}^- V_S^-$)
10^{18} cm^{-3} excess S (p-type)		$4.56 \cdot 10^{-5}$	1.22		Atoms along interstices or neutral vacancies in Pb sublattice

*The diffusion annealing was carried out in a sulfur atmosphere.

The diffusion coefficient D is found from Fick's differential equation

$$\frac{\partial n}{\partial t} = D \frac{\partial^2 n}{\partial x^2}$$

(n is the concentration of particles). The theory and methods for measuring the diffusion coefficient are given in Boltaks' monograph [76]. When the temperature is increased, the diffusion coefficient D varies in accordance with the law

$$D(T) = D_0 e^{-\mathscr{E}_D / k_0 T},$$

where $\mathscr{E}_D$ is the diffusion activation energy.

TABLE 1.3. Diffusion Coefficients of Impurities in PbS, PbSe, and PbTe

Diffusant	PbSe*			PbTe*			PbS*		
	D_0, cm²/sec	$\mathscr{E}_D$, eV	Ref.	D_0, cm²/sec	$\mathscr{E}_D$, eV	Ref.	D_0, cm²/sec	$\mathscr{E}_D$, eV	Ref.
Pb	$4.98 \cdot 10^{-6}$	0.83*	[80]	$2.9 \cdot 10^{-5}$	0.6*	[76, 81]	—	—	—
Se	$2.1 \cdot 10^{-5}$	1.2*	[82]	—	—	—	—	—	—
Te	—	—	—	$2.7 \cdot 10^{-6}$	0.75*	[76, 82]	—	—	—
Sb	0.34	2.0*	[82]	$4.9 \cdot 10^{-2}$	1.54*	[76, 82]	—	—	—
Cd	17.6	2.0	[83]	—	—	—	—	—	—
Zn	37.0	2.1	[84]	—	—	—	—	—	—
Fe	0.64	1.62	[85]	—	—	—	—	—	—
In	$9 \cdot 10^{-5}$	1.35	[85]	—	—	—	—	—	—
Sn	$1.2 \cdot 10^{-6}$	0.81	[86]	$3 \cdot 10^{-2}$	1.56*	[76, 82]	—	—	—
Ag	$7.4 \cdot 10^{-1}$	0.35	[87]	—	—	—	—	—	—
Au	$5.6 \cdot 10^{-2}$	0.75	[88]	—	—	—	—	—	—
Cu	$2 \cdot 10^{-5}$	0.31	[65]	—	—	—	$\begin{cases} 4.6 \cdot 10^{-4} \\ 5 \cdot 10^{-3} \end{cases}$	0.357 0.310	[65] [45]
Cl	$1.6 \cdot 10^{-8}$	0.45	[89]	—	—	—	—	—	—
Gd	$7.4 \cdot 10^{-6}$	0.30	[88]	—	—	—	—	—	—
Ni†	—	—	—	—	—	—	17.8	0.96	[46]

*The carrier density in p- and n-type PbSe was 10^{18} cm⁻³; in p-type PbTe, it was 10^{17} cm⁻³ and in p-type PbS, it was 10^{18} cm⁻³.
†Data on the diffusion of Ni in PbS are also given in [90].

The available results on the diffusion of impurities in lead chalcogenides can be divided into three groups:
1) diffusion in undoped single crystals;
2) diffusion in doped single crystals;
3) diffusion in undoped and doped polycrystalline samples.

A. Diffusion in Undoped Single Crystals

As pointed out in §1.3, the highest melting points of all lead chalcogenides correspond to compositions with an excess or deficiency of one of the components. Therefore, all crystals grown without special precautions from a stoichiometric melt are always nonstoichiometric. We shall call them "quasistoichiometric" samples.

Experiments on the self-diffusion of Pb and S in single crystals of PbS [77-79] have demonstrated that the diffusion activation energy is sensitive to deviations from stoichiometry. This is particularly noticeable in the self-diffusion of Pb (Table 1.2). The sixth column of Table 1.2 lists possible mechanisms of the diffusion of Pb and S in nonstoichiometric samples of PbS.

The dependences of the self-diffusion coefficients of S (Fig. 1.13a) and Pb (Fig. 1.13b) in PbS on the partial pressure of sulfur have been investigated at 700°C [78, 79]. The minimum in Fig. 1.13a and the kinks in Fig. 1.13b correspond to the pressure of sulfur which is necessary to produce stoichiometric PbS at 700°C.

Most of the results on the diffusion coefficients in lead chalcogenides have been obtained using "quasistoichiometric" samples. Many of the published results seem to be inaccurate (some of them are denoted by an asterisk in Table 1.3). This is evidently due to the poor quality of the investigated single crystals. However, in spite of this, the experimentally obtained values of the parameters D_0 and $\mathscr{E}_D$ for various diffusing impurities (Table 1.3) can be useful in calculations concerned with practical problems.

The value of the activation energy $\mathscr{E}_D$ can be used approximately to deduce diffusion mechanism. It is usually assumed [76] that impurities diffuse along the interstices if $\mathscr{E}_D \leqslant 1$ eV and that they diffuse along lattice sites (vacancies) if $\mathscr{E}_D > 1$ eV.

An investigation of the diffusion of Na in PbSe [87, 91] has established an anomalous distribution of the concentration through the depth of a sample, which departs from the standard Fick distribution. This anomaly has been explained by postulating "fast" (D_f) and "slow" (D_s) diffusion fluxes. It has been found experimentally that at low temperature the value of D_f is three orders of magnitude higher than D_s. At high temperatures the difference between these two diffusion coefficients becomes less. The temperature dependences of D_f and D_s for Na in PbSe can be written in the form

$$D_f(T) = 5.6 \cdot 10^{-6} e^{-0.4/k_0 T}, \quad D_s(T) = 15 e^{-1.74/k_0 T}. \tag{1.2}$$

B. Diffusion in Doped Single Crystals

The self-diffusion coefficients of Pb in PbSe and of Pb and S in PbS have been investigated [78, 80] in the presence of Bi_2Se_3, Ag_2Se, and Bi_2S_3, Ag_2S dopants. Usually the presence of dopants reduces $\mathscr{E}_D$, compared with its value for stoichiometric samples. Thus, for example, in the self-diffusion of Pb in PbS with 0.5 mol.% Bi_2S_3, the value of $\mathscr{E}_D$ decreases from 1.52 eV for a stoichiometric sample (Table 1.2) to 1.07 eV. In the self-diffusion of S in a sample with the same amount of Bi_2S_3, the diffusion activation energy decreases from 1.38 eV for a stoichiometric sample to 0.69 eV for a doped crystal.

C. Diffusion in Polycrystalline Samples

In contrast to single crystals, the hot-pressed samples exhibit, in addition to volume diffusion (D_v), diffusion along the grain boundaries (D_b). At temperatures $T > (0.8-0.9)T_{mp}$, it is found that $D_v > D_g$, but at $T < (0.4-0.5)T_{mp}$, $D_v < D_g$ [92]. If a continuous exchange of diffusing atoms takes place between the grain boundaries and the interior, we can introduce an "effective" diffusion coefficient D_{eff} [93]:

$$D_{eff} = \frac{D_{0v} n_v + D_b n_b}{n_v + n_b} \tag{1.3}$$

(n_b and n_v are the diffusing impurity concentrations in the grain boundaries and in the interior of a crystal).

An investigation of the diffusion of Sb and Ni in hot-pressed samples of p-type PbSe with an admixture of 0.5 mol.% Na_2Se

($p \approx 10^{20}$ cm^{-3}) was reported in [91]; the self-diffusion of Pb in PbS was studied in [94]. In these investigations, D_V and D_b were identified by varying the temperature and duration of the annealing. After annealing for a short time, the impurity distribution is governed mainly by the grain-boundary diffusion (D_b). As the annealing duration is increased, the role of the volume diffusion (D_V) increases. The higher the temperature, the greater is the contribution of the volume diffusion to the total flux. At temperatures above 600°C, the volume diffusion predominates, whereas at temperatures below that value the grain-boundary diffusion makes the principal contribution. The temperature dependence of the diffusion coefficient in polycrystalline samples can be written in the form

$$D(T) = D_{0v} e^{-\mathscr{E}_{Dv}/k_0 T} + D_{0b} e^{-\mathscr{E}_{Db}/k_0 T} . \tag{1.4}$$

An investigation of PbSe with 0.5 mol. % Na$_2$Se has yielded the following parameters for the diffusion of Sb and Ni, respectively: $D_{0v} = 16$ and $16 \cdot 10^2$ cm^2/sec; $D_{0b} = 6.3 \cdot 10^{-7}$ and $1.6 \cdot 10^{-7}$ cm^2/sec; $\mathscr{E}_{Dv} = 1.8$ and 1.9 eV; $\mathscr{E}_{Db} = 0.25$ eV and 0.30 eV [91]. These results show that the activation energy of the grain-boundary diffusion is very low.

CHAPTER II

OPTICAL AND PHOTOELECTRIC PROPERTIES

Investigations of the optical and photoelectric properties of semiconductors are among the most important sources of information on the band structure, the frequencies of the optical modes of the phonon spectrum, the permittivity, the recombination parameters, and other properties of semiconductors. It is also essential to know the optical and photoelectric properties of lead chalcogenides for the applications of these materials in semiconductor electronics.

2.1. Absorption and Reflection Light

The absorption and reflection of electromagnetic waves in lead chalcogenides have been investigated in detail in a wide range of frequencies [64, 95-129]. Before we present the results of these investigations, we shall first review briefly the principal optical parameters of a solid, the methods used to measure these parameters, and the microscopic processes responsible for their values. More detailed information on the optical properties of semiconductors are given in various papers, as well as in special monographs and reviews [130, 131].

A. Optical Parameters and Measurement Methods

It is convenient to describe the propagation of a plane electromagnetic wave in a solid by a frequency-dependent (the frequency is denoted by ω) complex refractive index $\mathcal{N}^* = \mathcal{N} + i\mathcal{K}$, where the real part of the refractive index $\mathcal{N}$ is related to the velocity of propagation of the wave and the extinction coefficient

$\mathscr{K}$ represents the decay of the amplitude of oscillations of the electric field:

$$\mathbf{E}(x, t) = \mathbf{E}_0 \exp\left[i\omega\left(\mathscr{N}^*\frac{x}{c} - t\right)\right] =$$
$$= \mathbf{E}_0 \exp\left[i\omega\left(\mathscr{N}\frac{x}{c} - t\right)\right] \exp\left(-\mathscr{K}\frac{\omega x}{c}\right). \qquad (2.1)$$

The optical properties of a crystal such as a lead chalcogenide one is governed by the interaction between the crystal and the electric field of the electromagnetic wave. The optical parameters $\mathscr{N}$ and $\mathscr{K}$ are related to the frequency-dependent electrical conductivity σ and electric susceptibility χ. In a static electric field, the conductivity represents the value of the current of free electric charges and the electric susceptibility represents the displacement of bound charges. In an alternating electric field, the charges oscillate and the distinction between free and bound charges disappears at the microscopic level. However, the current excited by an alternating electric field can still be represented formally as the sum of a conduction current, varying in phase with the electric field, and a displacement current, lagging in phase behind the field by $\pi/2$:

$$\mathbf{j} = \sigma\mathbf{E} + \chi\frac{\partial \mathbf{E}}{\partial t}, \qquad (2.2)$$

where the frequency-dependent real coefficients of proportionality σ and χ represent the conductivity and electric susceptibility in an alternating electric field.

If, instead of the current, we consider the polarization and the associated electric induction, we can introduce a complex permittivity $\varepsilon^* = \varepsilon_1 + i\varepsilon_2$ as a coefficient of proportionality between the electric induction and the field. The relationships between the three pairs of quantities we have introduced ($\mathscr{N}, \mathscr{K}$; σ, χ; $\varepsilon_1, \varepsilon_2$) are given by the following formulas:

$$\varepsilon_1 = 1 + 4\pi\chi = \mathscr{N}^2 - \mathscr{K}^2, \qquad (2.3)$$

$$\varepsilon_2 = \frac{4\pi}{\omega}\sigma = 2\mathscr{N}\mathscr{K}. \qquad (2.4)$$

The reflection coefficient of a perfectly smooth surface, which represents the reflected part of the energy of a normally incident beam, can be expressed in terms of the optical parameters $\mathcal{N}$ and $\mathcal{K}$:

$$\mathcal{R} = \frac{(\mathcal{N}-1)^2 + \mathcal{K}^2}{(\mathcal{N}+1)^2 + \mathcal{K}^2}. \tag{2.5}$$

The transmission coefficient of light through a plate of thickness d is given by the following formula, which takes account of multiple internal reflections:

$$\mathcal{T} = \frac{(1-\mathcal{R})^2}{1 - \mathcal{R}^2 e^{-2\alpha d}} e^{-\alpha d}, \tag{2.6}$$

where α is the absorption coefficient, which can be expressed in terms of the extinction coefficient $\mathcal{K}$:

$$\alpha = \frac{2\omega}{c} \mathcal{K}. \tag{2.7}$$

The absorption coefficient is equal to the reciprocal of the depth of penetration of light into a bulk sample, i.e., it is equal to the depth at which the energy of a light wave decrease by a factor e.

It is evident from Eqs. (2.5)-(2.7) that independent measurements of the reflection and transmission coefficients make it possible to determine the two optical parameters $\mathcal{N}$ and $\mathcal{K}$, which are related to the electrical parameters σ and χ. The calculation of σ and χ is a problem in the microscopic theory, whose solution depends on the nature of the electron and phonon spectra and on other microscopic properties of a solid. This method of determination of the refractive index $\mathcal{N}$ and the extinction coefficient $\mathcal{K}$ based on the simultaneous measurement of the reflection ($\mathcal{R}$) and transmission ($\mathcal{T}$) coefficients has been employed in several investigations [100, 109, 117, 119].

In the range of frequencies in which absorption is weak $[\mathcal{K}^2 \ll (\mathcal{N}-1)^2]$, measurements of the reflection coefficient, given by

$$\mathcal{R} = \left(\frac{\mathcal{N}-1}{\mathcal{N}+1}\right)^2, \tag{2.8}$$

yield the refractive index without the need to measure the transmission coefficient. This method of determining the refractive index was employed by Dixon and Riedl [111-113, 116].

The refractive indices of lead chalcogenides have been determined also from the positions of interference maxima observed during the transmission of light through a thin plate [104, 108, 109]. The wavelength λ, at which an m-th order maximum is observed, is given by the formula

$$m\lambda = 2\mathcal{N}d. \tag{2.9}$$

Walton and Moss [106] determined the refractive index from the angle of refraction of light in a prism made of the investigated material.

Near the fundamental absorption edge, the absorption coefficient varies rapidly with frequency, while the reflection coefficient varies weakly with frequency and can be regarded as constant. This makes it possible to determine the detailed form of the absorption spectrum near the edge by measuring solely the transmission coefficient, as was done by Scanlon [98, 99] and other investigators [64, 96, 97, 101, 104, 107, 118, 120, 122, 125].

The values of $\mathcal{N}$ and $\mathcal{K}$ are not completely independent. The relationship between them is given by integral dispersion equations, which are independent of the actual properties of the substance under investigation but are related to the causality condition and are perfectly general [115, 132]. One form of the dispersion equations relates the real and imaginary components of the permittivity:

$$\varepsilon_1(\omega) = 1 + \frac{2}{\pi} \int_0^\infty \frac{\varepsilon_2(\omega')\, \omega'\, d\omega'}{\omega'^2 - \omega^2}, \tag{2.10}$$

$$\varepsilon_2(\omega) = \frac{2\omega}{\pi} \int_0^\infty \frac{\varepsilon_1(\omega')\, d\omega'}{\omega^2 - \omega'^2} \tag{2.11}$$

(the integrals in the above equations are understood to represent the principal values). The use of the dispersion relationships makes it possible to find both $\mathcal{N}$ and $\mathcal{K}$, by measuring only one quantity (for example, the reflection coefficient) in a wide range of frequencies.

If we consider not only the amplitude but also the phase of the reflected wave, we find it convenient to describe the reflection by a complex quantity

$$\mathscr{R}^* = \mathscr{R} e^{i2\vartheta} = \left(\frac{\mathscr{N} + i\mathscr{K} - 1}{\mathscr{N} + i\mathscr{K} + 1}\right)^2. \tag{2.12}$$

By measuring the value of $\mathscr{R}$ in a wide range of frequencies, we can find the phase ϑ from the dispersion formula [115]:

$$\vartheta(\omega) = \frac{\omega}{\pi} \int_0^\infty \frac{\ln \mathscr{R}(\omega') \, d\omega'}{\omega^2 - \omega'^2}, \tag{2.13}$$

and then determine $\mathscr{N}$ and $\mathscr{K}$ from Eq. (2.12). The dispersion relationships have been employed in [109, 110, 119] to determine the absorption coefficient or to check the results obtained by direct measurements.

To illustrate the application of the dispersion relationships, let us consider Dixon's investigation [110] of the contribution of free carriers to the absorption and dispersion by the method of measuring the reflection coefficient in samples with low and high free-carrier densities at frequencies below the fundamental absorption edge. The phase difference for pure and heavily doped samples is

$$\vartheta(\omega) - \vartheta_0(\omega) = \frac{\omega}{\pi} \int_0^\infty \frac{\ln \mathscr{R}(\omega') - \ln \mathscr{R}_0(\omega')}{\omega^2 - \omega'^2} \, d\omega'. \tag{2.13a}$$

In the fundamental absorption region, the relative importance of free carriers is small and $\mathscr{R} \approx \mathscr{R}_0$. In the range of frequencies below the fundamental absorption edge ($\omega < \omega_g$), the absorption coefficient of a pure sample is small and it follows from Eq. (2.12) that $\vartheta_0 = 0$. Thus, in the frequency range $\omega < \omega_g$, we have the formula

$$\vartheta(\omega) = \frac{\omega}{\pi} \int_0^{\omega_g} \frac{\ln \mathscr{R}(\omega') - \ln \mathscr{R}_0(\omega')}{\omega^2 - \omega'^2} \, d\omega', \tag{2.13b}$$

which can be used to find ϑ and then $\mathscr{N}$ and $\mathscr{K}$ without measuring the absorption, and avoiding the determination of any parameters in the fundamental absorption region. Another method for the determination of the refractive index and the extinction coefficient from measurements of the reflection alone was employed by Avery [95], who varied the angle of incidence in measurements of the reflection coefficient and used light polarized parallel and perpendicular to the plane of incidence of a ray.

B. Microscopic Processes of Interaction between Light and a Semiconductor

Light interacts with a semiconductor by exciting electron transitions and atomic vibrations. If, under the influence of light, an electron is transferred from a state of energy $\mathscr{E}_0$ to a state with a higher energy $\mathscr{E}_j$ (or if a phonon is generated), we observe the absorption of a light wave in a crystal. These real transitions in the electron (or phonon) system govern, to a great extent, the absorption coefficient (more exactly, the imaginary part of the permittivity). Real transitions should satisfy the laws of conservation of energy and momentum; therefore, the frequency of the absorbed light ω is equal to the frequency of the transition $\omega_j = (\mathscr{E}_j - \mathscr{E}_0)/\hbar$. The electric susceptibility (or the real part of the permittivity) represents the deformation of the electron (or ion) system by the electric field of the wave. Calculations of the electric susceptibility (related to polarizability) in the second approximation of the perturbation theory yield formulas which can be conveniently considered using the concept of "virtual" transitions. According to this concept, a system is transferred to an excited state and immediately returns to its initial state. Virtual transitions do not obey the law of conservation of energy and, therefore, the electric susceptibility (polarizability) includes the contributions from transitions corresponding to frequencies different from the frequency of light but the relative importance of such transitions decreases as the difference $\omega - \omega_j$ increases.

The effect of an electromagnetic wave on a solid can be illustrated conveniently by means of a very simple model consisting of n independent neutral atoms, fixed in space and having one electron each. The complex permittivity for this model is

$$\varepsilon^*(\omega) = 1 + \frac{4\pi n e^2}{m_0} \sum_j f_j \left\{ \frac{1}{\omega_j^2 - \omega^2} + \frac{i\pi}{2\omega} \delta(\omega - \omega_j) \right\}. \qquad (2.14)$$

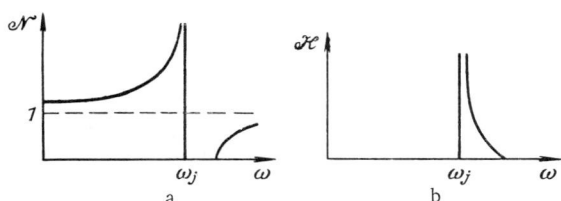

Fig. 2.1. Frequency dependences of the refractive index (a) and the extinction coefficient (b) in the absence of damping.

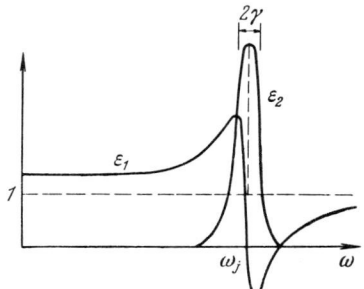

Fig. 2.2. Frequency dependences of the real and imaginary components of the permittivity when damping is taken into account.

The transition frequencies ω_j are then the natural frequencies of oscillators in the classical theory; the δ-shaped imaginary component of the permittivity gives rise to discrete absorption bands at the natural frequencies; the coefficients of proportionality f_j are the oscillator strengths, whose sum is equal to unity.

Using Eqs. (2.3), (2.4), and (2.14), we can obtain the spectral dependences of the refractive index and the extinction coefficient, shown in Fig. 2.1, for the case of a single natural frequency ω_j.

The broadening of the discrete bands is usually allowed for phenomenologically by the inclusion, in Eq. (2.1), of a damping factor $\exp(-\gamma t/2)$. Then, instead of Eq. (2.14), we obtain

$$\varepsilon^*(\omega) = 1 + \frac{4\pi n e^2}{m_0} \sum_j \frac{f_j}{(\omega_j^2 - \omega^2) - i\omega\gamma_j}. \qquad (2.15)$$

The frequency dependences of the real and imaginary components of the permittivity for a single natural frequency ω_j are shown in Fig. 2.2.

We shall now consider specific microscopic interactions between a light wave and a real semiconductor, which are important in various parts of the spectrum.

1. Electron transitions from the upper part of the valence band to the lower part of the conduction band are responsible for the shape of the absorption spectrum and for the dispersion near the fundamental absorption edge. Electron transitions can be direct, without phonon participation and without a change in the crystal momentum of an electron; or they can be indirect, in which the interaction with a phonon produces a considerable change in the crystal momentum if the wavelength of that phonon is short (indirect transitions involving long-wavelength optical phonons take place, like direct transitions, almost without any change in the crystal momentum of the electron). The various types of transition give rise to different frequency dependences of the absorption coefficient near the fundamental absorption edge. For the direct transitions, we have

$$\alpha \propto (\hbar\omega - \mathscr{E}_g)^n, \qquad (2.16)$$

where $n = 1/2$ if the transitions between the extrema of the conduction and valence bands are allowed by the selection rules, and $n = 3/2$ if such transitions are forbidden. The indirect transitions give rise to

$$\alpha \propto (\hbar\omega - \mathscr{E}'_g)^{n'}, \qquad (2.17)$$

where $\mathscr{E}'_g = \mathscr{E}_g \pm \hbar\omega_{ph}$ (ω_{ph} is the phonon frequency), and $n' = 2$, $n' = 3$, respectively, for the allowed and forbidden transitions. In the fundamental absorption range, light generates electron – hole pairs and this gives rise to several photoelectric effects.

2. Electron transitions from deep states in the valence band or to states high up in the conduction band govern the optical properties in the short-wavelength (ultraviolet) part of the spectrum. If the energy acquired by an electron is sufficiently high to overcome the surface barrier (work function), photoelectric emission (external photoeffect) is observed.

3. The interaction of light with free carriers involves two types of electron transition: within one band and between bands of the same type, for example, between valence bands, one of which (the upper) contains holes. Transitions within one band cannot be direct because the absorption associated with such transitions is possible solely due to scattering.

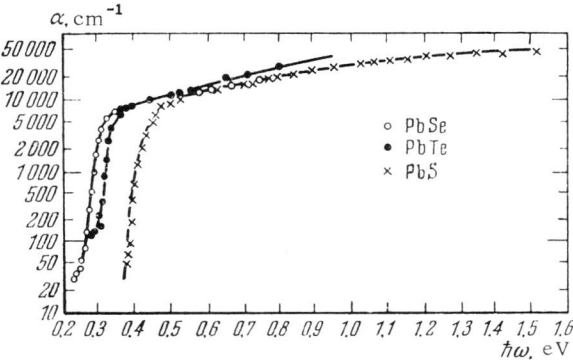

Fig. 2.3. Absorption coefficients of lead chalcogenides, plotted as a function of the photon energy near the fundamental absorption edge at T = 300°K.

4. The interaction of light with the lattice vibrations governs the absorption and dispersion in the far-infrared region. Electromagnetic radiation interacts effectively only with the transverse optical phonons and the natural frequency of the transitions is equal to the energy of these phonons.

C. Fundamental Absorption Edge

When the photon energy is close to the forbidden band width (in a nondegenerate semiconductor), the absorption increases rapidly with increasing frequency. The fundamental absorption edge of lead chalcogenides has been investigated by many workers [64, 95-101, 104, 108]. The dependence of the absorption coefficient of lead chalcogenides on the photon energy near the fundamental absorption edge was studied in detail by Scanlon [98, 99] at room temperature. Scanlon's curves are given in Fig. 2.3. In spite of the small areas of the very thin (down to 1 μ) crystals that were investigated, high-resolution absorption results were obtained by the infrared microscopy method.

If the top of the valence band and the bottom of the conduction band are located at the same point in the **k** space, electromagnetic radiation may induce direct electron transitions from one band to the other. In such transitions, a photon is absorbed and there is practically no change in the crystal momentum of an electron. If the matrix element of the momentum operator for wave functions

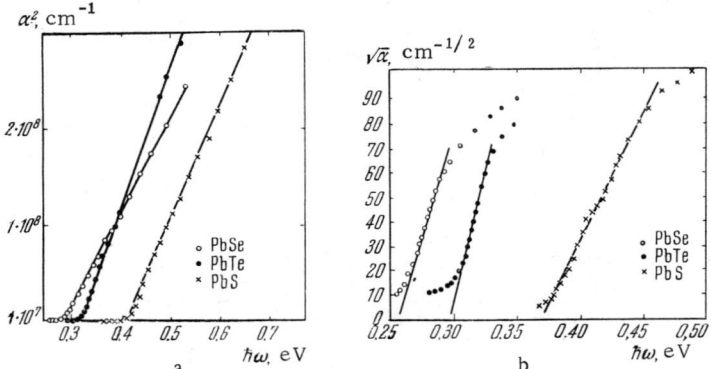

Fig. 2.4. Absorption coefficients, plotted as a function of the photon energy, in the region of direct (a) and indirect (b) transitions [99].

corresponding to the valence and conduction band edges is not equal to zero (allowed transitions), the dependence of the absorption coefficient on the energy of light quanta near the fundamental edge should be given by the formula

$$\alpha \propto (\hbar\omega - \mathscr{E}_g)^{1/2}. \qquad (2.16a)$$

Plotting α^2 along the ordinate axis and $\hbar\omega$ along the abscissa, Scanlon obtained straight lines for values of the absorption coefficient higher than 3000 cm^{-1} (Fig. 2.4a), whose intercepts on the abscissa are equal to the forbidden band width of each of the three chalcogenides. The values of $\mathscr{E}_g$, found in this way, are 0.32 eV for PbTe, 0.29 eV for PbSe, and 0.41 eV for PbS.

Generally speaking, the direct-transition threshold in ionic crystals may be somewhat higher than the energy corresponding to the forbidden band width because the polarization of the crystal lattice does not change in the short time represented by the interaction of a photon with an electron (cf., for example, [133]). The difference between the forbidden band width and the direct-transition threshold is of the order of the polaron energy. Estimates of the polaron energies give 0.002 eV for PbTe, 0.004 eV for PbSe, and 0.008 eV for PbS, i.e., these energies are smaller than the error in the determination of $\mathscr{E}_g$.

The indirect allowed transitions, in which a phonon is absorbed, should satisfy the following relationship near the absorption edge

$$\alpha \propto (\hbar\omega - \mathscr{E}_g')^2. \qquad (2.17a)$$

The dependence of $\alpha^{1/2}$ on $\hbar\omega$ is given by straight lines for values of the absorption coefficient lower than 3000 cm^{-1} (Fig. 2.4b). The indirect-transition threshold $\mathscr{E}_g'$ is only 0.03-0.04 eV smaller than the forbidden band width $\mathscr{E}_g$ and is 0.29, 0.26, and 0.37 eV for PbTe, PbSe, and PbS, respectively [98, 99]. The small difference between the direct and indirect transition thresholds indicates that the conduction and valence band edges are located at the same point in the **k** space and that indirect transitions involve the absorption of the long-wavelength optical phonons, whose energy is 0.02-0.03 eV.

The fundamental absorption edge shifts in the direction of shorter wavelengths when the temperature is raised. This is because, in contrast to the majority of semiconductors, the forbidden bands of lead chalcogenides broaden when the temperature is increased. The temperature dependence of the forbidden band width in the 77-300°K range is linear and fairly strong: $\partial \mathscr{E}_g/\partial T = 4 \cdot 10^{-4}$ eV/deg for all three chalcogenides [64, 95, 97, 123].

At temperatures above 400°K, Gibson [64] found that the temperature dependence of the forbidden band width is no longer linear. An investigation of the fundamental absorption edge, reported in [123], has indicated that the forbidden band width of PbTe above 400 °K is approximately constant and equal to 0.36 eV. Similar results have been obtained from an analysis of the thermal conductivity (cf. §4.2B). The nonlinearity of the temperature dependence of the forbidden band width of PbTe is due to the presence of two valence bands, whose mutual positions vary with temperature. This point will be discussed in greater detail in connection with the Hall effect in p-type PbTe (cf. §3.2B).

Investigations of the fundamental absorption edge under hydrostatic pressure have indicated that the forbidden band width decreases with increasing pressure and the derivative $\partial \mathscr{E}_g/\partial P$

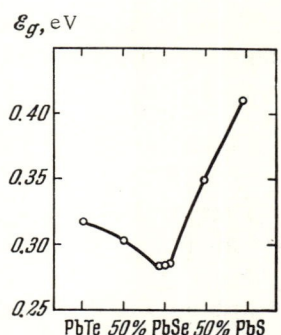

Fig. 2.5. Forbidden band width as a function of composition of PbTe – PbSe and PbSe – PbS alloys [99].

is $-(8\text{-}9.6)\cdot 10^{-6}$ eV·cm²·kg⁻¹ for PbS [125, 129] and $-(7.6\text{-}9.0)\cdot 10^{-6}$ eV·cm²·kg⁻¹ for PbTe [134], in satisfactory agreement with the results of measurements of the electrical properties under pressure (cf. §3.3A).

The fundamental absorption edge of heavily doped samples changes its shape because of the influence of free carriers. One effect of free carriers is an additional absorption (cf §2.2D). Since the coefficient representing the absorption by free carriers usually exhibits a power law dependence on the wavelength (known from measurements of the absorption at frequencies below the edge), we can subtract from the measured value of the absorption coefficient that component which is associated with the free carriers and thus calculate the fundamental absorption. Another effect, observed in heavily doped degenerate samples (to make the case specific, we shall consider an n-type semiconductor), is the filling with electrons of that part of the conduction band which lies below the Fermi level so that electron transitions between the band edges become impossible because of the Pauli exclusion principle. This shifts the fundamental absorption edge in the direction of shorter wavelengths. The absorption coefficient can be calculated, allowing for this effect, multiplying the coefficient of a pure sample by a factor

$$\left\{1 + \exp\frac{\left(1 + m^*_{d_n}/m^*_{d_p}\right)\zeta - (\hbar\omega - \mathscr{E}_g)}{\left(1 + m^*_{d_n}/m^*_{d_p}\right)k_0 T}\right\}^{-1},$$

where ζ is the Fermi level, measured from the bottom of the conduction band; $m^*_{d_n}$ and $m^*_{d_p}$ are the density-of-states effective masses of electrons and holes, defined by Eq. (4.9).

Analysis of the shift of the fundamental absorption edge due to the filling of the conduction band in PbS has been reported in [122, 127]. A comparison of the theoretical and experimental dependences has made it possible to find the number of electron constant-energy ellipsoids. The number (four) obtained indicates that the extrema in PbS are located at the boundary of the Brillouin zone along the <111> directions. The temperature coefficient of the fundamental absorption edge of degenerate semiconductors is smaller than $\partial \mathscr{E}_g/\partial T$ of pure materials. Similar investigations of PbSe and PbTe were carried out by Ukhanov [122].

The forbidden band widths of lead chalcogenides, determined from the absorption band edge, depend nonmonotonically on the atomic number of the group VI element. This would not be surprising were the band structures of the chalcogenides basically different, but a large number of results indicates a similarity of the band structure for all three compounds. The similarity of the energy band structure was also pointed out by Scanlon [99] on the basis of the absorption edge data for PbTe – PbSe and PbSe – PbS alloys. It is evident from Fig. 2.5 that pure PbSe has the narrowest forbidden band and that the band depends approximately linearly on the composition of these alloys. The curve in Fig. 2.5 shows no kinks, which would be expected if the nature of the band structure were to change with composition, as is the case in germanium – silicon alloys.

The fundamental absorption edge has been investigated also for alloys of lead chalcogenides and other $A^{IV}B^{VI}$ compounds (germanium and tin chalcogenides) [135, 136]. Interesting results have been obtained for $Pb_xSn_{1-x}Te$ alloys [136, 137]. It has been found that the forbidden band width is a linear function of the composition x:

$$\mathscr{E}_g(x) = |x\mathscr{E}_g(\text{PbTe}) - (1-x)\mathscr{E}_g(\text{SnTe})|$$

(cf. Fig. 6.8). The forbidden band width of SnTe at room temperature has been determined in [138] from the tunnel effect; its value is 0.18 eV. The linear relationship between the forbidden band width and the composition indicates that the band width of PbTe decreases when lead atoms are replaced with tin and that, at some value $x = x_{cr}$, this band becomes equal to zero and then again increases. The alloys investigated had compositions in the range $x > x_{cr}$. A similar relationship has been obtained for the same alloys at low temperatures from a study of the laser effect (cf. §2.2D). The temperature dependence of the forbidden band width of these alloys has been found to be the same as the dependence of PbTe, in contrast to the forbidden band width of SnTe, which decreases with increasing temperature [138]. The conclusions drawn by Dimmock et al. [137] from these investigations are discussed in Chap. VI.

Investigations of the dispersion of lead chalcogenides near the fundamental absorption edge [108, 128] have indicated that the refractive index, plotted against a function of frequency, has a maximum near $\hbar\omega = \mathscr{E}_g$. When the temperature is increased, this maximum shifts in the direction of higher photon energies. The forbidden band width and its temperature dependence, determined from the maximum of the refractive index [128], are in agreement with the results found from the absorption measurements.

Thus, investigations of the fundamental absorption edge have made it possible to determine the forbidden band width as well as its pressure and temperature dependences and to draw the following main conclusions about the nature of the energy band structure of lead chalcogenides:

1) the top of the valence band and the bottom of the conduction band are at the same point in the Brillouin zone;
2) the matrix element of the momentum operator for states corresponding to the band edges is not equal to zero.

D. Dispersion and Absorption at Long Wavelengths

The absorption and dispersion at long wavelengths, far from the fundamental absorption edge, are governed by the interaction of light with free carriers and lattice vibrations [107, 108, 110-114, 116-118, 120-122, 128]. The expression for the complex permittivity in this range of wavelengths can be represented thus [116]:

$$\varepsilon^*(\omega) = \varepsilon_\infty - \varepsilon_\infty \frac{\omega_p^2}{\omega} \frac{\omega - i\gamma}{\omega^2 + \gamma^2} + \frac{(\varepsilon_0 - \varepsilon_\infty)\omega_t^2}{\omega_t^2 - \omega^2 - i\gamma'\omega}. \qquad (2.18)$$

The first term, ε_∞, on the right-hand side of Eq. (2.18) represents the real and frequency-independent contribution of interband (band – band) electron transitions to the long-wavelength permittivity. The second term in the above equation is due to transitions within one

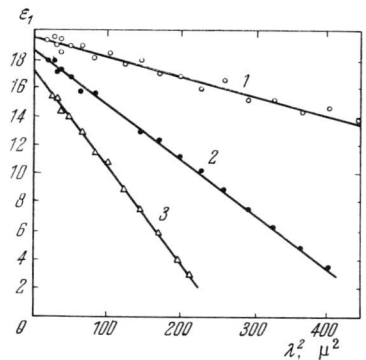

Fig. 2.6. Experimental dependence of the real component of the permittivity of n-type PbS on the wavelength [116] for samples with different electron densities: 1) $n = 4.0 \cdot 10^{17}$ cm^{-3}; 2) $n = 3.4 \cdot 10^{18}$ cm^{-3}; 3) $n = 9.2 \cdot 10^{18}$ cm^{-3}.

band and it includes the plasma frequency

$$\omega_p = \left(\frac{4\pi n e^2}{m^* \varepsilon_\infty}\right)^{1/2}. \quad (2.19)$$

The second term is proportional to the free-carrier density n and inversely proportional to the effective carrier mass m*. The damping constant γ is equal to the reciprocal of the relaxation time, which can be estimated from the carrier mobility. The third term in Eq. (2.18) describes the interaction of light with the transverse optical phonons of frequency ω_t. The damping constant of these phonons (vibrations) is denoted by γ'.

At frequencies which satisfy the inequality $\omega^2 \gg \gamma^2$ the contribution of free carriers to the electric susceptibility is, according to Eqs. (2.3) and (2.18), proportional to ω^{-2} and equal to

$$\chi_{fc} = -\frac{ne^2}{m^* \omega^2}. \quad (2.20)$$

This quantity can be determined experimentally because it represents the difference between the electric susceptibilities of pure and heavily doped semiconductors. The electric suscepbility in the region where $\mathscr{K}^2 \ll \mathscr{N}^2$ can be found from measurements of the reflection coefficient, as shown in §2.2A. The frequency dependence of χ_{fc}, described by Eq. (2.20), has been observed in the wavelength range 5-25 μ for PbTe [111-113] and PbS [108, 116].

It is evident from Fig. 2.6 that the dependence of the real component of the permittivity ε_1 on the square of the wavelength λ^2 is linear. Extrapolating this linear dependence to $\lambda = 0$ ($\omega = \infty$), we can find the high-frequency permittivity ε_∞. Such a procedure, as well as other extrapolation methods, all give values of ε_∞ which agree to within 10% [108, 113, 116, 122, 128]. The value of ε_∞ decreases slowly with increasing carrier density and temperature. When the temperature is increased from 77 to 300°K, the high-frequency permittivity decreases from 38 to 33 for PbTe, from 27 to 24 for PbSe, and from 19 to 17 for PbS [128]. The temperature dependence of ε_∞ is due to the variation of the forbidden band width (cf. §2.2E) and is in agreement with the empirical relationship of Moss [130]:

$$\varepsilon_\infty^2 / \mathscr{E}_g = \text{const.} \qquad (2.21)$$

In pure samples, at wavelengths equal to several microns, the contribution of free carriers to the polarizability is small and the refractive index depends weakly on the frequency and is close to the value $\mathscr{N}_\infty = \varepsilon_\infty^{1/2}$. Measurements of the refractive index have been reported in [95, 104, 106, 108, 113, 128]. Lead chalcogenides have relatively large refractive indices. At room temperature and $\lambda \approx 3\mu$, Avery's measurements of the reflection coefficient [95] yielded values of the refractive index $\mathscr{N}$ equal to 5.35, 4.59, and 4.10, for PbTe, PbSe, and PbS, respectively. These values are in good agreement with the results of measurements by an interference method [104] but are somewhat lower than the values obtained using a prism [106]. Since the refractive index depends weakly on the frequency, the reflection coefficient is approximately constant at frequencies below the fundamental absorption edge.

It follows from Eq. (2.20) that the measurements of χ_{fc} in the range $\omega^2 \gg \gamma^2$ make it possible to determine the effective mass of free carriers. This method, suggested by Spitzer and Fan [139], was used by Dixon and Riedl to determine the effective mass of electrons and holes in PbS [116] and of holes in PbTe [111-113]. For an ellipsoidal band, this methods give the electric-susceptibility effective mass m_χ^*, which is a combination of the longitudinal and transverse components of the effective mass and is given by the formula

$$\frac{3}{m_\chi^*} = \frac{2}{m_\perp^*} + \frac{1}{m_\parallel^*}. \qquad (2.22)$$

The same effective mass m_χ^* occurs in the expression for the electrical conductivity and, therefore, it is also called the conductivity effective mass.

Sec. 2.1]   ABSORPTION AND REFLECTION OF LIGHT   59

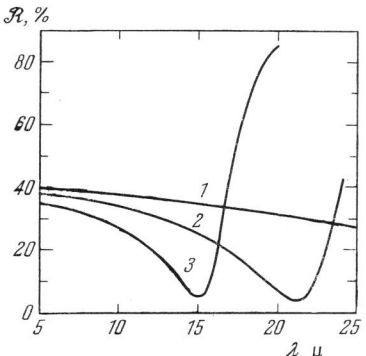

Fig. 2.7. Experimental dependence of the reflection coefficient of n-type PbS [116] on the wavelength in the near-infrared region at T = 85°K. Curve numbers 1-3 have the same meaning as in Fig. 2.6.

If the band is spherical (isotropic) but the effective mass depends on the energy (the band is nonparabolic), the effective mass in the degenerate case is

$$m^*_\chi = \left(\frac{1}{\hbar^2}\frac{1}{k}\frac{d\mathscr{E}}{dk}\right)^{-1}_{\mathscr{E}=\zeta}. \quad (2.23)$$

In more complex cases, the formulas for the electric-susceptibility effective mass m_χ^* will be found in the paper of Dixon and Riedl [113] for the Cohen model and in Appendix A for the Kane model (these models are discussed in §6.2B).

The advantage of this method of finding the effective mass is that we need not know the scattering mechanism or the carrier mobility. A different method for the determination of the effective mass from the infrared reflection data, put forward by Lyden [140], consists of the measurement of the position of the reflection minimum. If the reflection is weak ($\mathscr{K}^2 \ll \mathscr{N}^2$), the frequency is high compared with the reciprocal of the relaxation time ($\omega^2 \gg \gamma^2$), the dispersion is governed by free carriers and it follows from Eqs. (2.3), (2.8), and (2.18) that the reflection coefficient vanishes at a frequency

$$\omega = \omega_p \left(\frac{\varepsilon_\infty}{\varepsilon_\infty - 1}\right)^{1/2}, \quad (2.24)$$

which is close to the plasma frequency ω_p, given by Eq. (2.19), which includes the effective mass. In a more general case, a definite reflection minimum (Fig. 2.7) is observed near the plasma frequency; the position of this minimum is related to the value of the effective mass by a more complex expression [140]. Several workers have used this method to determine the effective mass [114, 122, 123].

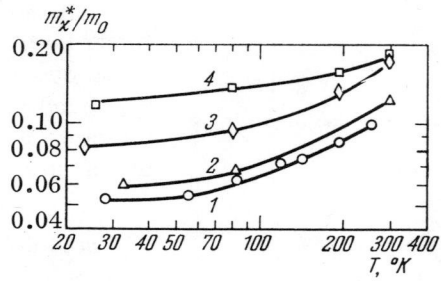

Fig. 2.8. Temperature dependence of the electric-susceptibility effective mass of holes in PbTe [113]: 1) $p = 3.5 \cdot 10^{18}$ cm^{-3}; 2) $p = 5.7 \cdot 10^{18}$ cm^{-3}; 3) $p = 1.5 \cdot 10^{19}$ cm^{-3}; 4) $p = 4.8 \cdot 10^{19}$ cm^{-3}.

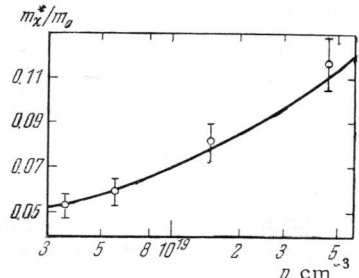

Fig. 2.9. Dependence of the electric-susceptibility effective mass of holes in PbTe [113] on the carrier density at $T = 30°K$.

The effective masses of electrons and holes have been found from the infrared reflection measurements for all three lead chalcogenides [113, 114, 116, 122]. When the temperature is raised from 77 to 300°K, the effective masses of electrons and holes increase from $0.05m_0$ to $0.10m_0$ for PbTe [113, 114] and from $0.10m_0$ to $0.15m_0$ for PbS [116]. The temperature dependence of the effective mass is nonlinear and the rate of rise increases considerably at higher temperatures (Fig. 2.8).

The dependence of the effective mass of holes in PbTe on the hole density is fairly strong [113] (Fig. 2.9), which indicates a strong nonparabolicity of the valence band. Dixon and Riedl [113] explained this dependence, in the temperature range 30-150°K, by the valence band nonparabolicity using the Cohen model (cf. §6.2B). However, in an analysis of the data obtained at higher temperatures it has been necessary to assume that, as the temperature is increased, some of the holes are transferred to a second valence band with a larger effective mass, lying deeper than the upper valence band. This second valence band has been deduced first from the Hall effect measurements (cf. §3.2B). Since the second valence band has little effect at low temperatures, it follows that at these temperatures the energy gap between the two valence bands $\Delta\mathscr{E}$ is not less than 0.19 eV. An agreement between a theoretical curve and the experimental data obtained at room temperature is reached assuming a smaller gap $\Delta\mathscr{E} = 0.08$ eV. This value has been obtained from an analysis of the absorption of light [107] (this will be discussed later). Thus, the energy gap between the valence bands decreases rapidly when the temperature is increased.

PbS also exhibits a strong dependence of the effective mass on the carrier density [116, 122]. This is difficult to explain because the gap between the Fermi level and the band edge in the investigated samples is usually small (0.06 eV) for a carrier density of 10^{19} cm^{-3} and therefore an appreciable nonparabolicity should not be expected, as confirmed by magneto-optical investigations (cf. §5.4C).

It follows from Eqs. (2.4), (2.7), and (2.8) that the absorption coefficient due to the interaction of light with free carriers at frequencies $\omega^2 \gg \gamma^2$ is

$$\alpha = \frac{4\pi n e^3}{c \mathscr{N} \mu m^{*2} \omega^2}, \qquad (2.25)$$

where μ is the carrier mobility, equal to $e/\gamma m^*$. The experimentally obtained dependences of the absorption coefficient on ω are described satisfactorily by a power function for all lead chalcogenides, with the exception of p-type PbTe, although the power exponent differs somewhat from -2 in some cases [104, 107, 118, 122].

Equation (2.25) has been deduced by a classical calculation; the same relationship is obtained by a more rigorous quantum-mechanical calculation only in some special cases. A quantum-mechanical calculation gives Eq. (2.25) only in the case of the energy-independent relaxation time, but for more realistic cases it yields formulas whose form (including the frequency dependence) is governed by the scattering mechanism. Comparison of the experimental curves for PbS with the results of quantum-mechanical calculations has been carried out only in a few investigations [122, 124], while in the majority of studies [104, 107, 118, 122] the classi-

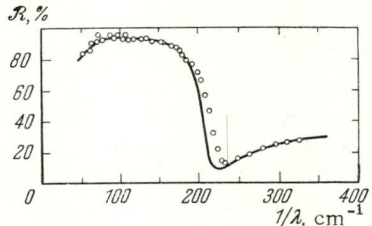

Fig. 2.10. Reflection spectrum of a PbS sample at room temperature in the far-infrared region [116].

cal formula (2.25) has been employed, and the effective mass obtained for all three lead chalcogenides has been found to be in satisfactory agreement with the results of the reflection and Faraday effect measurements (cf. §5.4A).

The proportionality $\alpha \propto \omega^{-2}$ is observed for p-type PbTe only at wavelengths considerably longer than 8 μ. In the intermediate part of the spectrum, between the fundamental absorption edge and the region where this dependence is observed, there is an additional absorption which has been attributed by Riedl [107] to electron transitions from the second to the first (upper) valence band. An additional absorption maximum is observed at 7.5 μ (at 300°K), which corresponds to an energy gap of 0.1 eV between the two valence bands. When the temperature is increased the additional absorption becomes stronger, which indicates that transitions between the valence bands are indirect.

Measurements of the absorption and dispersion in the far-infrared region (up to $\lambda = 2000$ μ) have been carried out for PbS [102, 103, 116, 117, 120], for PbSe [121], and for PbTe [126]. In this part of the spectrum, the optical properties are governed mainly by the interaction of light with the lattice vibrations, particularly in pure samples, in which the influence of free carriers is weak. The interaction with the lattice vibrations gives rise to a maximum in the reflection spectrum (Fig. 2.10), whose position (the wavelength of reststrahlen) is governed by the energy of the transverse optical phonons. An analysis carried out using Eq. (2.18), assuming a harmonic oscillator of frequency ω_t, yields the reststrahlen wavelength of 320 μ for PbTe [126], 230 μ for PbSe [121], and 150 μ for PbS [116, 117, 120]. This corresponds to transverse optical phonon energies of 3.9, 5.4, and 8.2 meV for PbTe, PbSe, and PbS, respectively, in agreement with the results obtained by neutron spectroscopy [141].

Knowing the optical phonon energies and the high-frequency permittivity, we can find the static permittivity ε_0 from the well-known Lyddane – Sachs – Teller relationship [142]:

$$\frac{\varepsilon_0}{\varepsilon_\infty} = \left(\frac{\omega_l}{\omega_t}\right)^2, \qquad (2.26)$$

where ω_l is the frequency of the longitudinal optical phonons. The frequencies of the longitudinal optical phonons have been determined from the tunnel effect (cf. §3.4A), recombination radiation (cf. §2.2C), and neutron spectroscopy [141]. The application of the Lyddane – Sachs – Teller relationship (2.26) gives the following values of the static permittivity: 400 for PbTe [126], 250 for PbSe [128], and 175 for PbS [128] at room temperature. These values of ε_0 are unusually high and they give rise to several effects which are manifested in the electrical properties of lead chalcogenides (cf. Chap. III). The value obtained for ε_0 for PbTe is in excellent agreement with the result deduced from measurements of the barrier capacitance of a p – n junction (cf. §3.4B). Measurements of the surface impedance [143] have yielded $\varepsilon_0 \leq 200$ for PbTe, which is only in order-of-magnitude agreement with the value just quoted. An analysis of the reflection spectrum of pure PbS between 500 and 2000 μ [117] has yielded $\varepsilon_0 = 150$, which is in satisfactory agreement with the value quoted above.

E. Dispersion and Absorption at Short Wavelengths

The reflection and absorption in lead chalcogenides have also been investigated at short wavelengths right down to the wavelength corresponding to a photon energy of 25 eV [64, 95, 105, 119]. The room-temperature reflection spectra obtained by Cardona and Greenaway [119] are shown in Fig. 2.11. The spectra of these compounds are similar and they each have several maxima. The main peak is observed at an energy $\mathscr{E}_2$, which is 2.5 eV for PbTe, 3.1 eV for PbSe, and 3.7 eV for PbS. An investigation of this peak and liquid-nitrogen temperature has revealed a fine structure. In addition to the main peak, there is a peak at energies higher than the fundamental absorption edge, on the long-wavelength side of $\mathscr{E}_2$, as well as several peaks at shorter wavelengths. In contrast to the forbidden band width, the photon energies at which these peaks are observed increase monotonically along the series PbTe → PbSe → PbS.

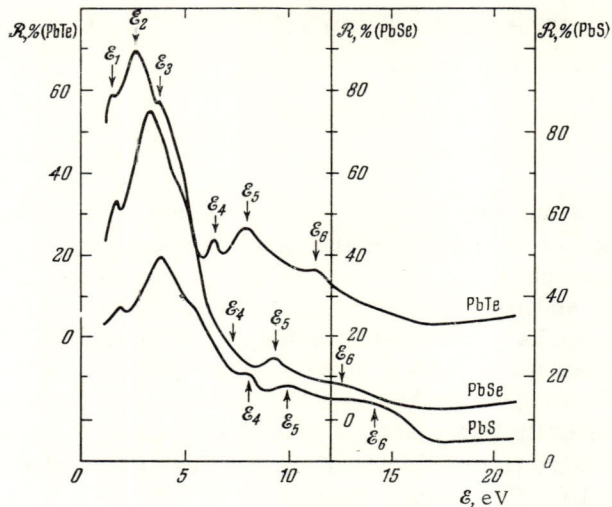

Fig. 2.11. Reflection spectra of PbTe, PbSe, and PbS samples at room temperature in the ultraviolet part of the spectrum.

The temperature dependences of the positions of the peaks $\mathscr{E}_1 - \mathscr{E}_3$ are weak compared with the temperature dependence of the forbidden band width. The other peaks have been investigated only at room temperature. The energy $\mathscr{E}_2$ of all three lead chalcogenides increases with increasing temperature at a rate of $\partial \mathscr{E}_2 / \partial T \approx 10^{-4}$ eV/deg. The positions of the $\mathscr{E}_1$ and $\mathscr{E}_3$ peaks of PbTe and of $\mathscr{E}_1$ of PbS depend in a similar manner on temperature. Changes in the positions of the peaks $\mathscr{E}_1$ of PbSe and $\mathscr{E}_3$ of PbS and PbSe are smaller and lie within the limits of the experimental error.

The absorption spectra, deduced from the transmission or calculated using the dispersion relationships, are similar.

The reflection spectra of SnTe and GeTe [119] resemble those of lead chalcogenides but the energies of the peaks $\mathscr{E}_1 - \mathscr{E}_6$ decrease when Pb is replaced with an element of a lower atomic number.

According to theoretical treatments (cf., for example, [144]), the reflection and absorption maxima appear near energies corresponding to vertical interband transitions of electrons at critical points. These critical points for a pair of branches of the energy spectrum of electrons are found as follows. Let $\mathscr{E}_{ij}(\mathbf{k}) = \mathscr{E}_i(\mathbf{k}) - \mathscr{E}_j(\mathbf{k})$

be the difference between the electron energies in two bands (filled and empty) for a given crystal momentum **k**. By definition, at the critical point, we have

$$\nabla_k \mathscr{E}_{ij}(\mathbf{k}) = 0 \tag{2.27}$$

and the density of states, considered as a function of $\mathscr{E}_{ij}$, has a singularity (a discontinuity in the derivative). The critical points are usually located at the center or at the edge (boundary) of the Brillouin zone, where nondegenerate bands always obey the relationships $\nabla_k \mathscr{E}_i = 0$.

If electron transitions at a given critical point are allowed by the selection rules, a reflection maximum is observed near the energy corresponding to this point. More exactly, the critical points correspond to maxima of the function $\varepsilon_2 \mathscr{E}^2$, which are close to the reflection coefficient maxima.

Investigations of the ultraviolet reflection can be used to detect the critical points and to measure the corresponding energy differences. Thus, such investigations are one of the very few methods for studying electron states which are far from the edges of the forbidden band. Unfortunately, this method does not permit us to plot even partly the energy band structure but requires knowledge of this structure. It is usually employed to check the results of theoretical calculations of the band structure. The relationship between the optical measurements in the ultraviolet region and the theoretically obtained energy bands is discussed later (cf. §6.1D). All the experimental values of the quantities $\mathscr{E}_1 - \mathscr{E}_6$ are given in that subsection.

Knowledge of the absorption in the range of energies above the fundamental absorption edge makes it possible to calculate the infrared permittivity ε_∞ [116, 119, 145]. The corresponding dispersion formula is

$$\varepsilon_\infty^{1/2} = 1 + \frac{\hbar c}{\pi} \int_{\mathscr{E}_g}^{\infty} \frac{\alpha(\mathscr{E}) d\mathscr{E}}{\mathscr{E}^2}. \tag{2.28}$$

The calculated values of ε_∞ and its temperature dependence are in reasonable agreement with those found from the optical measurements in the infrared range. It is found that the main contribution to the high-frequency permittivity is made by transitions corresponding to the $\mathscr{E}_2$ peak. However, the weak temperature dependence of this energy does not make it possible to explain the observed temperature dependence of ε_0 by means of the $\mathscr{E}_2$ transitions alone; this temperature dependence is found to be due to transitions near the fundamental absorption edge because of the strong temperature dependence of the forbidden band width.

2.2. Photoelectric Effects. Carrier Recombination and Recombination Radiation

A. Photoelectric Effects and the Determination of the Recombination Parameters

Photoelectric effects are changes in various electrical properties or the appearance of emf's due to illumination. Photoelectric effects are of practical importance and are being investigated vigorously, particularly in the case of PbS. The photoelectric effects in lead chalcogenides are widely used in the detection and measurement of infrared radiation. Measurements of these photoelectric properties make it possible to determine recombination parameters of semiconductors, investigate recombination mechanisms, and find band structure parameters, particularly the forbidden band width.

Photoconductivity is an increase in the electrical conductivity of a sample due to illumination. As a rule, the photoconductivity in uniform samples is due to an increase of the density of free carriers during illumination, but there are other photoconductivity mechanisms, for example, the heating of carriers. In the fundamental absorption region, light generates electron–hole pairs. The photoconductivity of polycrystalline films exhibits a number of special features, which we shall consider later. The photoconductivity of lead chalcogenides has been investigated by many workers (cf., for example, Moss's review [146], which gives a detailed list of references to publications which appeared up to 1955; other reviews [147-160] are also worth consulting).

In a uniform, thick sample exhibiting strong absorption of light and weak surface recombination, the intrinsic photoconductivity integrated over the thickness of the sample is given by the formula

$$\sigma_{pc} = (1 - \mathcal{R}) \beta \mathcal{J}_0 e (\mu_n \tau_n + \mu_p \tau_p), \qquad (2.29)$$

where $\mathcal{R}$ is the reflection coefficient; β is the photoconductivity ("internal photoeffect") quantum yield; $\mathcal{J}_0$ is the number of quanta incident on a unit area per unit time; μ_n and μ_p are the mobilities and τ_n and τ_p are the lifetimes of electrons and holes.

The difference between the lifetimes of electrons and holes is due to the influence of capture centers (traps). Carriers may be captured and remain at traps until a trap captures a carrier of opposite sign, which results in recombination, or until the trapped carrier returns back to an allowed band. Carriers of the opposite sign, which give rise to recombination, are free and contribute to the photoconductivity. If the capture of minority carriers is a fairly frequent event and the residence time of carriers at traps is not short, then a considerable fraction of the minority carriers generated by illumination will be trapped. This makes recombination a difficult process and increases the density of free carriers and consequently increases the photoconductivity. Since the excess densities of free electrons and holes in the presence of traps are not equal and their lifetimes are different as well, we obtain,

$$\mathcal{R}_r = \frac{\Delta n}{\tau_n} = \frac{\Delta p}{\tau_p}, \qquad (2.30)$$

where $\mathcal{R}_r$ is the number of recombination events per unit volume and per unit time; Δn, Δp are the excess densities of electrons and holes.

The state of the surface also affects the photoconductivity. Surface recombination reduces the effective lifetime of carriers generated near the surface and, therefore, the photoconductivity falls when the wavelength is decreased in the fundamental absorption region [151]. Surface trapping levels, like those in the interior, may increase the effective lifetime of the majority carriers and thus increase the photoconductivity.

In the absence of traps, the electron and hole lifetimes are equal ($\tau_n = \tau_p = \tau_0$) and the measurement of the photoconductivity in the spectral region where the quantum yield is equal to unity should give us the lifetime τ_0 if we use Eq. (2.29). This method has been used to determine the carrier lifetime in PbS [147] and PbSe [160] crystals at relatively low carrier densities.

Another method for the determination of carrier lifetime is based on a study of the photoconductivity kinetics. When the illumination ceases, the photoconductivity decays with time in accordance with the law $\exp(-t/\tau_0)$, provided traps are absent and surface recombination is weak. In the case of strong capture of minority carriers, the decay of the photoconductivity is governed (instead of by τ_0) by the majority-carrier lifetime if the occupancy of traps varies linearly with the carrier density. The carrier lifetime in PbS was deduced from the photoconductivity decay by Moss [161].

The photoconductivity in thin polycrystalline films, used in infrared photoresistors, is considerably higher than in single crystals. Oxidation plays an important role in the increase of the photoconductivity in films. The surface of photoconducting films is usually covered by layers of lead oxides and other lead compounds. The photoconductivity in polycrystalline films can be explained on the basis of two possible mechanisms.

The first of these mechanisms consists of an increase of the carrier density during illumination and does not differ from the photoconductivity mechanism in uniform bulk crystals [155, 156]. Using the carrier-density theory, we find that oxygen acts as an impurity which produces traps in the interior and on the surface of a film and, moreover, reduces the dark conductivity by filling vacancies of the group IV element so that the photoconductivity becomes more pronounced.

The second photoconductivity mechanism is associated with the complex structure of films and consists of a change in the resistance due to a reduction in the height of potential barriers during illumination. These barriers may appear at boundaries between crystallites [154] or between p- and n-type regions in a nonuniform sample [153]. Oxygen favors the formation of potential barriers. Films which exhibit n-type conduction before contact with oxygen may either retain this conduction or be converted to p-type, depending on the amount of absorbed oxygen. The absorption of oxygen makes films strongly nonuniform.

The relative contribution of each of these two mechanisms to the photoconductivity of films has not yet been determined reliably. The nonlinearity of the current – voltage characteristics of films indicates that barriers do exist but it does not mean that barriers are modulated by illumination [157]. Woods [158] found that the relative changes in the Hall coefficient and resistivity of PbS are equal to within 6%; hence, we may conclude that the main photoconductivity mechanism is that governed by the carrier density. Rogachev and Chechurin investigated the field effect [159] and concluded that the main cause of the photoconductivity is an increase in the density of the majority carriers but that modulation of the intercrystallite barriers by light can also play some role. An additional support for the barrier theory is provided by a change in the mobility due to illumination, deduced by Gobrecht et al. [524] from measurements of the Hall coefficient and field effect in PbSe films.

Measurements of the photoconductivity at very high frequencies, of the order of 10^{10} cps, at which the barrier resistances are shunted by the barrier capacitances, so that the barrier photoconductivity cannot be observed, are of very great interest from the point of view of the determination of the nature of the photoconductivity of films. By comparing the photoconductivity at low and high frequencies, Erofeichev and Kurbatov [149, 150] demonstrated that both (carrier-density and barrier) photoconductivity mechanisms are important and that their relative importance is determined by the quality of the sample.

Another important photoelectric effect is the photovoltaic effect at potential barriers, i.e., the appearance of a photo-emf due to the separation of photoinjected electron – hole pairs by the electric field of a barrier. A barrier may be a p – n junction. The photovoltaic effect at p – n junctions in PbS crystals was observed by Starkiewicz et al. [162]; the same effect was observed in PbSe by Kimmitt and Prior [163] and in PbTe by Laff [164]. The photovoltaic effect in p – n junctions is employed in photodiodes (cf. §7.2B). Apart from p – n junctions, the photo-emf can also be observed at barriers formed near point contacts between metals and semiconductors (cf. Moss's review [146]). This effect was used by Moss [161] to determine the lifetime in PbS. Moss determined the diffusion length, and hence the carrier lifetime, from the dependence of the photo-emf at a point probe on the distance between the probe and an illuminated strip.

High-voltage photo-emf's are observed in films [165-171] prepared and treated in oxygen in the same way as photoconducting films. The values of these photo-emf's reach 2-3 V, i.e., they are greater than the value of $\mathscr{E}_g/e$. The photo-emf is observed in those cases when a dc current is passed through a film during deposition, or when a film is deposited at some angle to the surface of the substrate, or when light is not incident normally to the film surface. The cause of these high photo-emf's in films is the complex structure of the films, but details of the actual mechanism have not yet been fully determined. One of the possible mechanisms is the addition of the photo-emf's of a large number of p—n junctions which appear in a film because of its inhomogeneity [171]. Another possible mechanism is the addition of the Dember photovoltages in a film with a sawtooth-shaped surface [172].

The photomagnetic effect, also known as the photomagnetoelectric effect, will be considered next. If electrons and holes are generated by light near the surface of a plate, they diffuse across the plate but, under steady-state conditions in the absence of a magnetic field, the simultaneous diffusion of carriers does not give rise to an electric current. A magnetic field deflects the electron and hole fluxes in different directions and, consequently, a photomagnetic current begins to flow along the plate.

The short-circuit current in the photomagnetic effect, observed in a p-type sample under the same conditions which apply to the photoconductivity formula (2.29), is

$$I_{\text{pme}} = (1 - \mathscr{R}) \beta \mathscr{J}_0 e (\mu_n + \mu_p) \frac{H}{c} (\tau_n D_n)^{1/2}, \qquad (2.31)$$

where H is the magnetic field intensity; D_n is the diffusion coefficient of the minority carriers (electrons). When the quantum yield is known and the surface recombination velocity is low, measurements of the photomagnetic effect make it possible to find the minority-carrier lifetime. If the electron and hole lifetimes are equal, a simultaneous measurement of the photoconductivity and of the photomagnetic effect makes it possible to determine, by means of Eqs. (2.29) and (2.23), the lifetime of carriers and the quantum yield [161]. The lifetime is then found from the ratio of the photo-

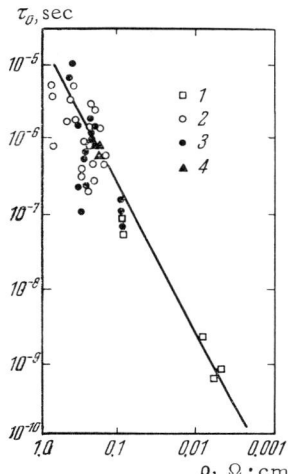

Fig. 2.12. Dependence of the carrier lifetime in natural PbS crystals on their resistivity [161]: 1) results obtained from measurements of the photomagnetic effect; 2) from the ratio of the photomagnetic effect and photoconductivity; 3) from the moving light spot method; 4) from the time of establishment of the photoconductivity.

magnetic effect and the photoconductivity, using the formula

$$\tau_0 = \left(\frac{\sigma_{pc}}{I_{pme}}\right)^2 \left(\frac{H}{c}\right)^2 D_n. \quad (2.32)$$

When this method is used the intensity of illumination and the quantum yield need not be known. Formula (2.32) is valid for any surface recombination velocity. The photomagnetic effect is particularly useful in the case of short carrier lifetimes when other methods for measuring these lifetimes are no longer applicable. It has been used as the main method in the measurement of lifetimes in PbS crystals [147, 161, 173]. Whenever possible the results have been compared with those obtained by other methods and the agreement has been found to be satisfactory. Lifetimes in PbSe crystals have also been determined from the ratio of the photomagnetic effect and photoconductivity [160].

The lifetimes obtained by Moss [161] for natural crystals of PbS at room temperature vary from $9 \cdot 10^{-6}$ sec to $6 \cdot 10^{-10}$ sec, proportionally to the square of the resistivity of the samples (Fig. 2.12) and the corresponding diffusion lengths lie within the range 1-100 μ. At high electron densities ($n > 2 \cdot 10^{18}$ cm^{-3}), Baryshev and Aver'yanov [147] found a weaker dependence of the lifetime on the carrier density, while at lower carrier densities ($n < 2 \cdot 10^{18}$ cm^{-3}) they found that $\tau_0 \propto n^{-2}$, in agreement with Moss's results. In heavily doped samples, the lifetime depends weakly on temperature but at $n < 2 \cdot 10^{18}$ cm^{-3} the lifetime decreases with increasing temperature. The lifetimes in p- and n-type samples with similar carrier densities are approximately equal.

Scanlon [173] observed a dependence of the carrier lifetime on the dislocation density in PbS samples with relatively low carrier densities. The lifetime is independent of the resistivity and inversely proportional to the dislocation density in the range of dislocation densities from 10^5 to 10^7 cm^{-2}. In crystals with the lowest dislocation densities, the carrier lifetime is equal to $2 \cdot 10^{-5}$ sec.

The time constant which governs the magnitude and kinetics of the photoconductivity of PbS films may be considerably higher than that of bulk crystals: it is of the order of $5 \cdot 10^{-4}$ to $5 \cdot 10^{-5}$ sec [1] but this is not due to a higher purity of the bulk material but due to the influence of trapping levels.

Surface recombination was found by Baryshev and Aver'yanov [147] to have no influence on the photoconductivity and photomagnetic effect of PbS samples. The agreement between the lifetimes, obtained from the photoconductivity and photomagnetic effect separately indicates the absence of trapping in PbS crystals [147].

The lifetimes in PbSe and PbTe crystals [160, 174] have not been investigated in as much detail as in lead sulfide. The lifetime in pure PbSe crystals [160] is $1.8 \cdot 10^{-6}$ sec, as found from the ratio of the photoconductivity and the photomagnetic effect. Measurements of the photoconductivity give $0.5 \cdot 10^{-6}$ sec, but this is the effective value which is not only due to the volume properties but is also governed by the surface recombination. Estimates of the surface recombination velocity from the dependence of the photoconductivity on the thickness of a sample have demonstrated that this velocity is less than 80 cm/sec for the cleanest etched surfaces. Mechanically polished surfaces exhibit surface recombination velocities not less than 220,000 cm/sec.

The lifetimes in polycrystalline PbSe and PbTe films are usually shorter than those in PbS films. The lifetime obtained for PbSe films from the photomagnetic effect [151] is $0.4 \cdot 10^{-6}$ sec for a surface recombination velocity of 200 cm/sec.

B. Recombination Mechanisms

In investigations of recombination mechanisms we must determine first whether the recombination effect takes place in one stage (interband recombination) or whether recombination involves intermediate energy levels associated with defects. Secondly, we must determine the mechanism of the loss of energy of the order of the forbidden band width, which is evolved during each recombination event.

In pure samples, only the interband recombination is possible, but in heavily doped samples recombination usually takes place via intermediate centers. There are several mechanisms of loss of the excess energy. In radiative recombination, this energy is dissipated in the form of a photon. The radiative recombination process is the converse of the fundamental absorption of light. This type of recombination can be direct, without a change in the crystal momentum of an electron being transferred from the conduction to the valence band, or indirect, with the participation of a phonon. If the radiative recombination is appreciable, a crystal can be seen to emit light. Radiative recombination is important in narrow-gap semiconductors, which include lead chalcogenides.

In impact recombination (Auger recombination) mechanism, the excess energy is transferred to a third carrier which takes part in a multiple collision. The impact mechanism is also important in lead chalcogenides. There are other nonradiative recombination mechanisms, involving the heating of the lattice or the excitation of plasma oscillations, but these processes are not observed in lead chalcogenides. Therefore, we shall discuss only the radiative and impact recombination mechanisms.

The interband (band – band) radiative recombination lifetime, τ_r, for a sample with intrinsic conduction is the limiting value of the lifetime which can be reached at a given temperature. The radiative recombination velocity is calculated from the data on the optical absorption and reflection using the Van Roosbroeck – Shockley formula, deduced from the principle of detailed balancing:

$$\mathcal{R}_r = \frac{n_i}{2\tau_r} = \frac{2c}{\pi^2}\left(\frac{k_0 T}{\hbar c}\right)^4 \int_{u_g}^{\infty} \frac{\mathcal{N}^3 \mathcal{K} u^3 \, du}{e^u - 1}, \quad (2.33)$$

where $u = \hbar\omega/k_0 T$; u_g corresponds to the fundamental absorption edge ($u_g \approx \mathscr{E}_g/k_0 T$).

Integration in Eq. (2.33) has been carried out numerically [98, 99, 176, 177]. Scanlon [98, 99] made calculations for room temperature and obtained values of τ_r, equal to $6.8 \cdot 10^{-6}$, $8.0 \cdot 10^{-6}$, and $63 \cdot 10^{-6}$ sec for PbTe, PbSe, and PbS, respectively. The value of $20 \cdot 10^{-6}$ sec, found experimentally for PbS samples with a very small number of defects [173], is quite close to the radiative limit. Baryshev's calculation [177] shows that the value of τ_r decreases rapidly with increasing temperature because of an increase in the photon density.

The relative contribution of the radiative recombination mechanism to the total recombination velocity was estimated by Washwell and Cuff [174] for PbTe and PbSe crystals from the recombination radiation intensity. The ratio of the number of emitted light quanta to the number of absorbed quanta (in the injection of the minority carriers by illumination) is equal to the ratio of the total and radiative lifetimes:

$$\frac{\mathcal{J}_r}{\mathcal{J}_0} = \frac{\tau_0}{\tau_r}. \tag{2.34}$$

This ratio reaches several percent for PbTe with a hole density less than $3 \cdot 10^{18}$ cm^{-3} at temperatures below 100°K. Extrapolation of this result leads to the conclusion that, at sufficiently low temperatures and carrier densities, the radiative recombination becomes the dominant mechanism.

The important role of the radiative recombination mechanism at low temperatures in all three lead chalcogenides has made it possible to observe stimulated (laser) recombination radiation [178-182], which will be described later in more detail. The high efficiency of the radiative recombination in lead chalcogenides is not only due to the narrow forbidden band, but also due to the relatively high (compared with InSb) density of states because of the larger effective mass and the many-ellipsoid band structure.

Estimates of the interband impact recombination velocity, obtained by Baryshev et al. [183, 184], gave room-temperature lifetimes of 10^{-4}, 10^{-5}, and 10^{-2} sec for intrinsic samples of PbTe, PbSe, and PbS, respectively. Hence, we may conclude that the impact mechanism is comparable with the radiative recombination process only in PbSe. The low efficiency of the impact recombination is due to the approximate equality of the effective electron and hole masses, which makes it difficult to satisfy the laws of con-

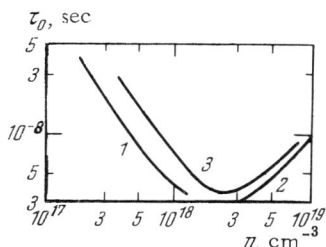

Fig. 2.13. Carrier-density dependence of the lifetime in synthetic n-type PbS crystals: 1, 2) theoretical curves for nondegenerate and degenerate samples, respectively [186]; 3) experimental curve [147].

servation of energy and momentum in triple collisions [178]. Moreover, the high permittivity reduces the interaction between carriers participating in such collisions.

Thus, in intrinsic samples as well as in extrinsic (doped) crystals, the dominant recombination mechanism at relatively low carrier densities and low temperatures is the interband radiative recombination.

Comparison of the results of theoretical estimates of the lifetime in the interband recombination case with the experimental values shows that the interband transitions cannot explain the lifetime values obtained for extrinsic samples with a sufficiently high carrier density and that, consequently, recombination must take place also via intermediate energy levels associated with lattice defects. Scanlon's experiments [173] show that recombination levels may be due to dislocations. However, in samples with low dislocation densities, recombination centers must obviously be of a different origin.

Baru and Khmel'nitskaya [185] calculated the lifetime as a function of the carrier density in PbS at room temperature on the assumption that the recombination takes place through local levels associated with deviations of lead sulfide from stoichiometry, i.e., the levels associated with vacancies. The activation energy of the levels has been assumed to be 0.03 eV, in agreement with the results of an analysis of the temperature dependence of the lifetime carried out by Baryshev and Aver'yanov [147]. It has been found that — in contrast to the interband recombination — the impact mechanism is more effective than the radiative recombination. The lifetime is inversely proportional to the square of the majority-carrier density and a value of $\tau_0 p^2 \approx 2 \cdot 10^{27}$ $cm^{-6} \cdot$ sec is obtained, compared with $\tau_0 p^2 \approx 3 \cdot 10^{27}$ $cm^{-6} \cdot$ sec obtained from measurements of Baryshev and Aver'yanov. Equally good agreement between the experimental data and the theoretical estimates of Pincherle was also reported by Moss [161] for natural PbS crystals, which exhibit a quadratic dependence of the lifetime on the resistivity.

In n-type PbS samples with carrier densities higher than $2 \cdot 10^{18}$ cm^{-3}, the quadratic law is not obeyed and the lifetime may even increase with the carrier density [147] (Fig. 2.13). Such a carrier-density dependence of the lifetime has been attributed by Baryshev and Uritsekii [186] to a reduction in the probability of transitions when energy levels merge to form an impurity band. Assuming that the impact recombination takes place via an impurity band, Baryshev and Uritskii obtained good agreement between a theoretical $\tau_0(n)$ curve and the experimental values in the electron density from $4 \cdot 10^{17}$ to $8 \cdot 10^{18}$ cm^{-3} (Fig. 2.13).

Thus, the experimental values of the lifetime in extrinsic samples of PbS can be explained satisfactorily by assuming that carriers recombine mainly via impurity levels or an impurity band associated with vacancies. However, this assumption is in conflict with the conclusions which follow from measurements of the Hall effect, and which show that, at the carrier densities considered, there are no localized levels in the forbidden band and there is no impurity band separated by a gap from an allowed band (cf. §3.2A).

Recombination in PbSe and PbTe has not been investigated experimentally as thoroughly as in lead sulfide, but the available results suggest that the selenide and telluride obey the same relationships as PbS.

C. Spectral Characteristics of Photoelectric Effects and of Recombination Radiation

The spectra of all the photoelectric effects considered so far have a threshold corresponding to the fundamental absorption edge. A correlation between the photoconductivity and optical absorption spectra was established in [97, 101]. The values of the forbidden band width and of its temperature coefficient, deduced from the edges of the photoconductivity spectra [97, 101, 146, 187], are in good agreement with those found from the optical data (cf. §2.1C). Moss [146] found that at room temperature the wavelength $\lambda_{1/2}$ (at which the photoconductivity falls to half its maximum value) is 3.9, 5.0, and 2.9 μ for PbTe, PbSe, and PbS, respectively. The photon energy, corresponding to the photocontuctivity edge, increases linearly with temperature at a rate $\partial(\hbar\omega)/\partial T = 4 \cdot 10^{-4}$ eV/deg [146].

When the photon energy is increased above the fundamental absorption edge, a reduction in the photoconductivity may be observed because electron – hole pairs are generated closer to the surface and the surface recombination reduces the effective lifetime [151]. Further considerable increase of the photon energy makes the effective quantum yield of the photoconductivity greater than unity and the photoconductivity increases. The multiplication of carriers in collisions results in a monotonic increase of the photoconductivity quantum yield of PbS when the photon energy is increased above 2 eV [188, 189].

In the long-wavelength part of the photoconductivity spectrum, the signal is due to carrier transitions from impurity levels. Such an impurity photoconductivity has been observed in PbS at wavelengths of the order of 10 μ [190], at which there is a corresponding impurity absorption band.

The edges (thresholds) of the photomagnetic effect spectrum [161] and of the spectrum of the photovoltaic effect at a p – n junction [162-164] also correspond to the forbidden band width determined from the optical absorption data. Measurements of the spectral distribution of the short-circuit current through a p – n junction in PbTe [164] have yielded the value $\mathscr{E}_g = 0.173 + 4.75 \cdot 10^{-4}T$ eV in the temperature range 40-300°K as well as a constant value $\mathscr{E}_g = 0.185 \pm 0.002$ eV at $T \lesssim 10°K$. The photomagnetic effect in PbS, calculated per one absorbed light quantum, is independent of the wavelength in the photon energy range above the absorption edge (in agreement with Moss [161]) and the quantum yield is 1.

The recombination radiation spectra have been investigated for all three lead chalcogenides. In investigations of the spontaneous recombination radiation, reported in [174, 191-195], the excess carriers were injected by illumination. The stimulated radiation (the laser effect) has been investigated at low temperatures [178-181] under conditions of strong excitation caused by passing a current through a p – n junction. When the current through a p – n junction is below the laser threshold, the spontaneous radiation is observed. Laser light can be also generated by carrier injection caused by electron bombardment [182].

Figure 2.14 shows the recombination spectrum of PbTe obtained by Washwell and Cuff [174]. This spectrum is in the form of a simple band about 0.04 eV wide. The long-wavelength edge of the band is in agreement with the forbidden band width, which is assumed to be $\mathscr{E}_g = 0.18 + 4.4 \cdot 10^{-4}T$ eV. The recombination radiation band edge corresponds to the forbidden band width even in

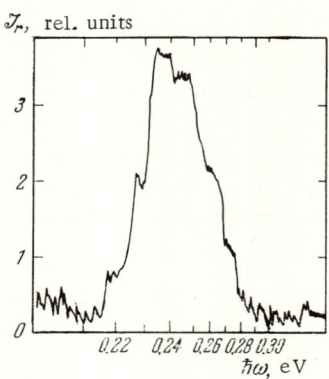

Fig. 2.14. Recombination radiation spectrum of p-type PbTe with a hole density $4.4 \cdot 10^{18}$ cm^{-3} at T = 90°K [174].

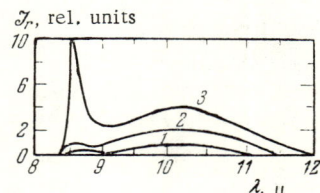

Fig. 2.15. Spontaneous recombination radiation spectrum of a PbSe p−n junction laser at 12°K [181]:
1) j = 75 A/cm^2; 2) j = 150 A/cm^2;
3) j = 300 A/cm^2.

degenerate samples, in contrast to the fundamental absorption edge, which is shifted by an amount which is of the order of the Fermi energy in such samples. The band edge is sharp without any density-of-states "tail" (such a "tail" indicates the merging of an impurity band with an allowed band); this lack of "tails" in lead chalcogenide is due to the high permittivity of these materials.

An examination of the detailed structure of the recombination radiation spectrum shows a periodic modulation of the spectral curve. The modulation period for PbTe is 0.012 eV, which is close to the energy of the longitudinal optical phonons $\hbar\omega_l$, determined from the tunnel effect (cf. §3.4A) and by neutron spectroscopy [141]. Thus, the fine structure of the spectrum is due to the influence of the longitudinal optical phonons. The minima in the spectral curve appear at photon energies given by $\mathscr{E}_g + m\hbar\omega_l$ where m = 1, 2, ...

The recombination radiation spectrum obtained by Washwell and Cuff for PbSe [174] is similar. The parameters found from this spectrum are: $\mathscr{E}_g = 0.13 + 4.4 \cdot 10^{-4}$T eV and $\hbar\omega_l = 0.018$ eV. Investigations of the recombination radiation spectra of PbS, reported in [194, 195], have shown no fine structure associated with phonons.

Washwell and Cuff [174] carried out a theoretical analysis of the recombination radiation spectrum. An assumption that the recombination is due to direct transitions of electrons from the

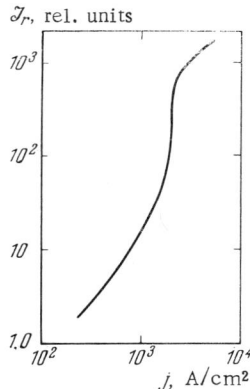

Fig. 2.16. Dependence of the intensity of the recombination radiation emitted by a PbSe p–n junction laser at T = 12°K on the density of the current through the junction [181].

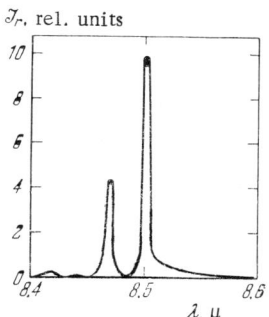

Fig. 2.17. Stimulated radiation spectrum of a PbSe p–n junction laser at T = 12°K and j = 4000 A/cm² [181].

conduction to the valence band was found to give good agreement between the calculated results and the experimental data.

Butler et al. [181] investigated the recombination radiation of p–n junctions in PbSe at low temperatures. They found that, at relatively low injection levels represented by a current of 100 A/cm² density flowing through the p–n junction, a wide band is observed at a wavelength of 10.1 μ as well as a smaller and narrower band at 8.5 μ (Fig. 2.15). The former band increases slightly with increasing current, while the latter increases rapidly (superlinearly) and becomes a sharp peak at 300 A/cm². Such a dependence of the radiation intensity on the current in the spontaneous radiation region (Fig. 2.16) differs from the dependence observed in lasers made of $A^{III}B^{V}$ materials, in which the spontaneous radiation increases linearly with the current and the superlinearity is associated with the laser effect.

D. Stimulated Recombination Radiation

The stimulated recombination radiation (the laser effect) is observed when a recombination event is not spontaneous but is caused by a photon and the emission of a recombination photon results in an increase in the total number of photons, which in their turn stimulate other recombination events so that the process becomes avalanche-like. The stimulated radiation appears at high injection levels when the electron distribution function shows an increase in the conduction band, i.e., when a band population inversion is observed, which may be represented by a negative temperature. The recombination radiation band becomes narrower at the laser threshold and the narrowing is so strong that the width is in fact limited by the resolving power of the monochromator (50 Å has been reported in [181]).

The stimulated radiation effect has been observed in p−n junction diodes made of PbTe [178, 179] and PbSe [180, 181], as well as in PbTe, PbSe, and PbS samples bombarded with fast electrons [182]; it has been found also in PbTe−SnTe alloys illuminated with light from a GaAs p−n junction laser [137]. The stimulated radiation spectrum of a laser made of PbSe, obtained by Butler et al. [181] at 12°K, is shown in Fig. 2.17. The laser photon energy is close to the forbidden band width. Laser radiation peaks have been observed at 6.5 μ in PbTe [178, 179] and at 8.5 μ in PbSe [180, 181] at 12°K. These wavelengths are longer than the wavelength of the radiation emitted by InSb lasers (5.2 μ). Heating to 110°K reduces the stimulated radiation wavelength of PbTe at 5.2 μ, which is due to the broadening of the forbidden band with increasing temperature. The observation of the laser effect confirms that transitions between extrema in the conduction and valence bands are direct.

The threshold value of the current density, at which the laser effect is observed, is of the order of 10^3 A/cm^2 (for example, 2000 A/cm^2 for PbSe [181]), which corresponds to a current of several amperes through a p−n junction. Hurwitz et al. [182] generated excess carriers by fast electron bombardment and have observed the laser effect beginning from a very low current density of 7.1 mA/cm^2, equivalent to 800 A/cm^2 through a p−n junction.

The intensity of the recombination radiation increases suddenly when the laser threshold value of the current is reached [181], as shown in Fig. 2.16. When the current is increased still further, the intensity increases linearly with the current, which reflects the fact that the quantum efficiency has reached its maximum value.

To generate stimulated radiation in lasers, the two opposite faces (the radiation emerges through one of them) should be polished to an optical-quality grade so that the sample acts as a Fabry – Perot resonator. The resonance condition for a line of the mth order is given by an interference formula such as Eq. (2.9), in which d is the distance between the polished (reflecting) faces. The principal mode of the laser radiation usually corresponds to a high value of m. It is evident from Fig. 2.17 that, apart from the principal mode, there are also auxiliary modes. The mode separation is given by the formula

$$\Delta\lambda = \frac{\lambda^2}{2d(\mathcal{N} - \lambda d\mathcal{N}/d\lambda)}. \tag{2.35}$$

The observed mode separation is equal to several tens or hundreds of angstroms, in agreement with estimates carried out using Eq. (2.35) [179-182].

The laser radiation wavelength and the threshold current decrease when a magnetic field is applied (cf. §5.4D).

Pratt and Ripper [196] carried out a theoretical analysis of changes in the laser effect in lead chalcogenides induced by deformation; this analysis shows that a uniaxial compression of 5000 kg/cm^2 along the [100] direction should increase the wavelength of the laser emission of PbSe from 8.5 to 12 μ due to the reduction in the forbidden band width. The change in the refractive index due to deformation should also modulate the wavelength because of the change in the resonance conditions. Apart from a change in the emission wavelength, Pratt and Ripper's theory predicts also a dependence of the polarization of the emitted radiation on the uniaxial deformation.

Experimental investigations carried out by Calawa et al. [197] confirmed a change in the wavelength and polarization of the laser emission of a PbSe diode subjected to a uniaxial compression perpendicular to the p – n junction plane.

Dimmock et al. [137] reported the generation of spontaneous and stimulated (coherent) radiation of about 15 μ wavelength, at 12°K, in lasers made of $Pb_xSn_{1-x}Te$ alloys. Carriers are generated in these lasers by light from a GaAs diode laser of 0.84 μ wavelength and the threshold value of the power of incident illumination

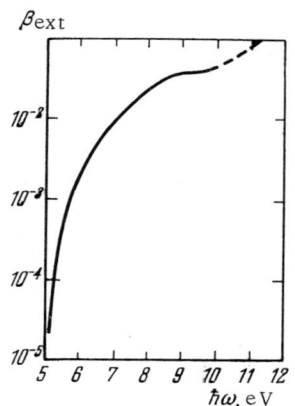

Fig. 2.18. Photoemission quantum yield (external photoeffect) of PbTe as a function of the photon energy [198].

is 1.5 W. The alloy with $x = 0.81$ emits its principal mode at a wavelength of 15.9 μ, which corresponds to a forbidden band width $\mathscr{E}_g = 0.078$ eV; the alloy with $x = 0.83$ emits its principal mode at a wavelength of 14.9 μ. The forbidden band width obtained from these measurements is in agreement with the optical data on $Pb_xSn_{1-x}Te$ alloys (cf. §2.1C). The dependence of the forbidden band width on the alloy composition is similar to that observed at 300°K: the forbidden band is 0.19 eV wide for PbTe ($x = 1$), it decreases linearly on the addition of Sn, vanishes at $x = 0.65$, and then increases linearly, reaching 0.3 eV for SnTe ($x = 0$), as found from measurements of the tunnel effect at 4.2°K [138] (cf. Fig. 6.8).

E. Photoelectric Emission

When the photon energy is sufficiently high, electrons may be ejected from a filled band into vacuum, giving rise to the effect known as photoelectric emission or photoemission (the external photoeffect). The quantum yield of the photoelectric emission and the dependence of the energy distribution of the emitted electrons on the wavelength of the incident light have been investigated for PbTe [198] and PbS [189, 199-202].

The dependence of the photoelectric quantum yield, β_{ext}, of PbTe on the photon energy has been obtained by Spicer and Lapeyre [198]; it is shown in Fig. 2.18. The threshold (minimum) photon energy $\mathscr{E}_{th}$, at which the photoelectric emission begins, is equal to the difference between the energy of an electron at rest in vacuum and the upper edge of the valence band. Thus, measurements of the position of the long-wavelength edge of the photoelectric emission spectrum make it possible to determine this difference.

The same quantity can be determined by investigating the energy distribution of the emitted electrons. The maximum photoelectron energy in vacuum is

$$\mathscr{E}_{max} = \hbar\omega - \mathscr{E}_{th}. \quad (2.36)$$

If the direction of the electric field between the photoemission cathode and the anode is such that it slows down the emitted electrons and returns them to the cathode and the absolute value of the potential difference (the applied voltage and the contact potential) is larger than $\mathscr{E}_{max}/e$, none of the photoelectrons can reach the anode and the current is zero. By measuring the minimum potential difference necessary to produce this effect, we can find $\mathscr{E}_{max}$ and consequently $\mathscr{E}_{th}$. Both methods give the same value of $\mathscr{E}_{th}$ = 4.9 eV for PbTe [198].

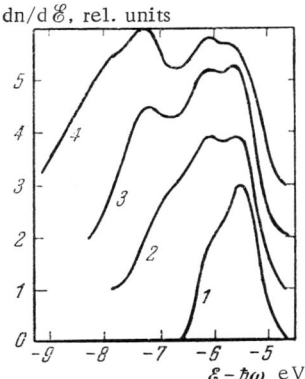

Fig. 2.19. Distribution of the energies of photoelectrons emitted from PbTe [198] as a function of the photon energy $\hbar\omega$ (eV): 1) 6.5; 2) 7.5); 3) 8; 4) 9.

The photoemission threshold depends strongly on the surface properties. Smith and Dutton [189] and Knapp [199] obtained a value of 5 eV, while Oman and Priolo [201] found a threshold of 3.8 eV. The latter value is in good agreement with the work function found from measurements of the contact potential difference [200].

When the photon energy is increased above the threshold value, the calculated photoemission yield increases, reaching a fairly high value of 15-20% for PbS at $\hbar\omega$ = 14 eV [189]. Investigations of the quantum yield and of the energy distribution of photoelectrons at high incident photon energies yield information on states located far from the forbidden band. Such investigations were carried out for PbTe by Spicer and Lapeyre [198] and by Sacks and Spicer [202].

The electron energy distributions for various incident photon energies are shown in Fig. 2.19 for PbTe.

The curves shown in Fig. 2.19 have several maxima. If the difference between the electron energy in vacuum ($\mathscr{E}$) and the photon ($\hbar\omega$) is plotted along the abscissa, as in Fig. 2.19, it is found that the positions of the maxima are independent of the photon energy.

The maxima in the photoelectron spectra correspond to the density-of-state maxima in the valence band, separated by $-(\mathscr{E} - \hbar\omega)_{max} - \mathscr{E}_{th}$ from the maximum of the upper valence band. The doublet, observed for PbTe at energies $\mathscr{E} - \hbar\omega$, equal to -5.6 eV and -6.1 eV, indicate the presence of bands located 0.7 and 1.2 eV, respectively, below the upper valence band. The doublet disappears rapidly when the photon energy is increased from 9 to 10 eV. Spicer and Lapeyre [198] suggest that the doublet is associated with direct transitions from the valence bands located at the point L of the Brillouin zone, to a higher-lying conduction band whose maximum is also at the point L, separated by about 8.3 eV from the maximum of the upper valence band; the curvature of this conduction band is considerably greater than that of the valence bands from which electrons are excited. A maximum at $\mathscr{E} - \hbar\omega = -7.3$ eV is probably associated with indirect transitions from a valence band located about 2.4 eV below the upper valence band.

Similar measurements, carried out for PbS [202], indicate the presence of maxima associated with valence bands lying 2.0 and 3.0 eV below the upper valence band. The former maximum (corresponding to 2.0 eV) is due to transitions from a valence band located at the point L and split by the spin-orbit interaction so that the gap is not more than 0.15 eV.

Comparison of the photoelectron emission spectra at high photon energies with the results of energy band calculations are given in §6.1D.

CHAPTER III

ELECTRICAL PROPERTIES

3.1 Mobility

A. Mobility for Various Scattering Mechanisms

The electrical conductivity of a solid with a carrier density n is given by the well-known formula

$$\sigma = e\mu n, \qquad (3.1)$$

where μ is the mobility equal to the average drift velocity of carriers under the action of a unit electric field. The mobility is governed by the scattering of carriers on imperfections (departures from the lattice periodicity), and in the case of elastic scattering the mobility in a semiconductor with a simple parabolic band is

$$\mu = \frac{e \langle \tau \rangle}{m^*}, \qquad (3.2)$$

where $\langle \tau \rangle$ is the average relaxation time; m^* is the effective mass.

In the simplest cases, the relaxation time is a power function of the carrier energy, temperature, and effective mass:

$$\tau \propto \mathscr{E}^r T^s m^{*t}. \qquad (3.3)$$

When carriers are scattered by the acoustical vibrations of the lattice, we have $r = -1/2$, $s = -1$, $t = -3/2$, provided one phonon is emitted or absorbed in each collision process. Analysis of the scattering by ionized impurities usually gives results which can be described approximately by the parameters $r = 3/2$, $s = 0$, $t = 1/2$. The single-phonon scattering by the polar optical vibrations of the lattice is characterized by a power dependence of the relaxation time on the carrier energy and lattice temperature only at temperatures considerably higher than the characteristic temperature $\theta_l' = \hbar\omega_l/k_0$, where ω_l is the frequency of the longitudinal long-wavelength optical vibrations. The quantity θ_l is usually called the Debye temperature. At temperatures $T \gg \theta_l$, we have in this case $r = 1/2$, $s = -1$, $t = -1/2$. At temperatures $T \ll \theta_l$, the relaxation time is independent of the carrier energy but its temperature dependence is stronger:

$$\tau \propto (e^{\theta/T} - 1). \tag{3.4}$$

In the intermediate case, the collision process is inelastic and the relaxation time concept cannot be used. In this and other inelastic scattering cases (for example, in the case of electron–electron collisions), the mobility is described by formulas more complex than Eq. (3.2).

Averaging of the relaxation time over the energies is carried out as follows:

$$\langle \tau \rangle = \frac{\int_0^\infty \tau(-\partial f/\partial \mathscr{E}) k^3 \, d\mathscr{E}}{\int_0^\infty (-\partial f/\partial \mathscr{E}) k^3 \, d\mathscr{E}}, \tag{3.5}$$

where $f(\mathscr{E}, \zeta)$ is the Fermi distribution function and k is the crystal momentum of carriers. In the case of classical statistics, the averaging process results in a power dependence of $\langle \tau \rangle$ on temperature and then

$$\mu_{cl} \propto T^{r+s}. \tag{3.6}$$

For example, in the scattering by the acoustical phonons $\mu_{cl} \propto T^{-3/2}$. In the degenerate case, the derivative $(-\partial f/\partial \mathscr{E})$ is approximately given by the delta function $\delta(\mathscr{E} - \zeta)$, where ζ is the Fermi energy and consequently $\langle \tau \rangle = \tau(\zeta)$. If the Fermi energy is independent of temperature, we then have

$$\mu_{deg} \propto T^s. \tag{3.7}$$

In the case of an arbitrary degree of degeneracy, the mobility in a semiconductor with a parabolic band can be expressed in terms of Fermi integrals. For example, for $r = -1/2$ (the acoustical scattering), we have

$$\mu = \mu_{cl} \frac{\sqrt{\pi}}{2} \frac{F_0(\zeta^*)}{F_{1/2}(\zeta^*)}, \qquad (3.8)$$

where the Fermi integrals are given by

$$F_n(\zeta^*) = \int_0^\infty f(x, \zeta^*) x^n \, dx \qquad (3.9)$$

and the reduced Fermi level is

$$\zeta^* = \zeta/k_0 T. \qquad (3.10)$$

We shall demonstrate later that none of these simple cases applies to lead chalcogenides and this is true of the majority of semiconductors. First, several scattering processes are usually acting at a given carrier density and temperature. In such a case, the reciprocals of the relaxation times, due to the i-th scattering mechanism, can be summed as follows:

$$\tau^{-1} = \sum_i \tau_i^{-1}, \qquad (3.11)$$

and the dependence $\tau(\mathscr{E})$ cannot be represented by a power law.

Secondly, the energy bands of lead chalcogenides are anisotropic; they have ellipsoidal constant-energy surfaces and two components of the effective mass: the longitudinal $m_\parallel^*$ and the transverse $m_\perp^*$. In this case, the denominator of Eq (3.2) should contain what is known as the effective conductivity mass, defined by Eq. (2.22). The dependence of the relaxation time on the direction of the electron crystal momentum relative to the ellipsoid axes is different for different scattering mechanisms. In particular, with the acoustical scattering, the relaxation time depends weakly on the direction (it is practically isotropic) and is inversely proportional to the effective density-of-states mass (taken to a power of 3/2):

$$m^*_{d1} = (m^*_\| m^{*2}_\perp)^{1/3}, \qquad (3.12)$$

while collisions with ionized impurities result in an anisotropic scattering.

Thirdly, the dependence of the carrier energy on the crystal momentum in lead chalcogenides is nonquadratic at sufficiently high energies. This implies band nonparabolicity, which is characterized by an energy-dependent effective mass. Then, the denominator of Eq. (3.2) has the following effective mass:

$$m^* = \left(\frac{1}{\hbar^2}\frac{1}{k}\frac{d\mathscr{E}}{dk}\right)^{-1}. \qquad (3.13)$$

The same quantity gives the density of states in the allowed bands. It is usually assumed that the relaxation time is

$$\tau \propto \frac{1}{|M|^2 \rho(\mathscr{E})}, \qquad (3.14)$$

where ρ is the density of states and M is the average matrix element of the interaction of an electron with a scatterer. The matrix element is assumed to depend on the value of the crystal momentum in the same way as in the case of a parabolic band. In particular, the matrix element for the acoustical scattering is independent of the crystal momentum and the quantity $\tau\rho$ does not vary with the carrier energy. Thus, the influence of the nonparabolicity on the momentum dependence of the relaxation time enters only through the density of states. This conclusion follows from a calculation [203, 204] in which carriers are described by plane waves. In fact, the wave functions are of the Bloch type:

$$\psi = u_k(\mathbf{r})\, e^{i(\mathbf{kr})}, \qquad (3.15)$$

and the dependence of the periodic part of the Bloch function $u_k(\mathbf{r})$ on the crystal momentum in the nonparabolicity region is very strong (cf., for example, [205]), which should result in an additional dependence of the matrix element on the crystal momentum.

A reliable calculation of this effect is difficult and, therefore, it is usually assumed that the nonparabolicity does not affect the value of the matrix element, but this approximation should be treated with caution.*

The situation is more complex in the simultaneous presence of the anisotropy and nonparabolicity of the energy bands. In this case, the longitudinal and transverse components of the effective mass may have different energy dependences. Models describing such special cases are considered in §6.2B. In the case of a nonparabolic band, the mobility is not expressed in terms of the Fermi integrals but by more complex formulas (cf. §6.2B and Appendix A).

Finally, the temperature dependence of the effective mass, which is fairly strong in lead chalcogenides, also contributes to the temperature dependence of the carrier mobility. This effect is particularly strong in the scattering by the acoustical vibrations, when the mobility is proportional to $(m^*)^{-5/2}$.

We shall now consider in more detail the results of investigations of the carrier mobility in lead chalcogenides. The involved temperatures can be divided into two regions: the low-temperature region, $T < 50\text{-}100°K$, where the mobility depends weakly on temperature; and the region of moderate and high temperatures, where the mobility decreases quite rapidly with increasing temperature.

---

*Recent calculations, carried out by the authors of the present book and their colleagues show that the mean square of the matrix element $|M|^2$ for lead salts decreases markedly (due to the nonparabolicity) with increasing energy in the case of scattering by the acoustical and optical phonons. For example, if a two-band model, described later (cf. §6.2B), is used and if it is assumed that the deformation potentials of the conduction and valence bands are approximately equal, it is found that in the case of scattering by the acoustical phonons

$$|M|^2 \sim 1 - \frac{8}{3} \frac{\mathscr{E}(\mathscr{E}_g + \mathscr{E})}{(\mathscr{E}_g + 2\mathscr{E})^2}$$

and therefore the effect considered here may increase the relaxation time by a factor of 2-3. The corresponding factor for the polar scattering is obtained by replacing the coefficient 8/3 in the above equation with 2.

B. Temperature Dependence of the Mobility at Moderate and High Temperatures. Scattering by Lattice Vibrations

The mobility in lead chalcogenides, determined from measurements of the Hall effect and electrical conductivity in the extrinsic (impurity) conduction region, varies with temperature — at sufficiently high temperatures — in accordance with the power

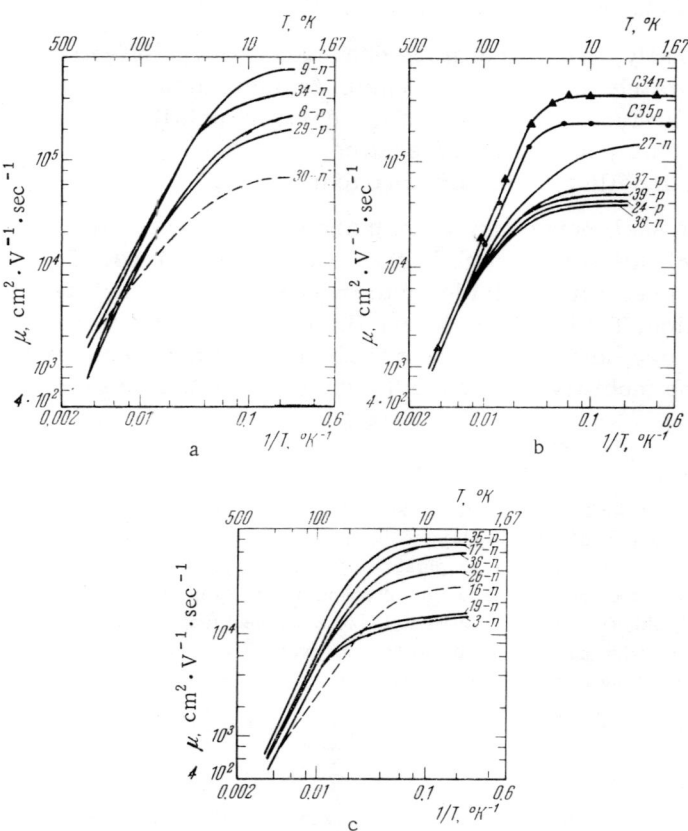

Fig. 3.1. Temperature dependences of the electron and hole mobilities in various n- and p-type samples of PbTe (a), PbSe (b), and PbS (c) [220]. The upper two curves for PbSe are taken from [38].

MOBILITY

law [206-224]:

$$\mu \propto T^{-\nu}, \quad (3.16)$$

where the power exponent ν is usually 2-3 and may vary slowly with temperature. The usual value is $\nu = 5/2$. The range of temperatures in which the power-law dependence is observed begins at temperatures of 20-50°K in samples with relatively low carrier densities and at higher temperatures in samples with greater carrier densities; kinks may be observed in the curves plotted on the logarithmic scale. The temperature dependences of the carrier mobility in PbTe, PbSe, and PbS, obtained by Allgaier and Scanlon [220], are given in Fig. 3.1.

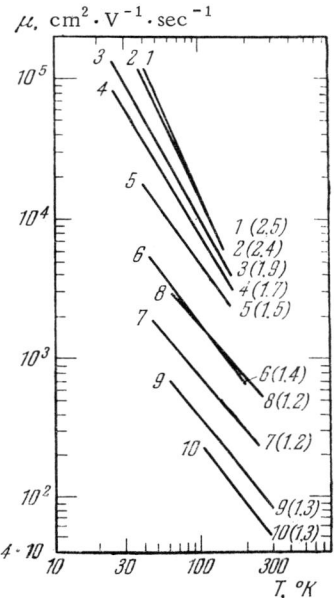

Fig. 3.2. Temperature dependences of the mobility of holes, governed by the scattering on the lattice vibrations in PbTe (1-7) and SnTe (8-10) [223]. The upper lines represent samples with relatively low carrier densities and the lower lines, those with higher densities [the quantity $\nu = -d\ln\mu/d\ln T$ is shown in parentheses].

The carrier mobility depends less strongly on temperature in degenerate than in nondegenerate samples [222, 223]. It is evident from Fig. 3.2 that the value of ν decreases with increasing carrier density. For example, $\nu = 1.3$ for PbTe with a hole density of $5 \cdot 10^{19}$ cm^{-3} at room temperature [223]. At high temperatures (of the order of 500-1000°K), the rate of fall of the mobility with increasing temperature is more rapid than at room temperature, as found by Stavitskaya et al. [224].

There is no basic difference between the temperature dependences of the Hall mobilities of all three chalcogenides, both of n- and p-type, with the exception of a stronger temperature dependence of the mobility found for p-type PbTe [223]. The mobility determined from the Nernst – Ettingshausen effect has the same temperature dependence as the Hall mobility [209, 225].

The formulas given earlier in the present section show that in the nondegenerate case the mobility varies with temperature as $T^{-3/2}$ in the acoustical scattering case, as $T^{3/2}$ in the case of collisions with ionized impurities, and as $T^{-1/2}$ in the case of polar scattering by the optical vibrations if the temperature is considerably higher than the Debye temperature θ_l. None of these theoretical results explains the observed rapid fall of the mobility with increasing temperature.

Below the Debye temperature, in the optical scattering case, the mobility should, as shown earlier in the present section, decrease exponentially with increasing temperature. Combining the expressions for the scattering by the optical and acoustical vibrations, some authors have obtained curves which are in agreement with the experimental data [212, 214, 217, 219]. However, such agreement is obtained by assuming that the Debye temperature of PbS [217] and PbTe [219] is 1100°K and that of PbSe is higher than 800°K [212]. These values of the Debye temperature disagree greatly with the values of 150-200°K, obtained from measurements of the specific heat [226], and with the energies of the longitudinal optical phonons (cf., the table in Appendix C).

An attempt to explain the observed temperature dependence of the mobility was also made by Stil'bans et al. [218, 228]. They assumed that the carrier scattering is dominated by the interaction with two acoustical phonons. Since the laws of conservation of energy and momentum of colliding quasiparticles impose much less rigorous restrictions on the energy of the phonons taking part in two-phonon collisions than in the case of one-phonon scattering, it would seem that the two-phonon processes could be more effective in the scattering of carriers than the one-phonon collisions. However, a rigorous calculation of the influence of the two-phonon processes on the mobility, carried out later in [229] by the density matrix method, has demonstrated that in a covalent semiconductor the relative change in the mobility due to the two-phonon processes at temperatures above the Debye temperature is of the following order of magnitude

$$\frac{\Delta\mu}{\mu} \sim -\frac{\lambda_{fc}}{l}-\frac{\overline{u^2}}{a^2}-2\frac{\Delta\Xi}{\Xi}-\frac{5}{2}\frac{\Delta m^*}{m^*}. \tag{3.17}$$

The first term in Eq. (3.17) is of the order of the ratio of the wavelength of free carriers λ_{fc} and the mean free path l. This ratio is approximately equal to 0.4 for n-type PbTe with a carrier mobility of 1000 cm$^2 \cdot$V$^{-1} \cdot$sec^{-1}, if the effective mass is assumed to be $0.05m_0$. Hence, it follows that the term associated with the emission or absorption of two phonons makes some contribution to the mobility above room temperature but it cannot be the dominant term. The second term in Eq. (3.17) is of the order of the ratio of the average of the square of the displacement of an atom from its equilibrium position to the square of the lattice constant. This value is always considerably smaller than unity. The third term is equal to double the relative change in the deformation potential, while the fourth term is proportional to the change in the effective mass due to the interaction of a carrier with the lattice vibrations.

When the relative change in the deformation potential and the effective mass, due to the thermal expansion of the lattice is added to the last two terms, the correction to the mobility includes all relative changes in these quantities caused by temperature variations. As demonstrated in §6.3B, the relative temperature change in the effective mass may be considerable in the case of semiconductors with a narrow forbidden band (which include lead chalcogenides), and it may considerably exceed the relative change in the deformation potential due to temperature variations.

Moizhes et al. [230] were the first to show that the temperature dependence of the effective mass of PbSe could be accounted for by a decrease in the carrier mobility with increasing temperature. The temperature dependence of the effective mass was deduced by Moizhes et al. from measurements of the thermoelectric power (cf. §4.1B) and was sufficiently large to explain the observed temperature dependence of the mobility. In the temperature range 100-400°K and at carrier densities up to $6 \cdot 10^{18}$ cm^{-3}, the effective mass of electrons varied as $T^{0.35}$ and that of holes as $T^{0.45}$. Such an unusually strong temperature dependence of the effective mass was confirmed by measurements of several other effects. It was found that the temperature dependences of the effective masses in PbTe and PbS were also fairly strong. The temperature dependence of the effective mass was in rough agreement with the temperature dependence of the forbidden band width [230].

The theory of nondegenerate semiconductors yields the following expression for the scattering by the long-wavelength acoustical vibrations:

$$\mu_{cl} \propto m^{*-5/2} T^{-3/2}. \qquad (3.18)$$

The first factor in the above expression is approximately inversely proportional to temperature if the temperature dependence of the effective mass is of the type given in the preceding paragraph. Consequently, we obtain the $T^{-5/2}$ temperature dependence, in agreement with the experimental observations.

At sufficiently high carrier densities (about 10^{19} cm^{-3} for PbTe at room temperature), the statistics of electrons and holes is degenerate. In the case of strong degeneracy

$$\mu_{deg} \propto m^{*-2} T^{-1}, \qquad (3.19)$$

and the temperature dependence of the mobility is weaker than in the absence of degeneracy. In the case of arbitrary degeneracy, we should use Eqs. (3.8) and (3.18). The value of μ_{cl}, obtained using Eq. (3.8) for PbTe (with an electron density 10^{19} cm^{-3}) from the experimental values of μ in the temperature range 750–850°K, varies with temperature as $T^{-3.5}$ [224]. This temperature dependence of μ_{cl} is in agreement with the temperature dependence of the effective mass m* $\propto T^{0.6}$ found for the same temperature range from measurements of the thermoelectric power. The observed temperature dependence of the effective mass may be misleading: at high temperatures, a large fraction of electrons has energies comparable with the forbidden band width and the conduction band nonparabolicity affects the average effective mass. This is due to an increase in the number of high-energy electrons with increasing temperature and, consequently, at high temperatures there are more electrons with a larger effective mass.

The carrier mobility in p-type PbTe at temperatures above 150°K, determined from the product $cR(77°K)\sigma(T)$, decreases with increasing temperature more rapidly than $T^{-5/2}$. This is due to the fact that when the temperature is increased, some holes are transferred to a deeper valence band in which the mobility is an order of magnitude lower than in the upper valence band [223, 231, 232]. The usual Hall mobility $cR(T)\sigma(T)$ does not decrease as rap-

TABLE 3.1. Mobility in Lead Chalcogenides
at 300°K [220]

Compound	μ_n, cm$^2 \cdot$V$^{-1} \cdot$sec^{-1}	$n \cdot 10^{-18}$, cm^{-3}	μ_p, cm$^2 \cdot$V$^{-1} \cdot$sec^{-1}	$p \cdot 10^{-18}$, cm^{-3}
Pb Te	1730	1.08	839	0.33
Pb Se	1045	2.38	995	4.28
Pb S	614	3.74	621	2.66

idly, because a reduction in the average mobility due to the transfer of holes to the lower band is compensated by an increase in the Hall coefficient (cf. §3.2B).

Table 3.1 gives the electron and hole mobilities at room temperature obtained by Allgaier and Scanlon [220]. The table lists the highest mobilities, together with the corresponding carrier densities.

It must be pointed out that samples of the same type with similar values of the carrier density may have mobilities differing by a factor of two, because of inhomogeneities, microscopic cracks, etc. Therefore, the values listed in Table 3.1 should not be regarded as the highest possible mobilities because in samples with fewer imperfections the mobility may be higher. The hole mobility listed in Table 3.1 for PbS is higher than the electron mobility, although measurements of the Hall effect in the mixed conduction region indicate that hole mobility is lower than that of electrons (cf. §3.2).

Knowing the mobility and assuming that it is entirely due to collisions with the acoustical phonons, we can estimate the value of the deformation potential constant, which governs the effectiveness of the scattering by the acoustical vibrations. The expression for the mobility, which allows for the ellipsoidal shape of the constant-energy surfaces and for the scattering anisotropy, is of the form [233, 234]:

$$\mu = \frac{2^{3/2} \pi^{1/2} \hbar^4 e c_l}{3 m_\chi^* m_{d1}^{*3/2} (k_0 T)^{3/2} \Xi^2}, \qquad (3.20)$$

where the effective masses m_χ^* and m^*_{d1} are defined by Eqs. (2.22) and (3.12); c_l is a combination of the elastic moduli which govern the average velocity of propagation of the acoustical vibrations; Ξ is some combination of the deformation potential constants Ξ_d and Ξ_u, which is defined by Eq. (3.54) and depends on the anisotropy coefficient K and on the elastic moduli. If the deformation potential constants Ξ_d and Ξ_u have very different absolute values, the constant Ξ is governed by the larger of the two.

Comparison of Eq. (3.20) with the highest values of the experimental mobilities (Table 3.1) yields deformation potential constants whose absolute values are 30-40 eV. However, these values require considerable refinement. First of all, as will be shown later, the polar scattering makes an appreciable contribution to the scattering at relatively low carrier densities. Therefore, the mobility which would be obtained in the case of the acoustical scattering is higher than the experimental value. Secondly, the nonparabolicity must be taken into account in an analysis of the experimental data. Thirdly, the values of the mobilities at relatively low carrier densities have a considerable scatter; therefore, more reliable estimates of the deformation potential constants are obtained using the data (corrected for the degeneracy) for samples with relatively high carrier densities (of the order of 10^{19} cm^{-3}). Calculations which take into account all the points just mentioned yield the following values of Ξ for electrons in n-type materials: 24 eV for PbTe, 23 eV for PbSe, and about 20 eV for PbS. The deformation potentials of holes are similar. We shall consider again the deformation constants in §3.3B and §6.3A.

Lead chalcogenides are partly ionic compounds and, therefore, one would expect the interaction of carriers with the polar optical lattice vibrations to have some effect on the mobility. The effective ionic charge, which occurs in the expression for the mobility governed by the interaction with the optical phonons, is given by Callen's formula [235]:

$$e^* = \left\{ \frac{1}{4\pi} M\Omega \omega_l^2 \left(\frac{1}{\varepsilon_\infty} - \frac{1}{\varepsilon_0} \right) \right\}^{1/2}. \qquad (3.21)$$

The use of this formula gives the following values of the effective charge: 0.18e for PbTe, 0.21e for PbSe, and 0.26e for PbS.

The interaction of electrons with the polar vibrations of the lattice is governed by the dimensionless polaron coupling constant [270]:

$$\alpha_p = \frac{e^2}{\hbar}\left(\frac{m^*}{2\hbar\omega_l}\right)^{1/2}\left(\frac{1}{\varepsilon_\infty} - \frac{1}{\varepsilon_0}\right). \tag{3.22}$$

Estimates obtained for lead chalcogenides show that the polaron coupling constants of these compounds are a few tenths and, therefore, we can use the perturbation theory in the calculation of the interaction of electrons with the polar vibrations.

Another quantity (which is important in calculations of the mobility) governed by the scattering of electrons on the polar vibrations is the Debye temperature for the longitudinal optical phonons $\theta_l = \hbar\omega_l/k_0$. Using the values of $\hbar\omega_l$, obtained from measurements of the tunnel current (cf. §3.4), we find that the values of the Debye temperature are 160°K for PbTe, 190°K for PbSe, and 300°K for PbS.

In contrast to the deformation potential constant, which governs the efficiency of the acoustical scattering, the constant which represents the efficiency of the polar scattering is expressed in terms of the parameters which can be found by independent methods. Consequently, the mobility due to the polar scattering can be calculated without the use of adjustable parameters. For example, in the case of a nondegenerate parabolic band at temperatures $T \gg \theta_l$, the mobility governed by the polar scattering can be described quite accurately by the formula:

$$\mu = \frac{2^{5/2}\,\hbar^2}{3\pi^{1/2} e m_\chi^* m_{d1}^{*1/2} (k_0 T)^{1/2} (\epsilon_\infty^{-1} - \epsilon_0^{-1})}.$$

Estimates based on this formula show that, in addition to the acoustical scattering, the polar scattering also affects the mobility at relatively low carrier densities. Quantitative theoretical results can be obtained by taking into account the following factors, which complicate considerably the above formula:

1. The relaxation time approximation, which is valid at relatively high temperatures $T \gg \theta_l$, is inapplicable to PbTe and PbSe below 300°K and to PbS below 600°K. At temperatures which are of the same order as the Debye temperature, the mobility and other transport coefficients are usually calculated by the variational method. Such a calculation was carried out by Howarth and Sondheimer [236] ignoring the nonparabolicity and screening. The variational method gives relatively simple results in the strongly degenerate case, when the formulas for the transport coefficients can be obtained in the analytic form. The strong degeneracy approximation can be used at 77°K for carrier densities higher than $2 \cdot 10^{18}$ cm^{-3} in PbTe and PbSe, and higher than 10^{19} cm^{-3} in PbS.

2. Screening of the polar vibrations by free carriers reduces the scattering probability and alters the phonon spectrum at low values of the phonon wave vector [238]. These two effects are taken into account in the published theoretical calculations. In particular, at $T \gg \theta_l$ the screening reduces the reciprocal of the relaxation time by the factor [237]:

$$\frac{1}{\tau} \sim 1 - \frac{1}{4(kr_{sc})^2} \ln\left[1 + 4(kr_{sc})^2\right],$$

where k is the crystal momentum of a carrier and r_{sc} is the screening radius corresponding to the permittivity ε_∞. In this range of temperatures the change in the crystal momentum kr_{sc} is of the order of 1 for lead chalcogenides; it follows that the screening increases the relaxation time (and consequently also the mobility) by a factor of the order of 2.

3. The nonparabolicity alters the effective mass and the density of states and it changes the energy dependence of the matrix element. The two effects are governed by the value of the ratio $\mathscr{E}/\mathscr{E}_g$ and they affect considerably the carrier mobility (cf. §3.1A, §6.2B, and Appendix A).

We shall not quote the fairly complicated formulas which are obtained when all the factors just listed are taken into account. We shall mention only that the new formulas still do not have any unknown parameters. Calculations based on these formulas yield, for example, the mobility governed by the polar scattering: its value is 43,000 cm · V^{-1} · sec^{-1} for a sample of PbTe with an electron density of $5 \cdot 10^{18}$ cm^{-3} at 77°K. The corresponding experimental value of the mobility is 21,000 cm^2 · V^{-1} · sec^{-1}. Thus, direct calculations show clearly that the polar scattering plays, like the acoustical scattering, an important role in these semiconductors.

When the carrier density is increased, the relative importance of the polar scattering decreases. The acoustical scattering becomes the dominant mechanism at carrier densities higher than $4 \cdot 10^{18}$ cm^{-3}, $9 \cdot 10^{18}$ cm^{-3}, and $1.1 \cdot 10^{19}$ cm^{-3} in n-type PbTe, PbSe, and PbS, respectively, at 77°K; at lower carrier densities the polar scattering predominates. The relative importance of each of the two scattering mechanisms varies quite slowly in the temperature range 77-300°K because, at these temperatures, both scattering mechanisms yield approximately the same temperature dependence of the mobility. At higher temperatures the relative importance of the polar scattering gradually decreases.

At temperatures below 77°K the relatively pure samples [with a carrier density of the order of $(2-4) \cdot 10^{18}$ cm^{-3}] exhibit the $T^{-5/2}$ law for the mobility. Since the effective carrier mass varies weakly below 77°K, the acoustical scattering does not explain the observed curve. Inclusion of the polar scattering mechanism produces a satisfactory agreement between the theory and experiment without introduction of any additional adjustable parameters.

The allowance for the polar scattering explains also the rise of the mobility when the carrier density is reduced below 10^{18} cm^{-3} at 77°K.

The relative importance of the polar scattering increases along the series PbTe – PbSe – PbS. A reduction in the mobility along the same series is mainly due to an increase in the polar scattering, because the mobility governed by the acoustical scattering is approximately the same for all three semiconductors (at the same carrier density).

In addition to the described mechanisms of the scattering by the long-wavelength acoustical and optical phonons, semiconductors with a complex energy band structure may exhibit scattering by the short-wavelength phonons, in which carriers are transferred from one extremum to another. However, in lead chalcogenides the transfer of a carrier between the equivalent extrema, located at the Brillouin zone boundary along the <111> directions, is not very likely because transitions between the L points are forbidden by the selection rules [239]. This theoretical conclusion was confirmed experimentally by Nanney [525] by measuring the relaxation time for intervalley transitions using an ultrasonic method. Nanney determined the relaxation time for intervalley transitions in PbTe with an electron density of $1.42 \cdot 10^{17}$ cm^{-3} at 65°K: the relaxation time was $1.2 \cdot 10^{-10}$ sec, which was approximately 80 times as high as the total relaxation time found from the mobility. Interband scattering accompanied by hole transitions between inequivalent maxima in the valence band is evidently important in PbTe, at temperatures of the order of 400°K [267], as well as in SnTe and in PbTe – SnTe alloys.

C. Mobility at Low Temperatures. Scattering by Ionized Impurities, Vacancies, and Other Lattice Defects

The temperature dependence of the mobility at low temperatures (Fig. 3.1) is weaker than at room temperature. The rate of change of the mobility with temperature gradually decreases as the temperature falls and at 4.2°K the mobility is practically independent of temperature in the majority of samples. Such a temperature dependence of the mobility is similar to the temperature dependence in metals exhibiting the residual resistance phenomenon.

The mobilities of lead chalcogenides at liquid helium temperatures are very high [220, 223, 240, 241] and reach values of the order of 10^5-10^6 cm$^2 \cdot V^{-1} \cdot $sec^{-1}. To explain the low-temperature mobility in lead chalcogenides, we must consider the scattering by ionized impurities. In order to calculate the value of the mobility governed by collisions with ionized impurities, we must know first the static permittivity ε_0. This permittivity of lead chalcogenides is unusually high and this explains why the scattering by impuritiy ions is weak at room temperature even in samples with impurity concentrations of 10^{20} cm^{-3}, and it also accounts for the high values of the mobility at low temperatures.

Kanai and Shohno [242] calculated the electron mobility in PbTe at liquid-helium temperature assuming $\varepsilon_0 = 400$, which had been found by measuring the barrier capacitance of a p – n junction [242, 243] and which agrees well with the values of the permittivity obtained by other methods (cf. §3.4B). The calculation was carried out using the Mansfield – Dingle formula [244], deduced for the scattering of electrons by a screened Coulomb center in a degenerate semiconductor:

$$\mu = \frac{e}{2\pi^2 \hbar} \left(\frac{3n}{\pi} \right)^{1/3} \frac{1}{n_d} \frac{x^2}{\ln(1+x) - \frac{x}{1+x}}, \quad (3.23)$$

$$x = \left(\frac{\pi\hbar}{e}\right)^2 \left(\frac{3n}{\pi}\right)^{1/3} \frac{\varepsilon_0}{m^*},$$

where n_d is the concentration of charged impurity centers which, in an n-type semiconductor with fully ionized donors, is equal to the electron density n.

The formula (3.23) was derived on the assumption that the potential energy of an electron at a distance r from a singly charged scattering center is

$$U = -(e^2/\varepsilon_0 r) \exp(-r/r_{sc}), \qquad (3.24)$$

where r_{sc} is the screening radius, whose value in the case of strong degeneracy is given by the expression

$$r_{sc} = \left[\frac{\varepsilon_0 \hbar^2}{4m^* e^2} \left(\frac{\pi}{3n}\right)^{1/3}\right]^{1/2}. \qquad (3.25)$$

It is found that Eq. (3.23) predicts satisfactorily the experimentally obtained values of the low-temperature mobility only if the carrier density is relatively low ($n < 10^{18}$ cm^{-3}). At carrier densities higher than 10^{18} cm^{-3}, the mobility is lower than that predicted by Eq. (3.23) and, as demonstrated by Kanai et al. [240], it decreases with increasing carrier density in accordance with a law close to $\mu \propto n^{-4/3}$. An approximately similar dependence of the mobility on the carrier density has been observed in p-type PbTe with hole densities higher than 10^{19} cm^{-3} [223].

Stil'bans et al. [221, 222] drew attention to the fact that at short distances from an impurity ion the approximation of a point-like Coulomb center in a uniform medium filled with a free-electron gas, used in the derivation of the Mansfield – Dingle formula, ceases to be valid. Only at distances exceeding the lattice constant can we use the screened potential of Eq. (3.24), which is not very effective in the case of a high permittivity. Under these conditions, the potential of an impurity ion can be divided into two zones: a near zone in the form of a deep potential well (having a volume of the order of a^3) in the direct vicinity of the ion and a far zone with a screened Coulomb potential.

If the permittiviity is sufficiently high, the field in the far zone can be neglected and the potential of an ionized impurity or vacancy becomes similar to the potential of a neutral atom. The potential of a neutral point defect cannot be given in its general form and, therefore, some simplifying assumptions on the energy dependence of the scattering cross section must be made in an analysis of the experimental data on lead chalcogenides. Stil'bans et al. [221, 222], assumed that the relaxation time is independent of the energy, as in the scattering by the hydrogen atom [245], i.e., the scattering cross section s is proportional to $\mathscr{E}^{-1/2}$. This assumption was found to agree with the results of an analysis of the experimental dependences of the mobility on temperature and electron density in PbTe at temperatures $T \geq 77°K$.

To explain the dependence of the mobility on the carrier density at liquid-helium temperatures, Kanai et al. [240] suggested that the scattering cross section is independent of the carrier energy. Then, bearing in mind that the electron velocity v close to the Fermi level in a degenerate semiconductor is proportional to $n^{1/3}$ and assuming that the electron density is equal to the donor concentration n_d, they obtained

$$\mu \propto \frac{1}{n_d v s} \propto n^{-1/3}, \qquad (3.26)$$

in agreement with the experimental dependence. Using the experimental values of the mobility, Kanai et al. [240] found that the scattering cross section is of the order of the square of the atomic radius ($s = 4 \cdot 10^{-16}$ cm^2). Estimates of the scattering cross sections [221, 240] show that carriers are indeed scattered mainly by the inner part of the potential of ionized impurities.

The scattering cross sections were also calculated by Shimizu [246] and Morita [247]. Shimizu [246] modeled the potential near an impurity ion by a rectangular well, whose depth is of the order of the atomic energy (several electron volts) and whose radius is of the order of the Wigner–Seitz cell radius (2 Å). The calculation was carried out by the method of partial waves in the Born approximation. The scattering cross section has been found to be independent of the energy and its value is in order-of-magnitude agreement with the experimental values. Morita [247] calculated the effective potential near an impurity ion allowing for the interaction

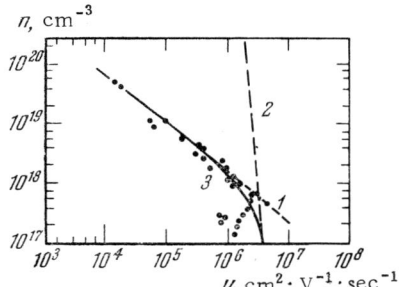

Fig. 3.3. Dependence of the electron mobility in n-type PbTe at 4.2°K on the carrier density. The curves are theoretical [247]: 1) for the scattering by the inner part of the potential of ionized impurities; 2) for the scattering by the screened Coulomb potential; 3) the combined curve. The experimental points are taken from [240].

of an electron with the longitudinal optical vibrations of the lattice. Using reasonable values of the parameters, Morita obtained the same scattering cross section for the inner part of the impurity ion potential as in the case of a rectangular well.

Figure 3.3 shows theoretical curves and experimental points for the dependence of the electron mobility in PbTe on the donor concentration at liquid-helium temperature. Curve 1 describes the mobility due to scattering on the inner part of the ionized impurity potential; curve 2 gives the mobility governed by the long-range part of the potential. The experimental points for electron densities 10^{18}-10^{20} cm^{-3} are closer to curve 1.

Thus, the scattering of carriers at low temperatures in lead chalcogenides takes place mainly at the inner part of the potential of ionized impurities, which are usually not foreign atoms but vacancies. Some differences in the assumed dependences of the scattering cross sections on the energy are unimportant particularly at high impurity concentrations, when the role of impurities in the scattering is very considerable and the theoretical expressions for the scattering cross section are distorted by the dependence of the carrier mass on the energy (the nonparabolicity of the energy bands); the influence of the nonparabolicity on the scattering cross sections is difficult to allow for in theoretical calculations because it requires the knowledge of the dependence of the form of the wave functions of carriers on the crystal momentum.

Carriers may be scattered also by inhomogeneities of the crystal lattice resulting from the deformation of the lattice by impurity ions. Estimates of the effectiveness of such scattering [246] have indicated that the cross sections for the scattering by inhomogeneities due to dilation are an order of magnitude smaller and those

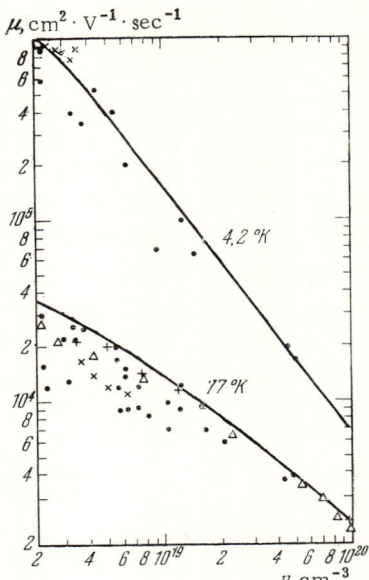

Fig. 3.4. Dependence of the electron mobility in PbTe on the electron density in the $2 \cdot 10^{18}$ to $1 \cdot 10^{20}$ cm^{-3} range at 77 and 2.4°K; the values are taken from various papers reviewed in [249].

due to shear are three orders of magnitude smaller than the experimentally observed cross sections.

Dislocations may scatter carriers and affect strongly the carrier mobility, particularly at low temperatures [220, 227]. Dislocations may capture electrons, holes, and ions and they can behave like charged lines. These lines are surrounded by cylindrical sheaths consisting of charge of opposite sign. The considerable scatter of the mobilities measured in different samples may be due to the influence of dislocations. Allgaier and Scanlon [220] increased the dislocation density in a particular sample. They have found that the carrier density had not been affected but the mobility had been reduced strongly.

In inhomogeneous samples, the lines of flow of the electric current are curved and this reduces the observed mobility [248]. Such curvature of the current flow lines may be due to dislocations surrounded by space-charge regions. The effect of such inhomogeneities may be manifested by a large scatter of the observed mobility values not only at low but also at high temperatures. Internal microscopic cracks in the samples may have a similar effect.

Using the conclusions reached in our discussion of the scattering mechanisms, we shall now discuss in detail (by way of example) the dependence of the mobility on the carrier density in degenerate n-type PbTe samples at liquid-nitrogen temperatures [249]. Figure 3.4 shows the mobilities at 4.2 and 77°K for samples with electron densities $2 \cdot 10^{18}$–10^{20} cm^{-3}. The experimental points, taken from various investigations, were joined by Moizhes and Ravich [249] by continuous curves which they then analyzed. The curves representing the mobility are the upper limits of fairly strongly scattered experimental values. The points lying considerably below the curves obviously apply to imperfect samples in which the mobility is governed by dislocations, cracks, inhomogeneities, etc.

The Fermi level in the cited range of carrier densities lies fairly high above the bottom of the conduction band and the nonparabolicity of this band has a considerable effect on the carrier-density dependence of the mobility. An allowance for the nonparabolicity can be made using one of the two simple models described in §6.2B. The results obtained using these two models do not differ greatly from one another and, therefore, we shall quote only the results obtained by means of the Kane model.

In the Kane model, the expression for the mobility in a strongly degenerate semiconductor with an arbitrary dependence of the relaxation time on the energy $\tau(\mathscr{E})$ is of the form [250]:

$$\mu = \frac{e\tau(\zeta)}{m^*_{\chi 0}(1 + 2\zeta/\mathscr{E}_{gi})}, \tag{3.27}$$

where ζ is the Fermi energy; $\mathscr{E}_{gi}$ is the effective width of the forbidden interaction band (cf. §6.2B); $m^*_{\chi 0}$ is the electric-susceptibility (or conductivity) effective mass near the band edge [cf. Eq. (2.22)].

The dependence of the Fermi energy on the electron density is found by analysis of the data on the thermoelectric power in a strong magnetic field and is given in Fig. 4.7 (cf. §4.1C). Using this dependence, we can find the relaxation time as a function of the energy at 77 and 4.2°K. Since, at helium temperatures, the mobility is entirely due to the interaction of electrons with defects, the quantity τ_{e-ph}, defined by

$$\frac{1}{\tau_{e-ph}} = \frac{1}{\tau_{77°K}} - \frac{1}{\tau_{4.2°K}}, \tag{3.28}$$

describes the interaction of electrons with phonons at 77°K.

Differentiating graphically the function $\tau(\zeta)$, we obtain the quantity

$$\frac{d \ln \tau}{d \ln \zeta} = \left(\frac{\partial \ln \tau}{\partial \ln \zeta}\right)_n + \left(\frac{\partial \ln \tau}{\partial \ln n}\right)_\zeta \frac{d \ln n}{d \ln \zeta}. \qquad (3.29)$$

If we bear in mind that τ_{e-ph} does not depend explicitly on the carrier density n and that $\tau_{4.2°K} \propto n^{-1}$, we can easily show that the partial derivative $(\partial \ln \tau / \partial \ln n)_\zeta$ is given by

$$\left(\frac{\partial \ln \tau}{\partial \ln n}\right)_\zeta = -\frac{\tau_{77°K}}{\tau_{4.2°K}}. \qquad (3.30)$$

The quantity $d \ln n / d \ln \zeta$ is calculated by differentiating a graph obtained from measurements of the thermoelectric power in a strong magnetic field (Fig. 4.7). Then, using Eqs. (3.29) and (3.30), we can find the quantity

$$r(\zeta) = \left(\frac{\partial \ln \tau}{\partial \ln \mathscr{E}}\right)_{n, \mathscr{E}=\zeta}. \qquad (3.31)$$

This quantity (which we shall refer to later as r_μ in order to indicate its origin) differs considerably from the quantity r_α deduced from the thermoelectric power (cf. §4.1E). This difference indicates that the scattering in lead chalcogenides is inelastic. Although our analysis of the relaxation time is not completely accurate (in fact, τ_{e-ph} depends weakly but explicitly on n because of the influence of screening on the polar scattering), the conclusion that the scattering is inelastic has been confirmed by other results. We shall consider the inelasticity of scattering in Chap. IV.

D. Role of Scattering of Carriers on Each Other

Collisions between carriers have much in common with the scattering of carriers by ionized impurities. Therefore, in nondegenerate covalent semiconductors, the carrier − carrier scattering mechanism is usually important only at those carrier densities and temperatures at which the scattering by the screened Coulomb potential of ionized impurities must be taken into account (cf., for example, [251]). In view of the high permittiviity of lead chalcogenides, the scattering by Coulomb centers is not very effective in these compounds. However, there are several important differences between the scattering of electrons on heavy ions at rest and on electrons moving rapidly across a crystal and participating in the transport phenomena.

PbTe-type crystals are partly ionic and the high-frequency permittivity ε_∞, due to the polarization of the rapidly responding electron system, is considerably smaller than the static permittivity ε_0, associated with the polarization of the electron and ion systems. For example, $\varepsilon_0 = 400$ but $\varepsilon_\infty = 35$ in the case of PbTe (cf. 2.1D and §3.4B). An electron, moving rapidly across a crystal, polarizes only the electron shells of the lattice ions, while the centers of these ions do not shift during the short time of the passage of the electron near a given ion. Estimates show [249] that at an electron density of $2 \cdot 10^{18}$ cm^{-3} in PbTe an electron at the Fermi level travels a distance equal to the screening radius in about 10^{-14} sec. The response of the lattice ions is represented by the period of the longitudinal optical vibrations, $3 \cdot 10^{-13}$ sec [252], which is more than an order of magnitude longer than the transit time of such an electron. Thus, an impurity ion can be considered to be at rest relative to the lattice and the lattice around it should be regarded as completely polarized.

It follows that the expression for the screened Coulomb potential, established by an impurity ion, contains the static permittivity, while the potential of the electron–electron interaction includes the high-frequency permittivity, which increases the effectiveness of the electron–electron scattering by more than two orders of magnitude. The electron–electron interaction potential can be divided into two zones, near and far, but since in this case the screened Coulomb potential is relatively large, it may play a more important role than in the scattering by ionized impurities.

On the other hand, there are factors which reduce the importance of the electron–electron collisions in the total scattering. Because the total momentum is conserved, the electron–electron collisions do not alter the electric current directly and their role reduces to a redistribution of the energy between the colliding particles. In calculations of the electrical conductivity and the Hall effect of a strongly degenerate semiconductor, we need include only the contribution to the current of the electrons having the Fermi energy, assuming that the distribution function is step-like. Therefore, a redistribution of the energy between carriers is unimportant and the electron – electron collisions do not have much influence on the mobility and Hall effect in degenerate semiconductors. The situation is different in the thermoelectric and thermomagnetic effects and in the thermal conductivity, where a redistribution of the energy is important even in degenerate samples. Measurements of these properties indicate that the electron – electron collisions play an important role in degenerate PbTe samples with electron densities of the order of 10^{18}-10^{19} cm^{-3} at liquid-nitrogen temperature. This problem will be discussed in greater detail in Chap. IV.

Collisions between carriers of opposite sign affect the current directly and therefore the scattering of the minority by the majority carriers may reduce the minority-carrier mobility. This effect was observed by Baryshev et al. [227] in PbSe. The minority-carrier mobility, determined by measuring the photomagnetic effect in a strong magnetic field, was found to depend less strongly on the temperature than the majority-carrier mobility and at 77°K the minority carriers had a much lower mobility than the majority carriers. Theoretical estimates given by Baryshev et al. [227] were in agreement with the experimental results.

E. Mobility in Solid Solutions

The mobility in solid solutions exhibits some special features. When atoms of one chalcogen are replaced by atoms of another, the band structure of the resultant solid solutions differs little from the structure of pure substances because the band structures of all three lead chalcogenides are similar. However, departures from the lattice periodicity give rise to an additional scattering mechanism, similar to the scattering by neutral impurities. When lead is replaced by atoms of other elements, for example, by Sn, the situation becomes more complex.

The carrier mobility has been investigated in the following solid solutions: PbTe – PbSe [216, 253, 254], PbTe – SnTe [253-256], PbTe – AgSbTe$_2$ [257], and several other systems. Stil'bans et al. [216, 253, 254] demonstrated that the carrier mobility in the PbTe – PbSe system is lower than that in pure telluride or selenide. The dependences of the electron and hole mobilities on the composition are shown in Fig. 3.5 for 300°K. They are in the form of curves

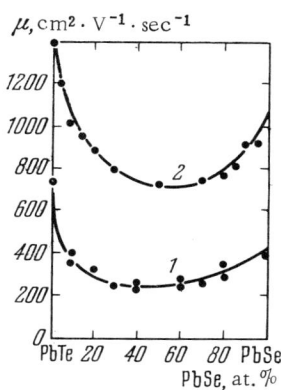

Fig. 3.5. Dependences of the hole (1) and electron (2) mobilities on the composition of PbTe – PbSe alloys [254].

with shallow minima. At low temperatures, there is a small maximum near a composition corresponding to equal numbers of Se and Te atoms; this maximum is due to some ordering of the lattice structure at this composition.

It is interesting to note that when one chalcogen is replaced with another, the mobility of holes decreases by a factor of 2.5 while the mobility of electrons decreases only by 25%, i.e., the decrease in the electron mobility is much smaller. This phenomenon has been observed also in other solid solutions and has been explained by Stil'bans et al. [253] by assuming that electrons travel mainly in the cation (Pb) sublattice, while holes travel mainly in the anion (chalcogen) sublattice. This phenomenon will be discussed later (cf. §6.1C) in the light of the results of theoretical calculations of electron wave functions for PbTe on the assumption of the total absence of ionicity.

The replacement of lead by tin reduces the electron mobility, but the results obtained for p-type PbTe – SnTe samples are more complex and contradictory [253-256]. In particular, the dependence of the hole mobility on the composition in the presence of a relatively small percentage of Sn is strongly nonmonotonic [254]. Since the energy bands of PbTe and SnTe differ considerably, the replacement of lead by tin alters the band structure and this affects the scattering mechanism.

The results obtained for solid solutions of $A^{IV}B^{VI}$ compounds should be supplemented by the results obtained by Efimova et al. [67] for PbTe doped with bismuth, which occupies the Pb sites, and with halogens (Cl, Br, I), which occupy the Te sites. The replacement of lead by bismuth is twice as effective in the electron scattering as the replacement of tellurium by halogens (the scattering cross sections have been compared). This indicates that the amplitudes of the electron wave functions in the conduction band are larger in the vicinity of Pb atoms than in the vicinity of Te atoms, which is in agreement with the results of theoretical calculations (cf. §6.1C).

F. Mobility in Thin Films

The electrical properties of thin films of lead chalcogenides are, in many cases, similar to the electrical properties of bulk single crystals [40, 128, 258, 259]. In less perfect films, the scattering by crystallites reduces the mobility at low temperatures [128]. In PbS films about a micron thick, investigated by Spencer and Morgan [260], the mobility is considerably lower than in bulk single crystals and is proportional to the film thickness. The effective value obtained for the mobility is explained, on the basis of Petritz's theory [154], by the influence of energy barriers which exist at the boundaries between crystallites, and the thinner the film the greater is the number of these boundaries. In very thin PbTe films, prepared by Zemel et al. [128], the mobility was found to vary with temperature in accordance with the law $T^{-1/2}$, which is explained by the surface scattering [261]. This slower change of the mobility with temperature than in bulk crystals has also been observed in thin single-crystal epitaxial films of PbSe and PbTe.

3.2. Hall Effect and Other Galvanomagnetic Phenomena

Charge carriers moving at a velocity **v** in a magnetic field **H** are acted upon by a force proportional to the vector product **v · H**. Consequently, the direction of the carrier drift in simultaneously applied electric and magnetic fields does not coincide with the direction of the electric field. In particular, in semiconductors with simple energy bands, carriers are deflected by a weak transverse magnetic field from the electric field direction by an angle $\tilde{\mu}H/c$, where the Hall mobility $\tilde{\mu}$ differs from the ohmic mobility by a factor A which is close to unity.

The deflection of the carrier drift direction by a magnetic field produces several galvanomagnetic effects. When a current in a sample flows along the x direction and a magnetic field is applied along the z direction, an electric Hall field appears in the y direction and this field is proportional to the current density and the

Sec. 3.2] HALL EFFECT 111

magnetic field:

$$E_H = RjH, \quad (3.32)$$

where R is the Hall coefficient, which depends on the carrier density, energy-band structure, carrier scattering mechanism, and degree of degeneracy. The ratio of the Hall field to an external electric field applied along the x axis in a sample with one type of carrier is equal to the tangent of the Hall angle $\tilde{\mu}H/c$ and hence we obtain

$$R = \frac{A}{enc}. \quad (3.33)$$

The numerical factor $A = \mu/\tilde{\mu}$ will be discussed later.

A change in the direction of the carrier drift causes such phenomena as the planar Hall effect and an increase of the resistance in the magnetic field. In the present section, we shall consider all these galvanomagnetic effects in nonquantizing magnetic fields, whereby we can assume that these fields do not alter the energy spectrum of electrons. The resistance in quantizing magnetic fields will be discussed in §5.2.

It is evident from Eq. (3.33) that measurements of the Hall effect in weak magnetic fields ($\mu H/c \ll 1$) make it possible to find the carrier density in the extrinsic (impurity) conduction region if the factor A is known or if it differs slightly from unity. Measurements of the Hall coefficient in strong magnetic fields ($\mu H/c \gg 1$) are even more convenient for the determination of the carrier density because then the coefficient is

$$R_\infty = \frac{1}{enc}. \quad (3.34)$$

Measurements of the Hall effect are among the most widely used methods for the determination of the carrier density in semiconductors. Since a change of the sign of carriers reverses the direction of the Hall field, the sign of the Hall effect reveals the type of conduction. Measurements of the temperature and carrier-density dependences of the Hall effect are used to detect bands whose edges lie a little below the top of the upper valence band or a little higher than the bottom of the lower conduction band.

Investigations of the Hall effect at high temperatures, when the minority carriers become important, make it possible to determine the intrinsic carrier density, forbidden band width, and ratio of the electron and hole mobilities in a given sample. Finally, the nature of a complex anisotropic structure of the energy bands can be determined by measuring the dependences of the galvanomagnetic effects (particularly that of the magnetoresistance) on the magnetic field strength and its orientation relative to the crystallographic axes.

A. Hall Effect in the Extrinsic Conduction Region

Before we present the results of investigations of lead chalcogenides in the extrinsic (impurity) conduction region, we shall consider the effects responsible for the carrier-density and temperature dependences of the Hall effect in weak magnetic fields. Equation (3.33) has two variables: n and A. The factor A is equal to unity if the following conditions are satisfied simultaneously:

1) there is only one type of carrier in a semiconductor;
2) the constant-energy surfaces are spherical;
3) the dependence of the energy on the crystal momentum is quadratic;
4) the relaxation time is independent of the energy.

(The last two conditions can be replaced by the condition of strong degeneracy.)

We shall now consider the case when conditions 2) and 4) are not satisfied. To take a specific case, we shall assume that the band structure in a cubic crystal is described by a many-ellipsoid model and that the scattering is isotropic, i.e., the relaxation time depends solely on the energy. The factor A is then a product of two quantities: the anisotropy factor

$$A_K = \frac{3K(K+2)}{(2K+1)^2} \qquad (3.35)$$

(where K is the ratio of the effective masses $m_{\parallel}^*/m_{\perp}^*$) and the statistical factor

$$A_\tau = \frac{\langle\tau^2\rangle}{\langle\tau\rangle^2}, \qquad (3.36)$$

where the angular brackets denote the statistical averaging described by Eq. (3.5). The value of A_K decreases as K increases and, when $K \gg 1$, it depends weakly on K. For example, $A_K = 0.82$ for $K = 10$; $A_K \to 0.75$ as $K \to \infty$; $A_K = 0$ when $K = 0$. For any deviation of the mass anisotropy coefficient K from 1, we find that $A_K < 1$, i.e., the anisotropy always reduces the Hall effect.

If the scattering is anisotropic, K in Eq. (3.35) is replaced with $K' = (m_\parallel^* / m_\perp^*)$ where $\tau_\parallel$ and $\tau_\perp$ are the relaxation times of electrons moving along and across the ellipsoid axis, respectively. The statistical factor A_τ depends on the degree of degeneracy and is given by

$$A_\tau = \frac{3}{2} \frac{2r + 3/2}{(r + 3/2)^2} \frac{F_{2r+1/2}(\zeta^*) F_{1/2}(\zeta^*)}{[F_{r+1/2}(\zeta^*)]^2}, \qquad (3.37)$$

where F_i are the Fermi integrals; r is the scattering parameter. For strongly degenerate semiconductors $A_\tau = 1$, while in the classical case

$$A_\tau = \frac{\Gamma(2r + 5/2) \Gamma(5/2)}{[\Gamma(r + 5/2)]^2}. \qquad (3.37a)$$

In general, the dependence of the relaxation time on the energy increases the Hall effect.

In particular, when carriers are scattered by the acoustical phonons ($r = -1/2$),

$$A_\tau = \frac{3\pi}{8} \approx 1.18;$$

and when $K = 10$

$$A = A_K A_\tau = 0.98.$$

In this typical case, the anisotropy and the energy dependence of the relaxation time compensate each other and the Hall factor becomes $A \approx 1$.

When carriers are scattered by Coulomb centers ($r = 3/2$), the statistical factor is larger (about 1.9) but in uncompensated samples this quantity is much smaller because of collisions between carriers (the statistical factor for covalent semiconductors is 1.2 and for polar semiconductors it is even smaller). We must also mention that the scattering of carriers by the screened Coulomb potential in lead chalcogenides is unimportant (cf. §3.1C).

Thus, in the cases considered so far the factor A differs little from unity and it is frequently ignored in the determination of the carrier density from the Hall effect data. In the case of a constant carrier density, the factor A can slightly alter (by 5-10%) the Hall effect when the temperature is increased, or when the degeneracy is lifted, or when the relative importance of various scattering mechanisms or the values of the mass anisotropy coefficient change.

Deviations of the statistical factor from unity may be caused also by the band nonparabolicity. This effect can be allowed for by replacing τ in Eq. (3.36) with (τ/m^*), where m^* is given by Eq. (3.13). Since the effective mass usually depends on the energy in accordance with a law which cannot be reduced to a power law, the statistical factor is then expressed not in terms of the Fermi integrals but in a more complex manner (Appendix A).

The presence of two or more types of carriers with different mobilities (for example, the presence of light and heavy holes) may alter the factor A considerably. This effect, which produces a fairly strong temperature dependence of the Hall effect in p-type PbTe, is discussed in the next subsection.

Investigations of the Hall effect in lead chalcogenides have been carried out by many workers [38, 44, 46-48, 62, 63, 66, 123, 209, 219, 220, 231, 232, 262-268]. As a rule, the temperature dependence of the Hall effect is very weak between low (liquid-helium) and room or higher temperatures: it amounts to 5-10% for all lead chalcogenides, with the exception of p-type PbTe. Such a weak temperature dependence, observed up to temperatures at which considerable numbers of minority carriers appear in a semiconductor, indicates that the carrier density is practically independent of temperature. A weak variation of the Hall coefficient may be due to a change in the factor A. The temperature dependences of the Hall coefficient and electrical resistivity of PbS, typical of all three lead chalcogenides, are shown in Fig. 3.6. In the region where the carrier density is constant, the observed increase of the resistivity with rising temperature is due to a reduction in the carrier mobility.

The Hall effect data yield information on the energy levels of electrons due to impurities or vacancies in semiconductors. Let us consider first how the carrier density should vary with temperature if the forbidden band has donor (or acceptor) levels separated by small gaps from the allowed bands ("shallow" levels).

At absolute zero, such levels are fully occupied by carriers, the free carrier band is empty, and the Fermi level lies in the middle between the energies of the impurity levels and the free band edge. As the temperature is increased, carriers are transferred from the impurity levels to the allowed band, the Fermi level shifts in the direction of the allowed band, and (provided the number of

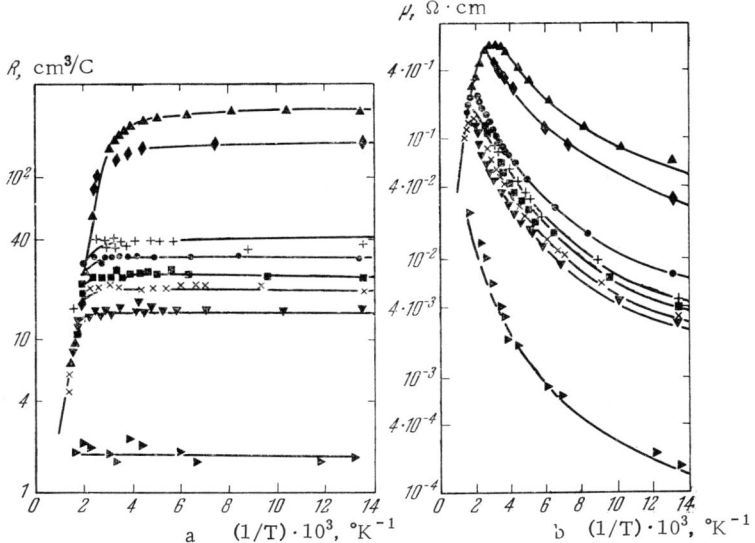

Fig. 3.6. Temperature dependences of the Hall coefficient (a) and resistivity (b) of PbS [48].

these impurity levels is sufficient) the Fermi level actually crosses into the band, i.e., the sample becomes degenerate. When the temperature is increased still further, the depletion of the impurity levels begins to have an effect, the degeneracy is gradually lifted, and the Fermi level shifts back to the forbidden band. When the Fermi level falls (rises) so that its gap from the donor (acceptor) levels becomes greater than k_0T, all the impurity levels become ionized, while the carrier density becomes independent of temperature and equal to the impurity concentration so that the carrier statistics approaches the classical case.

It is important to note that, even in the presence of very shallow impurity levels in the forbidden band, the carrier density is constant only at those temperatures at which there is no degeneracy. The Hall effect of lead chalcogenides remains constant at temperatures at which degeneracy is very strong. Hence, we may conclude that, below the Fermi level there are no donor or acceptor levels which could be responsible for the high carrier density in lead chalcogenides.

The depth of a level of a single charged impurity ion can be estimated roughly by assuming the potential of the ion to be of the Coulomb type [270]. It is found that the electron energy at an impurity level is less than that in the hydrogen atom because of the appearance, in a solid, of the factor $(m^*/m_0)(1/\varepsilon_0)^2$, which has a very small value for semiconductors with a small effective mass and high permittivity. For example, the ionization energy of an impurity center in PbTe ($m^* \approx 0.05 m_0$, $\varepsilon_0 = 400$) is $3 \cdot 10^6$ times smaller than the ionization energy of the hydrogen atom and is only 10^{-5}-10^{-6} eV.

The radius of a Bohr orbit in a solid, a_B, is $(m_0/m^*)\varepsilon_0$ times larger (in the example just given, it is 8,000 times larger) than in the hydrogen atom. At vacancy concentrations which are normally found in lead chalcogenides, the Bohr radius (in the example considered it is $4 \cdot 10^{-5}$ cm) is larger than the average distance between vacancies and, therefore, localized impurity levels merge into one impurity band. Moreover, when the condition $(na_B^3)^{1/3} \gg 1$ is satisfied, it is found that such an impurity band merges with an allowed band to form a single band; in this case, the properties of carriers far from the band edge differ little from the properties of carriers in lightly doped semiconductors (this is not true of car-

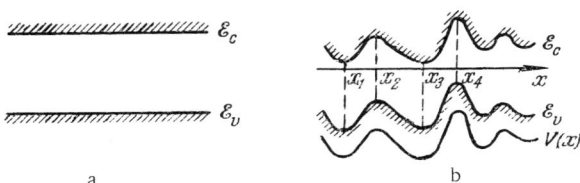

Fig. 3.7. Position of the bottom of the conduction band $\mathscr{E}_c$ and top of the valence band $\mathscr{E}_v$ in space [272]: a) in a pure semiconductor; b) in a heavily doped semiconductor. $V(x)$ is the electrostatic potential.

riers near the band edge). In our example, $a_B^3 = 0.8 \cdot 10^{-13}$ cm^3 and, therefore, the inequality just given is satisfied at carrier densities which are usually encountered in lead chalcogenides.

A theory of heavily doped semiconductors was developed by Bonch-Bruevich, Keldysh, and others [271, 272]. A change in the band structure of a semiconductor due to heavy doping can be explained qualitatively [272] by postulating that the charge of impurities distributed randomly in space gives rise to an electrostatic potential, which fluctuates from point to point and which shifts the conduction and valence band edges as shown in Fig. 3.7b. In such "distorted" bands, the fluctuations give rise to states in the conduction and valence bands which lie at energies corresponding to a forbidden band in a pure semiconductor. We thus obtain what are known as density-of-states "tails" extending into the forbidden band. This effect reduces the thermal width of the forbidden band but does not affect the edges of the optical recombination spectra since the local forbidden band width is still the same at each point. The situation is more complex when the effective electron and hole masses have different orders of magnitude [272]. Since the effective masses in lead chalcogenides are of the same order of magnitude, we can ignore this complication. Estimates show that the density-of-states "tails" of lead chalcogenides are negligibly short because of the high permittivity and small effective masses of these materials.

It must be mentioned that the theory [271, 272] predicts that at high impurity concentrations there are no impurity levels or impurity bands, either in the forbidden band or within the allowed bands, and that carriers are found only in the allowed bands. The role of impurities reduces simply to the establishment of a high carrier density in an allowed band, which is retained down to the lowest temperatures, so that lead chalcogenides can be regarded as impurity semimetals. This property makes it possible to observe superconductivity in PbTe [269] as well as several other effects in strong electric fields at liquid-helium temperatures (cf. Chap. V).

If the potential of an impurity atom is not of the Coulomb type, the impurity levels and bands, separated by an energy gap from an allowed band, may appear in the energy structure. Measurements of the Hall effect in PbS doped with copper and nickel in concentrations of about 10^{18} cm^{-3} were carried out by Bloem and Kröger [46, 47]; they found that the activation energies of copper and nickel are 0.02 and 0.03 eV, respectively. The existence of deep impurity levels in some PbTe samples has been reported in [62, 63, 209].

An increase of the Hall coefficient by up to 40% due to heating from room temperature to 850°K was observed by Alekseeva et al. [268] in heavily doped n-type PbTe samples ($n = 7 \cdot 10^{19}$-$2 \cdot 10^{20}$ cm^{-3}). Such behavior of the Hall coefficient in samples with high carrier densities is explained by the temperature dependence of the factor A, which is associated with the nonparabolicity of the conduction band of PbTe.

B. Special Features of the Hall Effect in p-Type PbTe

In contrast to other lead chalcogenides, p-type PbTe exhibits a considerable rise of the Hall coefficient above 150°K [223, 231, 232]. The ratio $R_{296°K}/R_{77°K}$ increases from 1.2 to 1.6 when the hole density is raised from $5 \cdot 10^{17}$ to $1.7 \cdot 10^{19}$ cm^{-3}. This increase in the Hall effect was explained by Allgaier [231] by electron transfer to a low-mobility valence band whose maximum (top) lies below the top of the upper valence band. In the presence of heavy holes, the expression for the Hall coefficient is of the form

$$R = \frac{1}{ec} \frac{A_{p1} b_p^2 p_1 + A_{p2} p_2}{(b_p p_1 + p_2)^2}, \qquad (3.38)$$

where p_1 and p_2 are the densities of the light and heavy holes, whose sum is independent of temperature and equal to the number of acceptors; $b_p = \mu_{p_1}/\mu_{p_2}$ is the ratio of the mobilities of the light and heavy holes. In view of the low mobility of holes in the second (heavy-hole) valence band, these holes make a relatively small

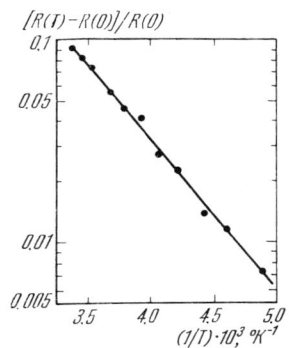

Fig. 3.8. Temperature dendence of the Hall coefficient of p-type PbTe with a hole density of $3.5 \cdot 10^{17}$ cm^{-3} [232].

contribution to the Hall effect and as holes are transferred to the second band, the effective hole density decreases, while the Hall coefficient increases in accordance with the formula

$$R = \frac{A_{p1}}{ec p_1}. \qquad (3.39)$$

When the carrier density increases, the Fermi level approaches the maximum (top) of the heavy-hole valence band and when the temperature is increased, an increasing number of holes may be transferred to the low-mobility (heavy-hole) band. This explains the increase of the ratio $R_{296°K}/R_{77°K}$ with increasing carrier density. When the Fermi level passes into the heavy-hole valence band, the ratio $R_{296°K}/R_{77°K}$ should decrease with increasing carrier density [232]. In view of the high density of states in the heavy-hole band, the Fermi level occupies a practically constant position when it reaches the edge of the second band. Such a dependence of the Fermi level on the carrier density is confirmed by the carrier-density dependence of the thermoelectric power [265] (cf. §4.1A).

An analysis of the temperature dependence of the Hall effect in PbTe samples with a hole density of $3.5 \cdot 10^{17}$ cm^{-3} [232] has been carried out using the formula [273]:

$$\frac{R(T) - R(0)}{R(0)} = \left(1 - \frac{1}{b_p}\right)^2 \left(\frac{m_{d2}^*}{m_{d1}^*}\right)^{3/2} e^{-\Delta \mathscr{E}/k_0 T}, \qquad (3.40)$$

which has been deduced for the classical statistics case. It follows from this formula that the slope of the straight line, giving the dependence $\ln \{[R(T) - R(0)]/R(0)\}$ on $1/t$ (Fig. 3.8), yields the value of the energy gap $\Delta \mathscr{E}$ between the two valence bands. If we assume that $\Delta \mathscr{E}$ varies linearly with temperature, it follows that the value of $\Delta \mathscr{E} = 0.14$ eV obtained in this way applies at absolute zero. Allowance for the temperature dependence of the effective mass alters

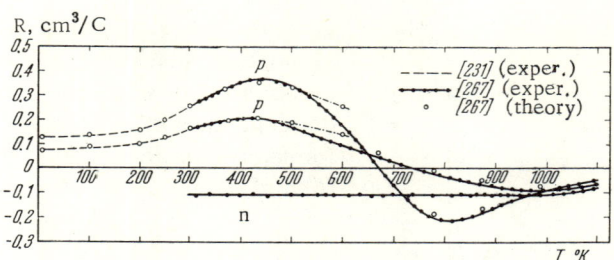

Fig. 3.9. Temperature dependences of the Hall coefficient of heavily doped PbTe samples.

the value of the gap to $\Delta\mathscr{E} = 0.16$ eV. The heavy-hole mobility is approximately an order of magnitude lower than the light-hole mobility but the effective mass of the heavy holes is larger than the free-electron mass.

Investigations of the temperature dependence of the Hall effect in p-type PbTe have been extended to temperatures up to 1100°K [265-267]. It has been found that $\Delta\mathscr{E}$ decreases quite rapidly with increasing temperature and vanishes altogether at some temperature.

The temperature dependence of the Hall coefficient of heavily doped p-type PbTe [223, 231, 267] is shown in Fig. 3.9. Between 4.2 and 150°K, the Hall coefficient retains its constant value; it then gradually increases, reaching a maximum at about 430°K [266, 267]. The position of the maximum is almost independent of the carrier density. A fall of the Hall coefficient at still higher temperatures is observed well before the intrinsic conduction region so that it cannot be attributed to the intrinsic carrier processes.

A theoretical analysis of the experimental temperature dependence of the Hall effect gives the following value for the energy gap between the valence bands: $\Delta\mathscr{E} = (0.17 - 4 \cdot 10^{-4} T)$ eV [267]. Consequently, at about 400°K, the edges of the two valence bands are at approximately the same level and at temperatures above 400°K the main contribution to the transport effects is made by the heavy-hole band. Changes in the band structure with temperature are shown schematically in Fig. 3.10, which is taken from a paper by Andreev and Radionov [267]. The forbidden band width at temperatures above 400°K depends weakly on temperature since the gap between the edges of the conduction and heavy-hole bands remains constant. Such a temperature dependence of the forbidden

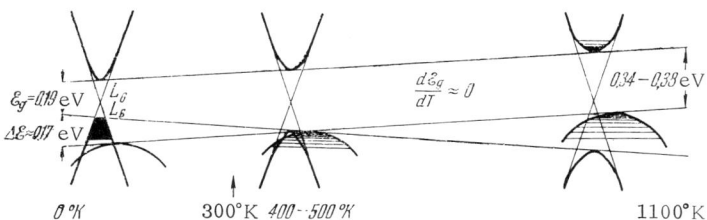

Fig. 3.10. Changes in the positions of the valence and conduction band edges of PbTe with increasing temperature [267].

band width has been found experimentally from the optical absorption [64, 123] and electronic thermal conductivity [274] (cf. Fig. 4.20). The value of the effective hole mass in the second (heavy-hole) valence band is $1.5m_0$ [267].

The presence of the heavy-hole valence band in PbTe affects also the carrier-density dependences of the Hall effect and thermoelectric power. Kolomoets et al. [265] compared the Hall density of electrons $n_H = 1/ecR$ at room temperature with the true electron density, which was assumed to be equal to the impurity concentration found by spectroscopic analysis. They found that the two values of the density differ considerably above $2 \cdot 10^{19}$ cm^{-3}. The energy gap $\Delta\mathscr{E}$, determined by Kolomoets et al. from the value of the Hall coefficient at a carrier density of $1.34 \cdot 10^{20}$ cm^{-3}, exceeds somewhat the value found from the temperature dependence of the Hall effect. The effective mass of the heavy holes found by Kolomoets et al. is $1.2m_0$. An analysis of the temperature dependence of the thermoelectric power shows that $\Delta\mathscr{E}$ decreases with increasing temperature [265] (cf. §4.1A), in agreement with the Hall-effect data.

Allgaier [231] investigated the valence bands of PbTe and measured the dependence of the Hall effect on the magnetic field intensity. He found that the Hall coefficient of PbTe with a hole density not less than $2 \cdot 10^{19}$ cm^{-1} does not change by more than 7% in a magnetic field at 77°K. The Hall coefficient in weak and strong fields differs by a factor which depends on the carrier scattering mechanism, degree of degeneracy, and anisotropy [as indicated by Eqs. (3.33) and (3.34)] so that a weak dependence on the magnetic field intensity can be explained by the influence of these effects. Hence, we may conclude that the properties of p-type PbTe at low temperatures and moderately high hole densities are governed solely by the light holes in the <111> ellipsoids, and that the heavy holes have no effect.

C. Hall Effect in Mixed and Intrinsic Conduction Regions

The minority carriers appear in considerable numbers when the temperature is sufficiently high and then the Hall effect of all three lead chalcogenides varies quite strongly with temperature. In the mixed conduction region, the Hall coefficient and electrical conductivity are

$$R = \frac{A_p \mu_p^2 p - A_n \mu_n^2 n}{(\mu_p p + \mu_n n)^2} \frac{1}{ec}, \tag{3.41}$$

$$\sigma = e\mu_p p + e\mu_n n. \tag{3.42}$$

The carrier densities n and p can be expressed in terms of the intrinsic density n_i and the donor density n_d:

$$np = n_i^2, \tag{3.43}$$

$$n - p = n_d \tag{3.44}$$

(for an n-type sample). By measuring the Hall coefficient and electrical conductivity at low temperatures, we can find the donor concentration n_d and the majority-carrier mobility μ_n. Assuming that the donor concentration is independent of temperature, extrapolating the temperature dependence of the mobility, and measuring R and σ at high temperatures, we find that we have four equations, (3.41)-(3.44), for the determination of four unknowns, n, p, n_i, and μ_p, so that we can find in this way the intrinsic carrier density n_i and the ratio of the electron and hole mobilities $b = \mu_n/\mu_p$.

These quantities can be found particularly conveniently at the temperature at which the Hall effect changes its sign in samples exhibiting p-type conduction at low temperatures. The change of sign of the Hall effect at the point of transition from p-type to intrinsic conduction is observed for all three lead chalcogenides; it indicates that the electron mobility is higher than the hole mobility. If we assume that the factors A_n and A_p do not differ from unity we find that, at a temperature defined by

$$b^2 n = b n_i = p, \qquad (3.45)$$

the Hall effect vanishes and the conductivity becomes equal to its intrinsic value. Having measured the intrinsic conductivity and the ratio of the carrier mobilities we can find the intrinsic carrier density at the temperature of the Hall effect inversion.

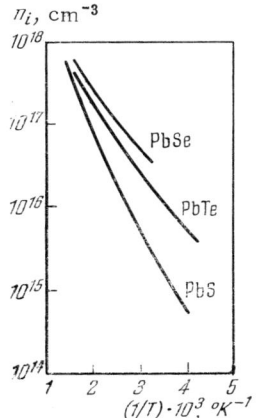

Fig. 3.11. Temperature dependences of the intrinsic carrier density in PbTe, PbSe, and PbS [5].

The temperature dependence of the intrinsic carrier density obtained from the Hall effect measurements is shown in Fig. 3.11 for PbTe, PbSe, and PbS. At room temperature, the intrinsic carrier density is $1.6 \cdot 10^{16}$, $3 \cdot 10^{16}$, and $2 \cdot 10^{15}\,\text{cm}^{-3}$ in PbTe, PbSe, and PbS, respectively. The temperature dependence of n_i is given by the formula [270]:

$$n_i = 2 \frac{\left(2\pi \sqrt{m_{dn}^* m_{dp}^*} k_0 T\right)^{3/2}}{h^3} e^{-\mathscr{E}_g/2k_0 T}. \qquad (3.46)$$

If we neglect the temperature dependence of the effective mass, we find that the dependence $\ln(n_i^2/T^3)$ on $1/T$ is a straight line whose slope can be used to find the forbidden band width. By assuming that the forbidden band width is a linear function of temperature in this temperature range $\left(\mathscr{E}_g = \mathscr{E}_{g0} + \frac{\partial \mathscr{E}_g}{\partial T} T\right)$ this method can be used to determine the value of $\mathscr{E}_{g0}$ by extrapolation of the linear dependence to absolute zero.

The first measurements of this kind yielded a greatly overestimated value for $\mathscr{E}_{g0}$ [208] because of irreversible changes in the composition of lead chalcogenides at temperatures above 500°K [3]. Measurements in the temperature range 300–500°K, carried out on PbTe [219] and PbS [48], gave values for $\mathscr{E}_{g0}$ which are in better agreement with the results of optical investigations but are still somewhat higher than the forbidden band width at 0°K. The main cause of the discrepancy is obviously the nonlinearity of the temperature dependence of the forbidden band width in the 300-500°K range (cf. §2.1C, §3.2B, and §4.2B).

An analysis of the temperature dependence of the Hall coefficient and electrical conductivity of PbTe in the intrinsic conduction region (500-900°K), reported in [123], has yielded $\mathscr{E}_{g0}$ = 0.36 eV, in agreement with the forbidden band width obtained in the same investigation by optical methods; the forbidden band width has been found to be independent of temperature above 400°K. The thermal values of the forbidden band widths of PbSe [38] and PbTe [264] are also in good agreement with the optical values.

The thermal values of the forbidden band width of PbTe and PbSe have been obtained also from measurements of the effect of hydrostatic pressure on the electrical properties (cf. §3.3A).

The ratio of the electron and hole mobilities, measured in n- and p-type samples, respectively, may not agree with the ratio of the electron and hole mobilities measured in the same sample because of a considerable scatter of mobilities in various samples. Therefore, it is interesting to determine this ratio from measurements at the temperature of inversion of the sign of the Hall effect. The values of b obtained in this way are 1.4 for PbS [48] and PbSe [212] and 2.5 for PbTe [219]. A more careful analysis has indicated [2] that the value of b increases with temperature.

The mixed conduction region of p-type PbTe begins at a lower temperature than that of n-type samples [267]; this is attributed to electron transitions to the second valence band and a consequent reduction of the average hole mobility, so that b = 9 at 700-800°K. When the value of b is so high, even a low density of electrons makes the electron contribution to the Hall effect quite appreciable, because this contribution is proportional to $b^2 n/p$.

D. Resistance in a Magnetic Field and the Planar Hall Effect

Because the carrier trajectories in a magnetic field are curved, the relationship between the electric field and current density is more complex than in the absence of a magnetic field:

$$j_i = \sum_{k=1}^{3} \sigma_{ik}(\mathbf{H}) E_k. \tag{3.47}$$

The conductivity is described by a tensor and each component of this tensor depends on the value and direction of the magnetic field.

When the coordinate axes in a cubic crystal are directed along the principal axes of this crystal, the relationship between the electric field and the current density in a weak magnetic field is given by the following formula, which is accurate up to terms quadratic in respect of the magnetic field,

$$E_i = \rho_0 \{j_i + a\,[\mathbf{jH}]_i + bj_i H^2 + cH_i(\mathbf{jH}) + dH_i^2 j_i\}, \qquad (3.48)$$

where ρ_0, a, b, c, and d are the parameters of a given material, and b, c, and d are equal to μ^2/c^2 (c is the velocity of light), multiplied by numerical factors which depend on the band anisotropy, scattering mechanisms, degree of degeneracy, etc.

The first term in Eq. (3.48) is related to the scalar resistivity of a crystal in the absence of a magnetic field. The second term represents the Hall field when $\mathbf{j} \perp \mathbf{H}$. The last three terms give rise to a magnetoresistance which is a quadratic function of the magnetic field intensity. The relative change in the resistivity in a weak magnetic field is [275]

$$\frac{\Delta \rho}{\rho_0} = \left\{ b + c\,\frac{(\mathbf{jH})^2}{j^2 H^2} + d\,\frac{\sum_{i=1}^{3} j_i^2 H_i^2}{j^2 H^2} \right\} H^2. \qquad (3.49)$$

In particular, when the current flows along the [100] direction, the magnetic field is made to rotate in a plane passing through the [100] direction and another cubic axis of the crystal [for example, when the magnetic field is made to rotate in a (010) plane], and when the direction of the magnetic field makes an angle ϑ with the direction of the current, the expression for the magnetoresistance simplifies considerably:

$$\frac{\Delta \rho}{\rho_0} = \{b + (c + d)\cos^2 \vartheta\} H^2. \qquad (3.50)$$

The term proportional to the parameter b is independent of the mutual directions of the current and magnetic field vectors and the sum c + d represents the difference between the longitudinal

($\vartheta = 0$) and transverse ($\vartheta = 90°$) magnetoresistances. In the special case, when the current flows along the [100] direction and the magnetic field vector is in a (010) plane, the ratio of the longitudinal and transverse magnetoresistances is

$$\frac{\Delta\rho^{[100]}_{[100]}}{\Delta\rho^{[001]}_{[100]}} = \frac{b+c+d}{b} \qquad (3.51)$$

(the subscript indicates the direction of the current and the superscript the direction of the magnetic field).

The terms proportional to b and c in Eq. (3.49) are independent of the current and magnetic field orientations relative to the crystallographic axes, while the term proportional to d is responsible for the magnetoresistance anisotropy. It is not equal to zero only for crystals with a complex anisotropic band structure, and then its value depends on the nature of the anisotropy.

The quantities b, c, and d can be determined separately by measuring the longitudinal and transverse magnetoresistances for two orientations of the current relative to the crystal axes, for example — using the current orientations [100] and [110]. Comparison of the values obtained with those calculated for specific band structure models makes it possible to select the appropriate model for a given semiconductor.

It follows from the theory that the longitudinal magnetoresistance vanishes in the case of isotropic bands or ellipsoid constant-energy surfaces when the band extrema are located along the <100> directions (this is known as the <100> model) and the magnetic field is directed along one of the principal axes of the crystals. The ratio (3.51) lies within the limits 0-2 for the <110> model and within the limits 0-4 for the <111> model; the highest value of this ratio is obtained for degenerate samples with strongly elongated ellipsoids ($K \gg 1$) (Fig. 3.12).

More detailed information on the nature of the anisotropy can be obtained using the following relationship between b, c, and d, which applies to the many-ellipsoid model [276]:

$$b + c + xd = 0, \qquad (3.52)$$

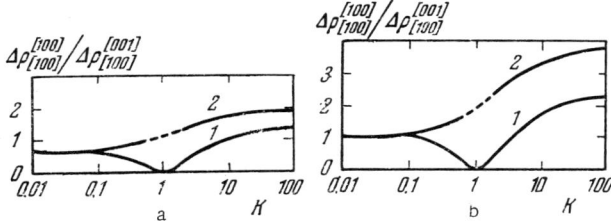

Fig. 3.12. Ratios of the longitudinal and transverse magnetoresistances as a function of the anisotropy coefficient. Theoretical curves: a) <110> model; b) <111> model; 1) nondegenerate semiconductors; 2) degenerate semiconductors.

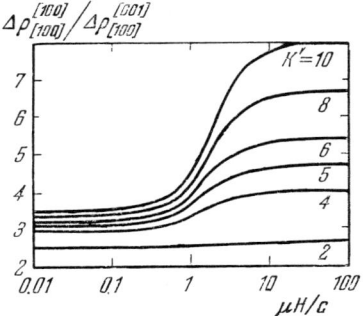

Fig. 3.13. Ratio of the longitudinal and transverse magnetoresistances of degenerate semiconductors in the <111> model as a function of the reduced magnetic field intensity for various values of the anisotropy coefficient K' (theoretical curves) [277].

where the constant factor is $x = 1$ for the <100> model, $x = -1$ for the <110> model, and $x = 0$ for the <111> model (in the last case, $d > 0$, in contrast to the spherical model for which $d = 0$).

Measurements of the magnetoresistance have played an important role in the determination of the nature of the anisotropy of the properties of Ge, Si, and other semiconductors. Measurements of the magnetoresistance of lead chalcogenides have been reported in [209, 262, 263, 277-280]. In contrast to Ge, Si, and InSb, the properties of n- and p-type samples of a given lead chalcogenide are similar, which indicates a similar structure of the edges of the conduction and valence bands in each of the lead chalcogenides.

In weak magnetic fields, the magnetoresistance is proportional to the square of the field intensity, in agreement with the theoretical predictions. In strong fields, the magnetoresistance increases more slowly with the field intensity but the predicted saturation is not observed, which may be due to the influence of inhomogeneities [248].

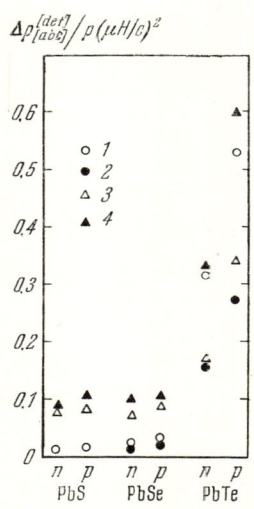

Fig. 3.14. Values of the magnetoresistance of lead chalcogenides at room temperature.

$1 - \Delta\rho_{[100]}^{[100]}/\rho(\mu H/c)^2$;

$2 - \Delta\rho_{[110]}^{[110]}/\rho(\mu H/c)^2$;

$3 - \Delta\rho_{[100]}^{[001]}/\rho(\mu H/c)^2 =$
$= \Delta\rho_{[110]}^{[001]}/\rho(\mu H/c)^2$;

$4 - \Delta\rho_{[110]}^{[1\bar{1}0]}/\rho(\mu H/c)^2$.

The angular dependence of the magnetoresistance for single crystals was obtained by Allgaier [277]; he used a current directed along the [100] axis and rotated the magnetic field in a (010) plane. His results agree with Eq. (3.50) and, in the case of PbTe, the ratio of the longitudinal and transverse magnetoresistances, given by Eq. (3.51), lies within the limits 2-5. Comparing this result with the theoretical predictions, we may conclude that the strongly elongated ellipsoids in PbTe are oriented along the <111> directions.

The correctness of the <111> model has been confirmed by measurements of the magnetoresistance of PbTe using a current flowing along the [110] direction [231, 262, 279]. Allgaier's measurements [231] yield the following values for the constant factor in Eq. (3.52): $|x| \leq 0.03$ for p-type and $|x| \leq 0.05$ for n-type samples.

The values of the anisotropy coefficient $K' = (m_\parallel^*/m_\perp^*)(\tau_\perp/\tau_\parallel)$ at 77°K are 2.5 and 4.2, respectively, for n- and p-type samples with a carrier density of about $2 \cdot 10^{18}$ cm^{-3} [280]. The corresponding values at 300°K are 3.3 and 4.7 [280].

We shall show later that the values of K' are 2-4 times smaller than the values of the anisotropy coefficient $K = m_\parallel^*/m_\perp^*$ found by measuring the Shubnikov–de Haas effect, cyclotron resonance, magneto-optical effects, and other properties (cf. Chap. V). This difference indicates a considerable anisotropy of the carrier scattering in PbTe at this carrier density. Estimates carried out on the basis of formulas of Herring and Vogt [233] show that the acoustical scattering cannot produce an appreciable deviation of the ratio $\tau_\parallel/\tau_\perp$ from 1. A calculation of the ratio in the polar scattering case, carried out by Gryaznov and Ravich [526], shows that the polar scattering exhibits an appreciable anisotropy. In particular, the value $\tau_{\|}/\tau_{\perp} \approx 3$ is obtained for $K=10$. Thus, allowance for the polar scattering makes it possible to obtain a semiquantitative agreement between the theory and the results of measurements of the magnetoresistance.

Allgaier [277] also measured the dependence of the ratio (3.51) on the magnetic field intensity in moderately strong magnetic fields at liquid-helium temperature. The agreement between the curves obtained and the theoretical dependences for various values of the anisotropy coefficient K' (Fig. 3.13) is satisfactory for $4 < K' < 6$.

Measurements of the magnetoresistance in nonquantizing magnetic fields have not given such definite information on the band structure of PbSe and PbS as the corresponding measurements carried out on PbTe. It is evident from Fig. 3.14 [280] that the longitudinal magnetoresistance of PbSe and PbS is much smaller than that of PbTe, which indicates a weaker anisotropy of the energy bands. Measurements of the magnetoresistance suggest that the constant-energy surfaces of PbS and PbSe are either spherical or ellipsoidal with an anisotropy coefficient differing little from unity. The information on the positions of the extrema in PbS and PbSe has been obtained from measurements of the magnetoresistance in quantizing magnetic fields (Shubnikov – de Haas effect) (cf. §5.2) and from other measurements.

Shogenji and Uchiyama [262, 263] measured the planar Hall effect in PbTe, which is a quadratic effect with respect to the magnetic field and exhibits a considerable anisotropy. In the observation of this effect, the current and measured voltage are mutually perpendicular, while the magnetic field is directed at an angle ϑ to the current. In weak magnetic fields, the measured effect is proportional to the square of the magnetic field and to $\sin 2\vartheta$, in agreement with theoretical results. The data obtained from the planar Hall effect have confirmed the <111> model of PbTe.

3.3. Influence of Deformation on Electrical Properties

Investigations of changes in the electrical properties of semiconductors due to deformation yield information on the energy band structure and carrier scattering. Moreover, an appreciable dependence of the resistance on the deformation makes it possible to use semiconductors as strain gauges. Methods for investigating semiconductors by deformation can be used in a wide range of temperatures and carrier densities.

It is usual to employ two convenient methods for establishing a uniformly stressed state in a semiconducting crystal: hydrostatic pressure and uniaxial deformation. The influence of a hydrostatic pressure or uniaxial deformation on the electrical, galvanomagnetic, thermoelectric, and optical properties of lead chalcogenides has been investigated quite thoroughly. In the present section, we shall consider the influence of the hydrostatic pressure on the electrical conductivity, Hall effect, and other galvanomagnetic effects (subsection A), as well as the piezoresistance effects produced by a uniaxial deformation (subsection B). The shift of the fundamental absorption edge and of the wavelength of stimulated recombination radiation has already been considered (§2.1C and §2.2D); changes in the magneto-optical properties of single-crystal films due to stretching along two directions will be discussed in §5.4C and the thermoelectric power under hydrostatic pressure will be considered in §4.1B.

A. Electrical Properties under Hydrostatic Pressure

An isotropic deformation — in particular, a hydrostatic compression — does not alter the symmetry of a crystal and does not produce any basically new effects which are not present in undeformed crystals. Changes in the parameters of a semiconductor (the forbidden band width, effective mass, etc.) due to compression produce changes in the effects which are governed by these parameters.

The influence of hydrostatic pressure on the electrical properties of lead chalcogenides was investigated by Averkin and others [129, 241, 281-289]. In the extrinsic (impurity) conduction region, the Hall effect is essentially independent of pressure for all three lead chalcogenides, with the exception of p-type PbTe; this is attributed to the constancy of the carrier density, while the pressure dependences of the mobility and thermoelectric power are governed mainly by changes in the effective mass during compression. When the intrinsic conduction region is approached the pressure dependence of the forbidden band width becomes important and this is manifested in various transport effects, including the Hall effect.

Methods for the determination of the parameters of semiconductors under hydrostatic pressure conditions have been developed directly from methods used in the absence of pressure (cf. §3.2C, §4.1B, etc.). Investigations of the pressure dependence of the forbidden band width and extrapolation of the results to zero pressure have made it possible to find more accurately the value of this width, as well as its temperature dependence in the absence of pressure.

Measurements of the thermoelectric power under pressure have shown that the effective masses of electrons and holes decrease in all three chalcogenides by about 2% at 1000 kg/cm^2 ([283, 286-289], cf. §4.1B). The same pressure increases the mobility by 5%, which is in agreement with the dependence of the mobility on the effective mass, $\mu \propto (m^*)^{-5/2}$, which applies when carriers are scattered by the acoustical vibrations. The quantity $S = d(\ln \mu)/d(\ln m^*)$, calculated from the pressure dependences of the mobility and effective mass, has been found to be close to -2.5 for PbSe [286] and PbTe [289]. In the case of degenerate samples, the value of S has been calculated using μ_{c1}, given by Eq. (3.8). The values obtained for S show that the relative change in the effective mass with pressure affects much more strongly the pressure dependence of the mobility than the relative change in the deformation potential constants.

An analysis of the Hall effect and electrical resistivity at high pressures gives the pressure dependence of the thermal value of the forbidden band width. The values of $\partial \mathscr{E}_g/\partial P$ found in this way, are $-(7-9) \cdot 10^{-6}$ eV·cm^2·kg^{-1} for PbTe [282, 283, 285, 289], $-(6-8) \cdot 10^{-6}$ eV·cm^2·kg^{-1} for PbSe [283, 284, 287], and $-(5.5-7) \cdot 10^{-6}$ eV·cm^2·kg^{-1} for PbS [129, 284]. The last value is in satisfactory agreement with the pressure dependence of the optical value of the forbidden band width of PbS, obtained from measurements of the fundamental absorption edge and photoconductivity (cf. §2.1C).

The value of the forbidden band width, found by extrapolation to zero pressure, is 0.29 eV for PbTe [289] and 0.27 eV for PbSe [287] at room temperature; the temperature coefficient of the forbidden band width of PbTe is $\partial \mathscr{E}_g/\partial T = 4 \cdot 10^{-4}$ eV/deg [289]. These results are in good agreement with those obtained by optical methods (cf. §2.1C).

The value of $\partial \mathscr{E}_g/\partial P$ indicates that the forbidden band width decreases approximately by 2-3% at 1000 kg/cm^2. The approximate agreement of the relative pressure-induced changes in the effective electron and hole masses and in the forbidden band width confirm that the extrema of the conduction and valence bands are located at the same point in the k space and that direct transitions between these extrema are allowed (for details see §6.3A).

It is interesting to note that the relative pressure-induced changes in the effective mass and the forbidden band width are an order of magnitude larger than the relative volume contraction. Such large relative changes in the effective mass and forbidden band width are typical of semiconductors with narrow gaps between the conduction and valance bands [286]. The order of magnitude of the shift of an electron energy level in a crystal due to deformation $\delta \mathscr{E}$ is $\mathscr{E}_A \delta a/a$, where a is the interatomic distance and $\mathscr{E}_A$ is the atomic energy, which is of the order of 10 eV. Hence, it follows that the relative change in the forbidden band width

$$\frac{\delta \mathscr{E}_g}{\mathscr{E}_g} \propto \frac{\mathscr{E}_A}{\mathscr{E}_g} \frac{\delta a}{a}$$

in narrow-band semiconductors, which include lead chalcogenides, is considerably larger than $\delta a/a$. The same is true of the relative change in the effective mass, which is approximately equal to $\delta \mathscr{E}_g/\mathscr{E}_g$ in these chalcogenides. The relative changes in the deformation potential constants, whose initial values are of the same order of magnitude as the atomic energy, amount to about $\delta a/a$ and are considerably smaller than $\delta \mathscr{E}_g/\mathscr{E}_g$.

The pressure dependences of the Hall coefficient, electrical conductivity, effect of weak magnetic fields on the magnetoresistance, in weak magnetic fields, and planar Hall effect of PbTe are reported in [285]. The constants b, c, and d, deduced from the magnetoresistance (cf. §3.2D), depend on the pressure as the square of the mobility, i.e., these constants are approximately doubled at the maximum applied pressure of 7000 kg/cm^2. This indicates that the anisotropy of the constant-energy surfaces depends weakly on pressure, i.e., the longitudinal and transverse components of the effective mass vary approximately identically with pressure.

The pressure dependences of the resistivity and Hall effect of p-type PbTe differ considerably from the dependences of the

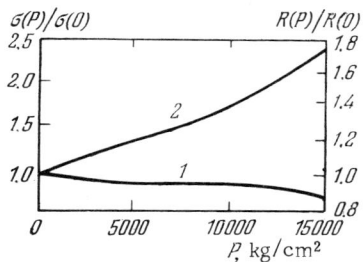

Fig. 3.15. Pressure dependences of the Hall coefficient (1) and electrical conductivity (2) of a PbTe sample with a hole density of $2 \cdot 10^{18}$ cm^{-3} [289].

same properties of the other lead chalcogenides because of the influence of the second valence band [285, 289]. In particular, the Hall coefficient of p-type PbTe depends on pressure even in the extrinsic (impurity) conduction region. The pressure dependences of the electrical conductivity and Hall coefficient are nonlinear (Fig. 3.15). This is explained by a redistribution of holes between the light- and heavy-hole bands (cf. §3.2B) due to an increase of the energy gap $\Delta\mathscr{E}$ between these bands when the pressure is increased. An analysis of the influence of pressure on the electrical properties of PbTe shows that $\partial\Delta\mathscr{E}/\partial P = (6-7) \cdot 10^{-6}$ eV $\cdot$ cm^2 $\cdot$ kg^{-1} [285, 289].

All these experimental investigations were carried out at pressures up to $(7-15) \cdot 10^3$ kg/cm^2. It is interesting to note that the extrapolation of the forbidden band width to higher pressures shows that at several tens of thousands of atmospheres the forbidden band width should disappear and the properties of lead chalcogenides should change radically.

The pressure dependence of the electrical resistance of lead chalcogenides has been investigated up to 200,000 kg/cm^2 [290]. As the pressure is increased, the resistance decreases at first and then rises rapidly at pressures of several tens of thousands of atmospheres, passes through a maximum, and decreases smoothly. This is attributed to polymorphic transformations in lead chalcogenides, which were deduced by Bridgman [291] from sudden changes in volume at pressures of this order of magnitude (25,000 kg/cm^2 for PbS, 43,000 kg/cm^2 for PbSe, and 42,000 kg/cm^2 for PbTe).

Phase transitions at high pressures have been investigated also by the x-ray diffraction method [527, 528].

B. Piezoresistance

A uniaxial pressure disturbs the cubic symmetry and its effect depends appreciably on the direction of deformation. Therefore, investigations of the electrical resistance under uniaxial pressure (piezoresistance) give information on the band structure of semiconductors which is additional to that obtained from measurements of the influence of hydrostatic pressure on the electrical properties.

A departure from the cubic symmetry in the case of a many-ellipsoid energy band structure may give rise to strong piezoresistance effects. Since different ellipsoids are oriented along different directions relative to the deformation axis, the displacements of the extrema during deformation are, in general, different. Changes in the positions of the band edges relative to one another cause a redistribution of carriers between extrema so that the contributions of different ellipsoids to the electrical conductivity are found to be different. In addition to a change of the resistivity along the direction of the applied pressure, a redistribution of carriers (carrier transfer effect) gives rise to components of the current transverse to the electric field, i.e., the nondiagonal components of the resistivity tensor are no longer equal to zero. The theory of this effect, which we shall now describe, has been developed by Herring and Vogt [233]. In particular, the carrier transfer effect is strongly temperature-dependent (inversely proportional to temperature in the classical statistics case) and proportional to the degree of anisotropy of the conductivity in each separate ellipsoid.

In addition to changes of the relative contribution of various ellipsoids to the conductivity, an anisotropic deformation alters the energies of the extrema near which possible final states of intervalley transitions are located. If the intervalley scattering is important, the deformation also alters the relaxation time. This effect is also inversely proportional to temperature. We have pointed out (§3.1B) that in lead chalcogenides the intervalley scattering should not be very important, at least at relatively low carrier densities, and therefore we shall not consider this form of scattering.

Apart from these major effects, a contribution is made to the piezoresistance by effects such as the dependence of the effective mass on the deformation. In the case of a strongly anisotropic band structure, such effects are considerably weaker than the carrier transfer effect; they depend weakly on the direction of the deformation relative to the crystal axes and do not exhibit a strong temperature dependence.

A theory of the piezoresistance, taking into account these "minor" effects for various types of band structure, was developed by Pikus and Bir [292].

The relationship between a change in the resistivity tensor $\delta\rho_{ik}$ and the deformation (strain) tensor ε_{ik} is given by a fourth-rank tensor m_{iklm}, known as the elastoresistance tensor:

$$\frac{\delta\rho_{ik}}{\rho_0} = \sum_{lm} m_{iklm}\varepsilon_{lm}. \qquad (3.53)$$

A relationship between the elastoresistance tensor and the piezoresistance tensor π_{iklm} (the piezoresistance tensor relates the change in the resistance to the stress tensor and is measured directly in experiments) is expressed in terms of elastic moduli.

The elastoresistance tensor for a cubic crystal can be expressed in terms of three independent quantities m_{11}, m_{12}, and m_{44}, known as the elastoresistance constants. Thus, in order to determine the piezoelectric properties of a crystal, we have to carry out three independent measurements. Uniaxial deformations along the principal axes of a crystal yield the values of m_{11} and m_{12}. When the deformation is directed so that it does not coincide with the edges of a cube, we obtain the value of m_{44}. Usually m_{11} and m_{12} are replaced by their combinations $m_{11}+2m_{12}$ and $m_{11}-m_{12}$. The first of these quantities represents the change in the resistivity due to an isotropic deformation and can be obtained from a hydrostatic pressure experiment. The elasticity and piezoresistance tensors are also expressed in terms of three independent quantities.

A shift caused by a uniform deformation of the edge of an ellipsoidal band located along the <100> or <111> direction in a cubic crystal can be expressed in terms of two deformation potential constants Ξ_d and Ξ_u, introduced by Herring and Vogt [233]:

$$\delta\mathscr{E} = \sum_{ik}(\Xi_d\delta_{ik} + \Xi_u u_i u_k)\varepsilon_{ik}, \qquad (3.54)$$

where u_1, u_2, u_3 are the components of a unit vector along the direction of the symmetry axis of the ellipsoid. The constant $\mathscr{E}_1$, which determines the shift of an extremum in the isotropic deformation case per unit relative change in volume, is

$$\mathscr{E}_1 = \Xi_d + \frac{1}{3}\Xi_u. \qquad (3.55)$$

The difference between the constants $\mathscr{E}_1$ for electrons and holes determines the pressure dependence of the forbidden band width:

$$\mathscr{E}_{1g} = \mathscr{E}_{1n} - \mathscr{E}_{1p} = -\frac{1}{\xi}\frac{\partial \mathscr{E}_g}{\partial P}, \qquad (3.56)$$

where $\xi = -\frac{1}{V}\frac{\partial V}{\partial P}$ is the compressibility of the crystal considered.

The carrier transfer effect depends considerably on the positions of ellipsoids in the **k** space. If the ellipsoids are elongated along the <100> directions, the quantities $m_{11} + 2m_{12}$ and m_{44} are equal to zero and the quantity $m_{11} - m_{12}$ in the classical statistics case is given by [233]:

$$m_{11} - m_{12} = \pm\frac{1}{3}\frac{\Xi_u}{k_0 T}\frac{\mu_\| - \mu_\perp}{\mu}. \qquad (3.57)$$

The upper sign applies to electrons, while the lower applies to holes; $\mu_\|$ and $\mu_\perp$ represent, respectively, the mobilities for the same ellipsoid when an electric field is applied along longitudinal and transverse directions. For the extrema directed along the <111> axes, the quantities m_{11} and m_{12} are equal to zero and the quantity m_{44} is given by the formula

$$m_{44} = \pm\frac{1}{9}\frac{\Xi_u}{k_0 T}\frac{\mu_\| - \mu_\perp}{\mu}. \qquad (3.58)$$

The piezoresistance of lead chalcogenides has been investigated fairly thoroughly [293-300]. To deduce the elastoresistance constants from the experimentally determined piezoresistance constants, we have to know the elastic moduli c_{11}, c_{12}, and c_{44}. The values of these moduli for PbTe and PbS were determined by Chudinov [301] from the velocities of ultrasound at temperatures between 100 and 600°K. The three elastic moduli have been deduced from measurements of the velocity of propagation of longi-

TABLE 3.2. Piezoresistance Constants at 300°K (in units of 10^{-12} cm^2/dyn) [307]

Compound	π_{11}	π_{12}	π_{44}
p-PbTe	+35	+40	+185
n-PbTe	+20	+25	−107
n-PbSe	+24	+19	+ 57
n-PbS	+12	+ 7	−11
p-PbS	+16	+16	−10

tudinal vibrations along the [100] direction and of torsional vibrations along the [100] and [110] directions. All three measured elastic moduli vary linearly with temperature [Appendix C lists the room-temperature values of c_{11}, c_{12}, and c_{44}, as well as the velocities of longitudinal ultrasonic waves and of the compressibility, which is equal to $3/(c_{11} + 2c_{12})$]. The values reported by Chudinov [301] differ somewhat from those reported for PbTe and PbS by other workers [295, 302, 303, 529]. Measurements of the compressibility of lead chalcogenides have been reported in [304-306]. Measurements of the three elastic moduli have not yet been carried out for PbSe but we may assume that their values are the arithmetic means of the values for PbS and PbTe, which is confirmed by the compressibility data [306].

Measurements of the piezoresistance in lead chalcogenides, reported in [293-300, 307], have yielded the results listed in Table 3.2.

Knowledge of the piezoresistance constants π_{ik} makes it possible to find the elastoresistance constants m_{ik} and to compare them with the theoretical predictions.

The absolute value of m_{44} for PbTe has been found to be considerably greater than $(m_{11} - m_{12})/2$, which is attributed to the carrier transfer effect when the extrema are located along the <111> directions [293, 295, 297-300]. The values of $(m_{11} + 2m_{12})/3$, related to the dependence of the effective mass on the isotropic deformation, has been found to be very large and comparable with m_{44}. An investigation of the temperature dependence of m_{44} has established that this quantity can be represented in the form of two terms, one of which is related to the changes in the effective mass and is weakly dependent on temperature, while the other is due to the carrier transfer effect and is inversely proportional to temperature. These two terms can be distinguished by their different temperature dependences.

Knowing the contribution of the carrier transfer effect to m_{44}, we can use Eq. (3.58) to find the value of the shear deformation potential constant Ξ_u. The values of Ξ_u, obtained from the piezoresistance measurements by various workers, are contradictory. The largest value of Ξ_u, obtained for relatively lightly doped samples of p-type PbTe, is 8 eV [298, 299]. Somewhat smaller values were obtained by Bir and Pikus [297] by analysis of the experimental data of Ilisavskii [295]: 6 eV for the valence band and 3 eV for the conduction band. A method based on the measurement of the absorption coefficient of ultrasound, applied by Nanney [525] to n-type PbTe, yielded Ξ_u, = 4.5 eV, which was in satisfactory agreement with the results obtained from the piezoresistance.

The absolute value of the constant Ξ_u is several times smaller than the value of Ξ, found from measurements of the mobility using Eq. (3.20). Hence, it follows that the mobility in lead chalcogenides is governed primarily by the deformation potential constant Ξ_d, whose absolute value is therefore several times larger than that of Ξ_u.

Investigations of changes in the effective mass during deformation and their relationship with changes in the forbidden band (cf. §6.3A) yield the constant $\mathscr{E}_{1g}$, which determines [according to Eq. (3.56)] the pressure dependence of the forbidden band width. The value 3-5 eV, obtained for PbTe [295, 297, 298], is in satisfactory agreement with the data quoted in the preceding subsection on the pressure dependence of the forbidden band width.

Burke [298] attributed the temperature dependence of $(m_{11} + 2m_{12})/3$, observed for p-type PbTe above 180°K, to the influence of a second valence band lying deeper than the ellipsoidal band (cf. §3.2B). An analysis of the temperature dependence of the elastoresistance indicates that the energy gap between the two valence bands increases under pressure (in agreement with the hydrostatic pressure data [285, 289]), but the coefficient which represents this increase is more than twice as large as that obtained by the hydrostatic pressure method.

The results obtained for PbSe and PbS are less clear and cannot be explained simply on the basis of the <111> model. The value of m_{44} of PbSe is considerably larger than the value of $(m_{11} - m_{12})/2$ but is weakly dependent on temperature and, therefore, it cannot be explained by the carrier transfer effect. The value of m_{44} of PbS is inversely proportional to temperature but is too small to be explained by the carrier transfer effect. The absence of an appreciable carrier transfer effect indicates that the anisotropy of the mobilities in PbSe and PbS is small or absent altogether. A large positive value of $(m_{11}+2m_{12})/3$ has been obtained for PbSe and PbS, which are similar in this respect to PbTe; this large value is due to a change in the effective mass caused by the variation of the forbidden band width induced by the isotropic deformation. The constant $\mathscr{E}_{1g}$, determined from the piezoresistance measurements, is 3-4 eV for PbSe [295, 297] and 3.5-4 eV for PbS [296], which is in agreement with the results of measurements of the forbidden band width under hydrostatic pressure as well as with the magneto-optical measurements in stressed PbS films.

Arsen'eva-Geil' [308-310] showed that inhomogeneous (compacted polycrystalline) samples of PbS have a very large piezoresistance, associated with deformation-induced changes in the heights of intercrystallite barriers. The strong strain sensitivity of lead chalcogenides makes them useful for materials for strain gauges (cf. §7.3).

3.4. Properties of p—n Junctions

The properties of p−n junctions have been investigated in connection with the use of semiconductors in the manufacture of diodes, transistors, photoelectric devices, lasers, etc., as well as for the purpose of determining the properties of semiconductors themselves. Investigations of p−n junctions can be used to find the value of the static permittivity and the energy of the longitudinal optical phonons, as well as to elicit information on the band structure, diffusion of impurities, etc.

The Fermi levels in an n-type sample and a p-type sample which are not in contact are at different distances from the level corresponding to the electron energy in vacuum. When two such samples are placed in contact, holes cross from the p-type to the n-type region and electrons travel in the opposite direction until the Fermi energy becomes the same for the whole system. Under equilibrium conditions, the junction region has a double electric layer depleted of carriers; the presence of this layer gives rise to a contact potential difference whose order of magnitude is $\mathscr{E}_g/e$. The space charge in the junction region is due to ionized impurities.

Usually, the n- and p-type regions are formed in the same sample because of the strong dependence of the donor and acceptor concentrations on the coordinate, i.e., in one part of the sample the donor concentration is higher than the acceptor concentration and the semiconductor exhibits n-type conduction, while in another part the relatively higher concentration of acceptors gives rise to p-type conduction. The form of the potential in the junction (transition) region is governed by the depth distribution of the impurity concentrations in the sample. Junctions of the p − n type may be formed also near a point contact between a metal and a semiconductor.

The most important property of p − n junctions is their rectifying action, which is due to a strongly nonlinear current − voltage characteristic.

Lead sulfide was one of the first materials in which the rectification of current was observed and a PbS detector was one of the first semiconducting devices. Rectification at a point contact was observed by Braun nearly a hundred years ago [311]. In the early stages of radio engineering, galenite was used to make crystal detectors, usually with tungsten "whiskers." The transistor effect [312, 313] was also first observed in lead salts and, in relatively pure p-type PbS samples, it reached a value comparable with the transistor effect in Ge. Lead sulfide transistors were never used on a wide scale because of the poor reproducibility of their properties, the brittleness of lead salts, and their easy oxidation.

Many workers have investigated p − n junctions in lead chalcogenides [163, 242, 243, 252, 314-324]. These junctions have been prepared by the alloying or impurity diffusion methods. Both n- and p-type materials have been used as the base. Thus, alloyed junctions have been prepared from PbTe with a carrier density of the order of 10^{17}-10^{19} cm^{-3} [242, 243]. PbTe and PbSe with a carrier density of $3 \cdot 10^{18}$ cm^{-3} have been used to make diffused diodes [324]. A junction in PbS, used to investigate the diffusion of sulfur, has been prepared by exposure of an n-type sample to sulfur vapor [317].

A. Current-Voltage Characteristic. Determination of Longitudinal Phonon Energy from Tunnel Effect Measurements

When a current flows through a p − n junction from the p- to n-type region (in the forward direction), the majority carriers in

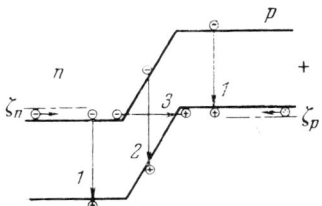

Fig. 3.16. Electron transitions in various mechanisms of the current flow through a p−n junction: 1) diffusion mechanism; 2) recombination − generation mechanism; 3) tunnel mechanism (the figure represents the case of degenerate p- and n-type regions). An external voltage V is applied in the forward direction. The difference between the Fermi levels in the n- and p-type regions is $\zeta_n - \zeta_p = eV$, where e is the electronic charge and V is the applied voltage.

each region move toward the junction and recombine there. There are several types of such recombination and corresponding mechanisms of current flow through the junction (Fig. 3.16).

Electrons may overcome the potential barrier of the p−n junction and cross over to the p-type region, where they are the minority carriers. Electrons injected into the p-type region are transferred to the valence band (where they recombine) in a layer whose thickness is equal to the diffusion length near the p−n junction. A similar process takes place in the case of holes. When the current is passed in the opposite direction, carriers are created by thermal generation in the same layer. This current-flow mechanism is usually known as the diffusion process. According to Shockley's theory [325], the current − voltage characteristic in this case is described by the formula

$$I = I_s(e^{eV/k_0T} - 1), \quad (3.59)$$

where I_s is a constant, equal to the saturation current in the reverse direction.

When carriers recombine or are generated within the space-charge layer, then Eq. (3.59) has to be modified somewhat and, according to Sah, Noyce, and Shockley [326], it becomes

$$I = I_s(e^{eV/\gamma k_0 T} - 1), \quad (3.60)$$

where the quantity γ assumes values ranging from 1 to 2; both γ and I_s depend weakly on the voltage, compared with the exponential law. This current-flow mechanism is known as the recombination − generation process.

The third mechanism represents direct electron transitions from the conduction band in the n-type region to the valence band

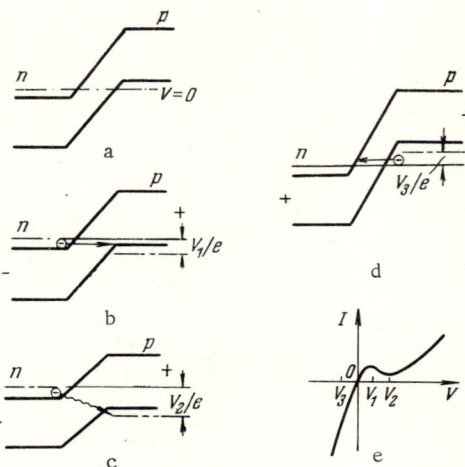

Fig. 3.17. Energy band diagrams (a-d) and the current – voltage characteristic (e) of a tunnel diode.

of the p-type region, or conversely. In this case, the potential barrier of the p – n junction is penetrated by the tunnel effect. Tunnel transitions are, in some respects, similar to recombination (or generation): in both cases, electrons disappear (appear) in the n-type region and holes behave similarly in the p-type region. In contrast to the ordinary recombination – generation transitions, the tunnel transitions take place without a change (or with a relatively small change) of the energy.

Let us consider the current – voltage characteristic of a tunnel diode [327] with degenerate n- and p-type regions (Fig. 3.17). In the absence of a voltage across the p – n junction, the Fermi level is the same throughout the sample (Fig. 3.17a). Under a forward voltage V_1, which is sufficiently small so that the bottom of the conduction band in the n-type region lies below the top of the valence band in the p-type region, the current through the diode is due to the direct tunnel transitions of electrons from the n- to the p-type region (Fig. 3.17b). The same process is responsible for a rapid rise of the current under a reverse voltage V_3 across the p – n junction (Fig. 3.17d).

When a sufficiently large forward voltage V_2 is applied to the diode, the bottom of the conduction band in the n-type region rises above the top of the valence band in the p-type region and the direct tunnel transitions that occur without a change of energy are impossible (Fig. 3.17c). Therefore, when the voltage is increased from V_1 to V_2, we observe a region in the current − voltage characteristic where the differential resistance is negative, i.e., the current decreases when the voltage increases (Fig. 3.17e). Because of other current-flow mechanisms through the junction, the total current through the diode does not fall to zero at V_2 but there is still some residual current (known as the excess current).

One of the possible mechanisms of current flow in this case is represented by the indirect tunnel transitions, accompanied by the emission of a phonon. Since the extrema of the conduction and valence bands of lead chalcogenides are located at the same point in the k space, the indirect tunnel transitions, like the indirect optical transitions, are accompanied by the emission of long-wavelength optical phonons. These indirect tunnel transitions increase the differential conductance of a diode at a positive voltage whose value is governed by the energy of the longitudinal optical phonons.

When the forward voltage is increased still further, the diffusion and recombination − generation mechanisms become important again and the current begins to rise once more when the voltage is increased.

Let us consider now the results of measurements of the current − voltage characteristics of diodes with p − n junctions prepared from lead chalcogenides. Since these chalcogenides have a narrow forbidden band, the contact potential difference between the n- and p-type regions is relatively small. This situation, together with the high carrier densities in the p- and n-type regions, gives rise to a narrow space-charge layer and a large tunnel current.

At low temperatures, the current-voltage characteristics in the forward direction can be divided into three regions. At moderate forward voltages (100-150 mV in alloyed PbTe diodes [243]), the current − voltage characteristic is described by the exponential formula of Eq. (3.60) [243, 324]; the values of the constant γ lie between 1 and 2, in agreement with the theory of the recombination − generation mechanism.

At higher voltages, the characteristics of p − n junctions investigated by Butler [324] are linear and are governed by the series resistance of the regions next to the junction.

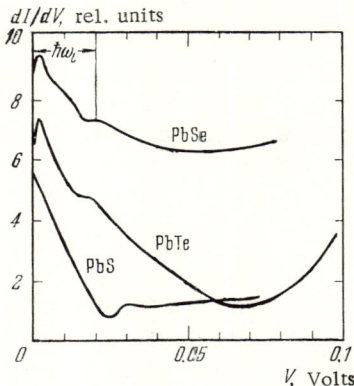

Fig. 3.18. Differential conductance, as a function of the voltage, of PbTe, PbSe, and PbS p−n junctions at 4.2°K [252].

At low forward voltages, the current is higher than could be expected from the theoretical formula. Since the n- and p-type regions are degenerate, the additional current may be due to the tunnel effect. This assumption has been confirmed by estimates made for the case of the direct tunnel transitions [243]. The tunnel effect is also responsible for the rapid rise of the reverse current.

Tunnel diodes, prepared from PbTe, PbSe, and PbS and exhibiting a negative resistance region at low forward voltages and low temperatures, have been investigated in [252, 321, 322]. The ratio of the peak to the valley current for these diodes is approximately 1.5.

The dependences of the differential conductance of the junctions on the voltage, obtained by Hall and Racette at 4.2°K [252], is shown in Fig. 3.18. Each of the three chalcogenides exhibits a narrow dip at $V=0$ and a threshold due to the interaction of electrons with the longitudinal optical phonons. Phonon-assisted tunnel transitions contribute to the current through the p−n junction and cause an increase in the conductance at some positive value of the voltage, from which we can find the energy of the long-wavelength longitudinal optical phonons $\hbar\omega_l$. The value of $\hbar\omega_l$ is deduced from the position of the point of the steepest positive slope of the current–voltage characteristic; it is found to be 13.6 ± 0.4 meV for PbTe, 16.5 ± 0.6 meV for PbSe, and 26.3 ± 0.4 meV for PbS [252]. These values are in good agreement with the results obtained for PbTe by neutron spectroscopy [141] and for PbTe and PbSe by investigating recombination radiation [174] (cf. §2.2C).

Hall and Racette investigated about ten samples of each chalcogenide and none of their measurements indicated the presence of the indirect phonon-assisted tunnel effect (with an appreciable change in crystal momentum) or a threshold in the forward tunnel effect.

This confirms the conclusion drawn from optical investigations that the conduction and valence band extrema of lead chalcogenides are at the same point in the k space. The same conclusion follows from an investigation of the tunnel effect in a magnetic field (cf. §5.3).

The nature of a minimum at $V = 0$ is not quite clear. One of the possible causes of this minimum is the polaron effect. Because of the polarization of the lattice in an ionic crystal, the energies of electrons and holes are lower than those in the absence of polarization. Electrons and holes are annihilated in tunnel transitions and, therefore, such transitions should yield an energy equal to the sum of the polarization energies. This energy is obtained by carriers from an external source and, therefore, the current is absent at voltages lower than a certain critical value. However, the value of the polaron energy in PbTe, estimated by Hall [321] from the width and depth of the minimum at $V = 0$ is $2 \cdot 10^{-4}$ eV, which is an order of magnitude smaller than the value obtained theoretically (cf. §2.1C).

A minimum at $V = 0$ can be due to other causes. For example, Vul et al. [328] explained a similar effect in GaAs by processes which occur at a contact between a semiconductor and a superconducting metal. At very low temperatures (of the order of 1°K), a complex fine structure has been observed in the current – voltage characteristics of lead chalcogenide tunnel diodes near $V = 0$ [329]; this fine structure has not yet been interpreted theoretically.

B. Barrier Capacitance. Determination of the Static Permittivity

When a small alternating voltage is applied to a p − n junction already held at a constant voltage V, the effects observed can be described by an equivalent capacitance connected in parallel with the junction. The inertia (delay) of the carrier injection into the neutral regions (governed by the carrier lifetime) results in the appearance of a diffusion capacitance. Charge transfer across the double electric layer in the junction corresponds to a barrier capacitance. Measurements of the barrier capacitance in PbTe have played an important role in studies of lead chalcogenides since Kanai and Shohno [242] used this method to obtain, for the first time, the value of the static permittivity ε_0.

Since there are practically no carriers within the space-charge layer, charge transfer is achieved by a change in the layer thickness. The charge density changes only at the boundaries between the space-charge layer and the quasineutral regions so that the resultant positive and negative charges are separated by a distance equal to the space-charge layer thickness d. Consequently, the barrier capacitance is

$$C = \frac{\varepsilon_0}{4\pi d}. \qquad (3.61)$$

Equation (3.61) cannot be used directly to determine the permittivity since the p−n junction thickness is difficult to measure by an independent method. This thickness depends on the distribution of donors and acceptors in the junction region, the total potential difference across the junction $V_c - V$, and the permittivity. Usually, two simple cases of the impurity distribution are considered: step-like and linear distributions.

Let us assume that the difference $n_a - n_d$ (the subscripts a and d denote acceptors and donors, respectively) changes suddenly from a positive to a negative value at some point but remains constant elsewhere. Let us also assume that the junction is strongly asymmetrical, i.e., in one region of a diode the absolute value of the difference between acceptor and donor concentrations is considerably smaller than in the other region. Then, the junction thickness is governed principally by the properties of the lightly doped region. Let us assume that the p-type region is lightly doped and that it has no donors, while the acceptor concentration in this region is n_a. In this case, the barrier capacitance is

$$C = \left[\frac{e\varepsilon_0 n_a}{2(V_K - V)}\right]^{1/2}. \qquad (3.62)$$

If the impurity distribution is not step-like but the acceptor concentration near the boundary between the space-charge region and the more heavily doped region remains constant, Eq. (3.62) retains its form except that the contact potential difference V_c is replaced by a different quantity, which is also independent of the voltage.

The acceptor concentration in a p-type base can be determined from the Hall effect. Knowing this concentration, we can use the slope of the dependence of C^{-2} on V for a given sample, or the slope of the line representing the dependence of C^2 on n_a, to find the permittivity ε_0.

If the difference $n_a - n_d$ depends linearly on the coordinate, $n_a - n_d = ax$, the capacitance is given by the formula

$$C = \left[\frac{e\varepsilon_0^2 a}{12 (V_0 - V)} \right]^{1/3}, \qquad (3.63)$$

where V_0 is some constant.

Under large reverse voltages ($|V| > 0.5$ V [243]), the space-charge layer is thick and the boundary of this layer lies in the region where the concentration of impurities is independent of the coordinate. The use of Eq. (3.62) has yielded $\varepsilon_0 = 400$ for PbTe [242, 243]. This anomalously high value of the static permittivity, confirmed later by other measurements (cf. §2.1D), is equal to the value suggested in [220] in order to explain the high carrier mobility at low temperatures. The use of diodes with n- and p-type bases in the measurement of the permittivity has increased the reliability of the results obtained and eliminated the possibility of the influence of surface barriers.

Under small reverse voltages, the space-charge layer is thin and, when the value of $n_a - n_d$ is expanded as a series in powers of the coordinate, we need retain only the linear term. Then, the experimental dependence of the barrier capacitance on the voltage is found to obey Eq. (3.63) [243].

Analysis of the results obtained in [243] has yielded the values of the donor concentration in the n-type region ($n_d = 10^{19}$ cm^{-3}) and the space-charge layer thickness in the absence of an external voltage ($d = 7 \cdot 10^{-6}$ cm) for a p−n junction with a p-type base ($n_a = 10^{18}$ cm^{-3}).

CHAPTER IV

THERMOELECTRIC AND THERMAL PROPERTIES

4.1. Thermoelectric Power and Thermomagnetic Effects

In the presence of a temperature gradient ∇T, the distribution of carrier velocities varies across a semiconductor. Consequently, even in the absence of an electric field, a macroscopic carrier flux is produced because the temperature gradient acts as an effective temperature field, which has some of the effects of the electric field. If the circuit is open and there is no current, an electric field is produced at each point in the sample and this field is found from the condition that the thermal current must be balanced by the ohmic current. The temperature gradient thus generates an electric field which gives rise to a thermo-emf or the Seebeck effect. Carriers with higher energies move from the hotter to the colder parts of the sample, and those with lower energies move in the opposite direction so that the total current is equal to zero.

The simplest thermoelectric circuit, consisting of a semiconductor and a metal, is shown in Fig. 4.1. The semiconductor-metal contacts are at different temperatures T_1 and T_2 ($\Delta T = T_2 - T_1$). In metals, the electron density and, in practice, their velocity distribution are independent of temperature and, therefore, the thermo-emf of metals is small compared with that observed in semiconductors. The potential difference ΔV, representing the thermo-emf, is due to processes taking place in the semiconductor and at the semiconductor-metal interfaces; it is independent of the properties of the metal. The thermoelectric power, which is the thermo-emf per unit temperature drop, is defined as

$$\alpha_0 = \lim_{\Delta T \to 0} \frac{\Delta V}{\Delta T}.$$

If we assume that the contact potential difference between the semiconductor and the metal is independent of temperature, the value of ΔV is equal to the linear integral of the electric field along the length of the semiconductor, except for the contact regions:

$$V^{(2)} - V^{(1)} = \int_{(1)}^{(2)} \nabla V \, dx = - \int_{(1)}^{(2)} E \, dx.$$

In fact, the contact potential difference V_c depends on the temperature and there is a finite difference $V_c(T_2) - V_c(T_1)$, which is also measured. The complication introduced by the contact potential difference can be avoided by calculating the integral not of ∇V but of the quantity $\nabla(V - \zeta/e)$, where ζ is the chemical potential measured from a band edge (cf., for example, [270]). The quantity $V - \zeta/e$

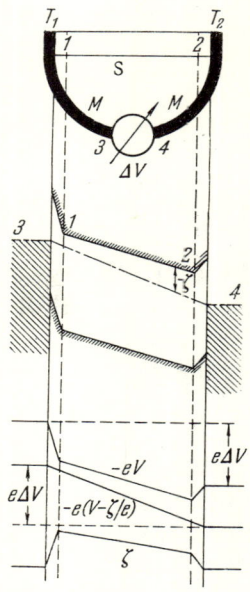

Fig. 4.1. Simplest thermoelectric circuit: S is a semiconductor; M is a metal. The middle part of the figure shows the energy band diagram; the lower part shows the variation of $-eV$, ζ, and $-e(V - \zeta/e)$ across the semiconductor and the metal – semiconductor contacts; here, ΔV is the measured potential difference.

does not vary when we pass through the contact, provided we neglect the temperature drop in the contact region; therefore, the integral of $\nabla(V - \zeta/e)$ in the contact region is equal to zero. Since $\zeta(^3) = \zeta(^4)$,

$$\Delta V = V^{(4)} - V^{(3)} = \int_{(3)}^{(4)} \nabla(V - \zeta/e) \, dx = \int_{(1)}^{(2)} \nabla(V - \zeta/e) \, dx.$$

It follows from this equation that the quantity $\nabla \zeta/e$ of a semiconductor is recorded as an additional electric field so that the total

Sec. 4.1] THERMOELECTRIC POWER AND THERMOMAGNETIC EFFECTS 151

electric field is $\nabla(V - \zeta/e)$. Therefore, the thermoelectric power is given by

$$\alpha_0 = \frac{\nabla(V - \zeta/e)}{\nabla T}. \tag{4.1}$$

For a homogeneous sample, the above definition is identical with that given earlier in this section; for an inhomogeneous sample, we have the relationship

$$\Delta V = \int_{(1)}^{(2)} \alpha_0 \nabla T \, dx,$$

where α_0 is defined by Eq. (4.1).

The α_0 expression for the thermoelectric power is directly related to the transformation of the thermal into electrical energy, and the processes in the thermoelectric circuit considered by us from the basis of the operation of thermoelectric power generators. Basically, the same quantities can be obtained by considering the isothermal effect, which is known as the Peltier effect and used in cooling units. When a current is passed through a contact between a metal and a semiconductor, the evolution of the Joule heat at the contact is accompanied by an additional absorption or evolution (depending on the direction of the current) of heat proportional to the amount of charge passing through the contact. The coefficient of proportionality between the amount of this additional heat and the electric charge is known as the Peltier coefficient.

The thermoelectric power and the Peltier coefficient of a semiconductor – metal contact are practically independent of the properties of the metal. The evolution or absorption of heat at a contact indicates that when a current flows in a semiconductor there is an energy flux **w**, directed toward or away from the contact and equal to

$$\mathbf{w} = -\Pi \mathbf{j}.$$

The Peltier coefficient is related to the thermoelectric power by the Thomson equation, which is independent of the actual properties of a semiconductor and applies generally:

$$\alpha_0 = \frac{\Pi}{T}.$$

Using the Thomson equation, we can determine the thermoelectric power ignoring the temperature gradient but calculating the energy flux under isothermal conditions (the Π-approach).

The thermoelectric power is affected by the application of a magnetic field, which bends the trajectories of the carriers. A change of the thermoelectric power in a magnetic field is, to some extent, similar to the magnetoresistance. In weak magnetic fields, the change in the thermoelectric power is proportional to the square of the field intensity. In strong transverse fields ($\mu H/c \gg 1$, but $\hbar\omega_c \ll k_0 T$) the thermoelectric power reaches saturation, tending to a limit α_∞.

In the presence of a temperature gradient and a transverse magnetic field, we observe an effect which is the analog of the Hall effect: a component of the electric field appears along a direction which is perpendicular both to the temperature gradient and to the magnetic field (this is known as the Nernst − Ettingshausen effect). In weak magnetic fields, the Nernst − Ettingshausen effect is a linear function of the field intensity. When the temperature varies along the x axis, the magnetic field is directed along the z axis, and the potential difference is measured along the y axis, the Nernst − Ettingshausen coefficient is given by the formula

$$Q = \frac{\nabla_y (V - \zeta/e)}{H \nabla T}. \qquad (4.2)$$

Like the thermoelectric power, the Nernst − Ettingshausen effect can be related to a corresponding isothermal effect in a magnetic field, namely, with the energy flux along a direction perpendicular to the current.

Carried out over a century ago, Stefan's measurements [330] of the thermoelectric power in natural crystals of galenite were the first experiments in which the semiconducting properties of one of the three lead chalcogenides were observed. Stefan found that the sign of the thermoelectric power varied from sample to sample,

Sec. 4.1] THERMOELECTRIC POWER AND THERMOMAGNETIC EFFECTS 153

a phenomenon which we now interpret as indicating n- and p-type conduction.

Thermoelectric and thermomagnetic effects in lead chalcogenides have been investigated quite thoroughly [23, 25, 44, 114, 122, 209, 216, 221, 224, 225, 228, 230, 237, 249, 265, 331-342]. These intensive investigations were stimulated both by practical applications of lead chalcogenides in thermoelectric devices, and by the possibility of determining the effective mass and the scattering mechanism by measuring the thermoelectric power and thermomagnetic effects.

A. Temperature and Carrier-Density Dependences

In the case of a quadratic dependence of the carrier energy on the crystal momentum and a power dependence of the relaxation time on the carrier energy, the formula for the thermoelectric power of an extrinsic semiconductor is [270]:

$$\alpha_0 = \frac{k_0}{e} \left[\frac{(r+5/2) F_{r+3/2}(\zeta^*)}{(r+3/2) F_{r+1/2}(\zeta^*)} - \zeta^* \right]. \qquad (4.3)$$

In the absence of degeneracy, the thermoelectric power is

$$\alpha_0 = \frac{k_0}{e} \left[(r+5/2) + \ln \frac{2(2\pi m_d^* k_0 T)^{3/2}}{nh^3} \right] \qquad (4.4)$$

and it depends logarithmically on the temperature and carrier density. If the carrier density n, the effective density-of-states mass m^*_d, and the scattering parameter r are all temperature-independent, the temperature dependence of the thermoelectric power of nondegenerate semiconductors is given by the formula:

$$\alpha_0 = \frac{3}{2} \frac{k_0}{e} \ln T + \text{const.} \qquad (4.5)$$

A dependence of the Eq. (4.5) type has been observed for n- and p-type lead selenide [230] (Fig. 4.2), but the coefficient in front of ln T has been found to be not 129 but about 200 μV/deg. This difference between the theoretical formula and experimental results has been attributed to the temperature dependence of the effective mass. Similar results have been obtained for the other two lead chalcogenides.

154 THERMOELECTRIC AND THERMAL PROPERTIES [CHAP. IV

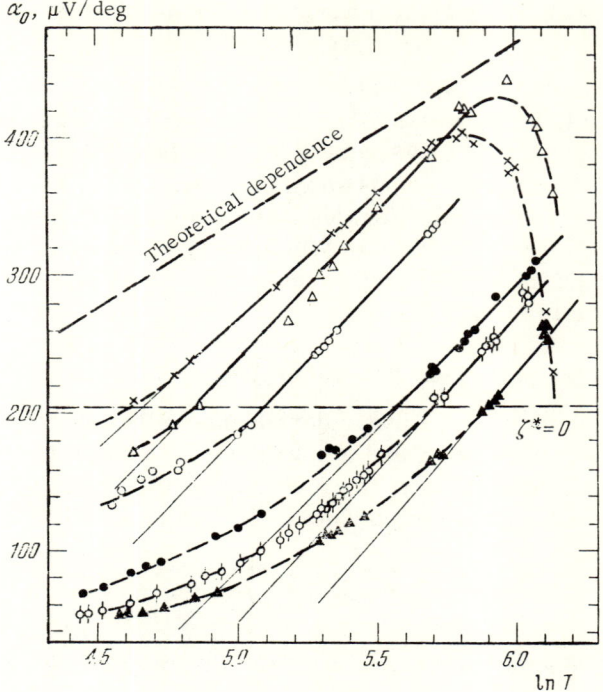

Fig. 4.2. Temperature dependence of the thermoelectric power of p-type PbSe [230].

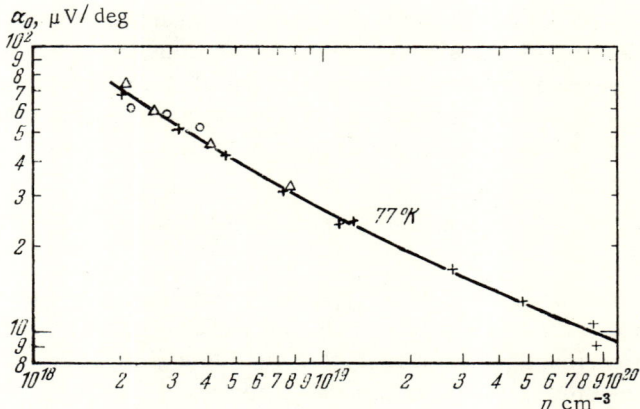

Fig. 4.3. Dependence of the thermoelectric power of n-type PbTe on the carrier density at T = 77°K [249].

Sec. 4.1] THERMOELECTRIC POWER AND THERMOMAGNETIC EFFECTS 155

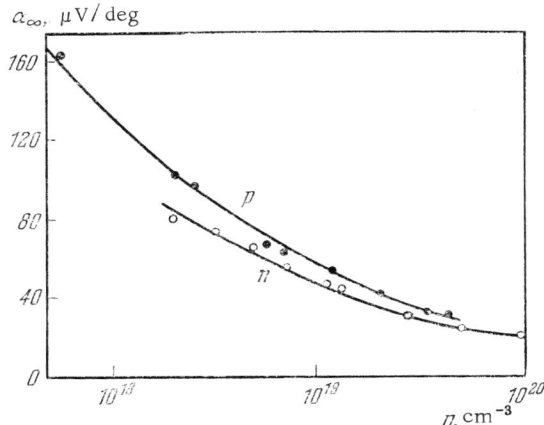

Fig. 4.4. Dependence of the thermoelectric power in a strong magnetic field on the carrier density in n- and p-type PbTe at T = 77°K [335, 336].

In the strongly degenerate case, Eq. (4.3) assumes the form

$$\alpha_0 = \frac{2\pi^{2/3} k_0^2 T m_d^*}{3^{5/3} e \hbar^2 n^{2/3}} (r + 3/2). \quad (4.6)$$

It follows from Eq. (4.6) that the thermoelectric power of degenerate semiconductors is proportional to the temperature in the case of weak temperature dependences of m_d^*, r, and n, and that it decreases with increasing carrier density (Fig. 4.3).

At high carrier densities or at high temperatures, when high-energy carriers contribute to the thermoelectric power, Eq. (4.3) should be modified in order to allow for the band nonparabolicity (Appendix A). In particular, the thermoelectric power of a strongly degenerate semiconductor which obeys Kane's model (cf. §6.2B) is given by

$$\alpha_0 = \frac{\pi^2 k_0^2 T}{3e\zeta} \left[\left(\frac{\partial \ln \tau}{\partial \ln \mathscr{E}}\right)_{\mathscr{E}=\zeta} + \frac{3}{2} - \frac{2\zeta/\mathscr{E}_{gi}}{1 + 2\zeta/\mathscr{E}_{gi}} + \frac{3}{2} \frac{\zeta/\mathscr{E}_{gi}}{1 + \zeta/\mathscr{E}_{gi}} \right]. \quad (4.7)$$

It must be pointed out that the formulas deduced for a parabolic band usually give quite satisfactory temperature and carrier-density dependences of the thermoelectric power even in the non-

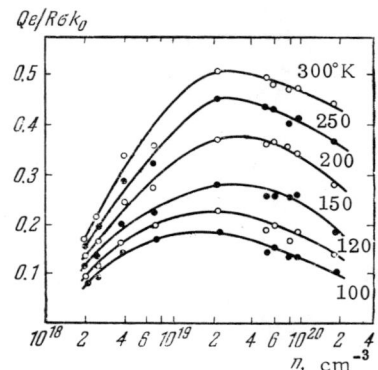

Fig. 4.5. Dependence of the Nernst – Ettingshausen coefficient on the carrier density in n-type PbTe at temperatures T = 100-300°K [334].

parabolic region. This is because the dependence of the effective mass on the energy, and the consequent dependence on the carrier density and temperature (we are talking here of the average effective mass), not only alters the temperature and carrier-density dependences of the Fermi level but, through the density of states, causes a reduction in the parameter r, whose role in Eq. (4.7) is assumed by the derivative $\partial \ln \tau / \partial \ln \mathscr{E}$. These two manifestations of the nonparabolicity compensate each other to a great degree and the carrier-density and temperature dependences of the thermoelectric power do not change greatly in the nonparabolic case. However, the use of the parabolic-band formulas in the determination of the parameters of a semiconductor with nonparabolic bands naturally gives rise to some errors.

Equation (4.3) should be modified also at high temperatures when carriers of opposite sign appear and the thermoelectric power decreases, because carriers of opposite sign make an opposing contribution to the thermoelectric power.

At very low temperatures, of the order of 10°K, the value of the thermoelectric power is higher than that predicted by Eq. (4.6) because of the drag of carriers by phonons. Greig [23] observed a definite maximum of the thermoelectric power below 20°K, which he attributed to the drag effect. This maximum appears only in those lead chalcogenide samples which have a high thermal conductivity, and the position of the maximum coincides with the position of the thermal conductivity maximum. A theoretical estimate of the drag effect has given a value which is in qualitative agreement with that obtained experimentally by Greig.

In a strong magnetic field, the thermoelectric power exhibits carrier-density and temperature dependences similar to those in

Sec. 4.1]   THERMOELECTRIC POWER AND THERMOMAGNETIC EFFECTS   157

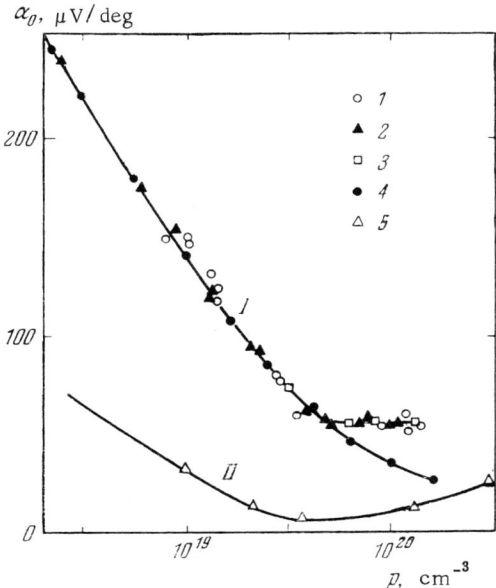

Fig. 4.6. Dependence of the thermoelectric power on the Hall carrier density in p-type PbTe [265]: I) T = 300°K [265]; II) T = 77°K (results obtained by one of the present authors); 1), 2) experimental points for samples prepared, respectively, by zone melting in hydrogen and by slow cooling; 3) calculated points for the two-band model; 4) calculated points for a single parabolic band; $r = -1/2$.

the absence of a magnetic field. The dependence of the thermoelectric power of PbTe on the carrier density in a strong magnetic field is shown in Fig. 4.4 at liquid-nitrogen temperature. We shall give later the formulas for the thermoelectric power in strong magnetic fields (cf. subsection C in the present section and Appendix A).

The expressions for the Nernst – Ettingshausen coefficient are somewhat more complex (cf., for example, [270] and Appendix A). In particular, the Nernst – Ettingshausen coefficient of a nondegenerate semiconductor with parabolic bands is proportional to the product of the mobility and some function of the scattering parameter r. The carrier-density dependences of the Nernst – Ettingshausen coefficient divided by the Hall mobility are shown in Fig. 4.5 for n-type PbTe at various temperatures.

The carrier-density and temperature dependences of the thermoelectric power of p-type PbTe have certain anomalies, which were discovered by Kolomoets et al. [265]. An analysis of these dependences gives information on the structure of the valence band of lead telluride. It follows from the Hall effect measurements (cf. §3.2B) that, in addition to prolate ellipsoids along the <111> directions, PbTe has a heavy-hole band lying deeper than the ellipsoidal bands. The effect of this band appears in the carrier-density dependence of the thermoelectric power. A theoretical curve, shown in Fig. 4.6, for a single-band model at 300°K is in good agreement with the experimental points only up to hole densities less than $3 \cdot 10^{19}$ cm^{-3}. At higher hole densities, the thermoelectric power remains constant (56 μV/deg). At 77°K and high carrier densities, the thermoelectric power increases with increasing carrier density.

This carrier-density dependence of the thermoelectric power can be explained as follows. In the two-band case, the thermoelectric power is given by the formula

$$\alpha_0 = \frac{\alpha_{p1} b_p p_1 + \alpha_{p2} p_2}{b_p p_1 + p_2},$$

where α_{p1} and α_{p2} are the thermoelectric powers observed in the presence of either just the light-hole band or just the heavy-hole band. The rest of the notation is the same as in Eq. (3.38). If the electrical conductivity due to the heavy holes is low compared with the conductivity due to the light holes, the quantity p_2 in the denominator can be ignored. If, moreover, $\alpha_{p2} \ll \alpha_{p1} b_p p_1$, the total thermoelectric power α_0 is approximately equal to the thermoelectric power of the light holes α_{p1}. According to Eq. (4.3), this thermoelectric power is a function of the Fermi level, which rises with increasing carrier density until it reaches the top of the heavy-hole band, after which it is practically unaffected by the carrier density because of the high density of states in the heavy-hole band. This makes the thermoelectric power independent of the hole density, as observed for high hole densities at room temperature (curve I in Fig. 4.6).

However, if the quantity $\alpha_{p2} p_2$ is comparable with $\alpha_{p1} b_p p_1$, but we still have $p_2 \ll b_p p_1$ because of the low mobility of the heavy holes, the total thermoelectric power is

$$\alpha_0 = \alpha_{p1}\left(1 + \frac{\alpha_{p2}p_2}{\alpha_{p1}b_p p_1}\right).$$

This quantity is larger than α_{p1} and the second term increases with increasing carrier density, which results in an increase of the total thermoelectric power at high carrier densities, as observed at 77°K (curve II in Fig. 4.6). When the density of the heavy holes is increased still further, we may reach a region in which $p_2 \gg b_p p_1$, $\alpha_0 \approx \alpha_{p2}$, and the thermoelectric power begins to decrease with increasing carrier density. Thus, the dependence $\alpha_0(p)$ should have a maximum.

Further information on the heavy-hole band is obtained from the temperature dependence of the thermoelectric power. This temperature dependence is found to be stronger than would follow for the single-band model. To make the experimental results agree with the theoretical predictions, we must assume that the energy gap between the two valence bands decreases with increasing temperature and that such a temperature dependence of the energy gap is found to be in agreement with that obtained from the Hall effect measurements [267] (cf. §3.2B).

The thermoelectric power of solid solutions based on lead chalcogenides has also been investigated. The replacement of one chalcogen atom with another produces no appreciable change in the thermoelectric power [343]. When lead is replaced by other elements of group IV, for example, by tin, the thermoelectric power changes in a complex manner because of changes in the band structure and associated scattering mechanisms [255].

B. Determination of the Effective Mass from Thermoelectric Power Measurements

We shall now consider a method for the determination of the effective mass which is based on measurements of the thermoelectric power. Knowing the parameter r, which represents the scattering mechanism, we can use the measured value of the thermoelectric power and Eq. (4.3) for a parabolic band to determine the reduced chemical potential ζ^* and then employ the formula for the electron density

$$n = \frac{\sqrt{2}(m_d^* k_0 T)^{3/2}}{\pi^2 \hbar^3} F_{1/2}(\zeta^*) \tag{4.8}$$

to find the density-of-states mass m^*_d, which is defined by the following expression in the case of a many-ellipsoid model:

$$m_d^* = N^{2/3}(m_\perp^{*2} m_\parallel^*)^{1/3}. \tag{4.9}$$

This method of measuring the effective mass has the advantages of great simplicity and validity in a wide range of temperatures and carrier densities; for these reasons, it is widely used by many workers. This is the method which was first used to observe the unusually strong temperature dependence of the effective mass in PbSe [230].

The effective density-of-states mass of PbTe has been determined from the thermoelectric power by many workers [114, 224, 337, 339, 341]. Since the conduction and valence bands of PbTe are strongly nonparabolic and analysis of the results of measurements in the majority of these investigations has been carried out using the formulas deduced for the parabolic band case, the average values obtained for the effective mass are found to vary appreciably with the carrier density. The effective density-of-states mass in the purest sample of PbTe (electron density of $3.6 \cdot 10^{17}$ cm^{-3}) has been found to increase approximately linearly from $0.16m_0$ to $0.22m_0$ [339] between 100 and 300°K. In a p-type sample with the same carrier density, the effective mass has increased from $0.18m_0$ to $0.28m_0$ in the same temperature range. The quantity

$$\Delta = \frac{d \ln m_d^*}{d \ln T}, \tag{4.10}$$

which gives the temperature dependence of the effective mass at electron densities below 10^{19} cm^{-3}, is reported to be 0.4 in the temperature range 100-400°K [339, 341] and 0.6-0.7 in the temperature range 300-650°K [340]. For holes at room temperature this quantity is $\Delta = 0.5$ [339].

The value of the effective mass as well as its temperature dependence depend considerably on the carrier density. The largest value $\Delta = 1$ has been obtained for a sample with an electron density of 10^{20} cm^{-3} at temperatures higher than 650°K [224]. Obviously, in this case we are dealing with the effect of the conduction-band nonparabolicity, which results in a temperature dependence of the average effective mass because of a redistribution of electrons in the nonparabolic band when the temperature is varied.

An analysis of the measurements of the thermoelectric power of PbTe, taking into account the band nonparabolicity, has been reported in [337, 342]. The effective density-of-states masses at the bottom of the conduction band and at the top of the valence band have been determined using the formulas of the Kane model (Appendix A). This analysis has yielded values of the effective mass which are practically independent of the carrier density. The effective mass of electrons, $m^*_d = 0.22 m_0$, reported in [337], is in satisfactory agreement with the results obtained by other methods, for example, from measurements of the Nernst − Ettingshausen effect (cf. subsection D of the present section). The effective mass near an extremum varies with temperature in the same way at all carrier densities and is approximately proportional to the forbidden band width, in agreement with the theoretical predictions (cf. §6.3B).

The value of Δ for PbSe with electron or hole densities of $3 \cdot 10^{17}$ to $6 \cdot 10^{18}$ cm^{-3} cm^{-3} is 0.45 for holes and 0.35 for electrons in the temperature range 100-400°K; the effective density-of-states mass at room temperature is $0.345 m_0$ for holes and $0.33 m_0$ for electrons [230]. An analysis of the results obtained for heavily doped PbSe has been carried out taking into account the band nonparabolicity in two investigations [332, 333].

The thermoelectric power of heavily doped PbS n- and p-type samples has been measured and the values are reported in [122]. By assuming the predominance of the scattering by the acoustical phonons, calculations based on the thermoelectric power give room-temperature effective mass ranging from $0.38 m_0$ at an electron density of $4 \cdot 10^{18}$ cm^{-3} to $0.44 m_0$ at an electron density of $7 \cdot 10^{19}$ cm^{-3}. For p-type samples, it has been found that $m^*_d = (0.45-0.48) m_0$ in the hole density range $(0.7-4) \cdot 10^{19}$ cm^{-3}. The effective mass increases with temperature and between 300 and 900°K this increase is represented by the coefficient $\Delta = 0.6-0.7$. Natural samples of PbS, with a relatively low electron density, have lower values of the effective mass [237, 338].

Measurements of the pressure dependence of the thermoelectric power yield the pressure dependence of the effective mass. Investigations of the thermoelectric power under hydrostatic pressure have been carried out for all three n- and p-type lead chalcogenides. It has been found that the effective masses of electrons and holes decrease approximately by 1.5-2.5% at a pressure of 1000 kg/cm^2 [282, 283, 286-289]. The pressure dependence of the effective mass becomes somewhat weaker at higher carrier densities. The effect of pressure on the effective mass results, in particular, in a corresponding change of the electrical conductivity (cf. §3.3A).

C. Investigations of the Band Edge Structure by Thermoelectric Power Measurements in Strong Magnetic Fields

The disadvantage of the method (described in the preceding subsection) for the determination of the effective mass from the thermoelectric power α_0 (in the absence of a magnetic field) is the need to know the carrier scattering mechanism (the scattering parameter r). The experimental data for PbTe and PbSe, reported in the investigations cited in the preceding subsection, have been analyzed on the assumption that the predominant scattering mechanism is the interaction of carriers with the acoustical vibrations of the lattice. The results for PbS have been analyzed using the same assumption as well as formulas for the polar interaction with the optical vibrations [237, 338].

It is now known that, in lead chalcogenides, carriers are scattered on the acoustical and optical phonons and, therefore, the energy dependence of the relaxation time is not given by a power law. At relatively low temperatures (of the order of 100°K) the scattering of carriers is inelastic and this also alters the effective scattering parameter r.

The problem of the value of r in the nonparabolic case is even more difficult. In this case, it is usual to assume [332-334] that the nonparabolicity affects the formula for the relaxation time only through the density of states and that the matrix element of the interaction of electrons with phonons or scattering centers depends on the crystal momentum in the same way as for a parabolic band. As pointed out in §3.1A, this approximation is fairly rough.

The methods for the determination of the effective mass based on the measurement of the thermomagnetic effects are free from this disadvantage. One of the methods is based on the measurement of the change of the thermoelectric power in a magnetic field (this is known as the longitudinal Nernst − Ettingshausen effect). This method was employed by Dubrovskaya et al. to study the nonparabolicity of the conduction and valence bands of PbTe, PbSe, and PbS [335, 336, 530].

Measurements of the thermoelectric power and of the Hall effect in strong magnetic fields have been carried out at a temperature close to that of liquid nitrogen using PbTe samples with electron densities ranging from $2 \cdot 10^{18}$ cm^{-3} to $1 \cdot 10^{20}$ cm^{-3}.

It follows from the formulas which we shall give later that, in order to determine the form of the electron spectrum, we need to know the thermoelectric power limit (saturation value) in strong magnetic fields, α_∞, which is reached when the condition $(\mu H/c)^2 \gg 1$ is satisfied. Since, in most of the samples, saturation of the thermoelectric power was not reached (because of low mobilities) in the maximum magnetic field of 32 kOe used in these investigations, the value of α_∞ had to be found by extrapolation. According to Rodot's calculation [344], the dependence of the thermoelectric power on the magnetic field intensity in the degenerate case is of the form

$$\Delta \alpha \equiv \alpha(H) - \alpha_0 = \frac{AH^2}{1 + BH^2}. \qquad (4.11)$$

The slope of the line representing the dependence $H^2/\Delta\alpha$ on H^2 gives the value of $\Delta\alpha_\infty = A/B$.

The relationships given later in the present subsection show that, by measuring the saturation value of the thermoelectric power, α_∞, in a strong (but not quantizing) magnetic field, we can find the density of states in a band as a function of the energy and the Fermi level as a function of the carrier density.

The saturation value of the thermoelectric power in a strong magnetic field $[(\mu H/c)^2 \gg 1]$ is given by the formulas obtained by Ravich [335] without making any assumptions about the actual form of the electron spectrum:

$$\alpha_\infty = \frac{k_0}{cn} \int_0^\infty \left(-\frac{\partial f}{\partial \mathscr{E}}\right) \frac{\mathscr{E} - \zeta}{k_0 T} \left(\int_0^{\mathscr{E}} \rho(\mathscr{E}') d\mathscr{E}'\right) d\mathscr{E}, \qquad (4.12)$$

where

$$n = \int_0^\infty \left(-\frac{\partial f}{\partial \mathscr{E}}\right)\left(\int_0^{\mathscr{E}} \rho(\mathscr{E}')\,d\mathscr{E}'\right) d\mathscr{E}. \tag{4.13}$$

Here, $\rho(\mathscr{E})$ is the dependence of the density of states on the energy.

Estimates show that the reduced chemical potential is $\zeta_1^* = 5.1$ for a lead telluride sample with the lowest electrical density $(2 \cdot 10^{18}$ cm$^{-3})$. Thus, at liquid-nitrogen temperature, all the lead telluride samples considered here should be sufficiently degenerate so that we can use the expansion in powers of $1/\zeta^*$. In the strongly degenerate case, Eqs. (4.12) and (4.13) assume the form

$$\alpha_\infty = \frac{\pi^2 k_0^2 T \rho(\zeta)}{3en}, \tag{4.14}$$

$$n = \int_0^\zeta \rho(\mathscr{E})\,d\mathscr{E}. \tag{4.15}$$

The first of these formulas allows us to determine ρ from the experimental values of α_∞. To calculate ζ, we differentiate Eq. (4.15) with respect to ζ and integrate it with respect to n. Consequently, we obtain

$$\zeta - \zeta_1 = \int_{n_1}^n \frac{dn'}{\rho(n')}. \tag{4.16}$$

The function $\rho(n)$ represents the density of states at the Fermi level, corresponding to a carrier density n, while ζ_1 is the Fermi level in the sample with the lowest carrier density n_1.

The function $\rho(n)$ has been calculated from the measured values α_∞ and n, using Eq. (4.14) and then numerical integration has been used to determine the value of ζ with an accuracy to within the constant term ζ_1. Since ζ_1 is governed completely by the density of states at energies $\mathscr{E} < \zeta_1$, Dubrovskaya et al. [335] were unable to find this quantity from their experimental data and they used two simple nonparabolic models, described in §6.2B. However, when $\mathscr{E} < \zeta_1$ the conduction band is almost parabolic and both models give similar values of ζ_1 for various reasonable values of the effec-

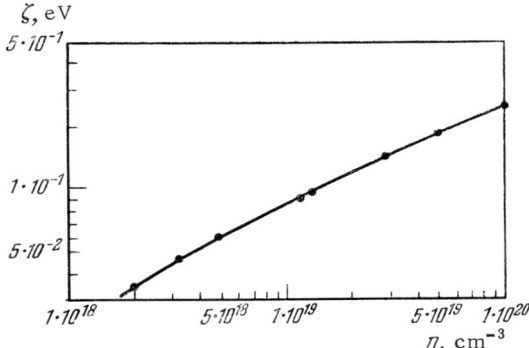

Fig. 4.7. Dependence of the Fermi level on the electron density in n-type PbTe at T = 77°K, obtained from measurements of the thermoelectric power in a strong magnetic field [335].

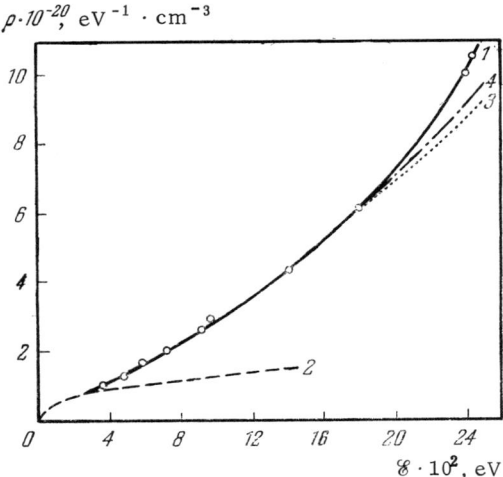

Fig. 4.8. Dependence of the density of states in the conduction band of PbTe on the energy [335]: 1) curve obtained by analysis of the experimental values of the thermoelectric power in a strong magnetic field; 2) theoretical curve for a parabolic band; 3) theoretical curve for the Cohen model and $\mathscr{E}_{gi} = 0.11$ eV; 4) theoretical curve for the Kane model and $\mathscr{E}_{gi} = 0.18$ eV.

tive width of the forbidden interaction band width $\mathscr{E}_{gi}$. In this way, the value $\zeta_1 = 0.034$ eV has been obtained for the sample with the lowest electron density of $2 \cdot 10^{18}$ cm^{-3}.

The dependence of the Fermi level on the carrier density is shown in Fig. 4.7 and the density of states is given as a function of the energy in Fig. 4.8. The shape of the $\rho(\mathscr{E})$ curve differs considerably from that predicted by the parabolic model.

The obtained function $\rho(\mathscr{E})$ has been compared with the dependence described by Eqs. (6.8) and (6.14), which follow from two simple nonparabolic models (cf. §6.2B). It has been found that by suitable selection of the parameters $\mathscr{E}_{gi}$ and m^*_{d0} the obtained function can be made to agree with either of these two models.

It follows from Kane's model that by plotting the quantity $n^{2/3}$ along the abscissa and using the ordinate to represent the square of the quantity

$$m^*_d = m^*_{d0}\left(1 + \frac{2\zeta}{\mathscr{E}_{gi}}\right) = \frac{3^{2/3} e a_\infty n^{2/3} \hbar^2}{\pi^{2/3} k_0^2 T}, \qquad (4.17)$$

we should obtain a straight line which has an intercept $(m^*_{d0})^2$ on the ordinate axis and whose slope has the tangent given by $2 \cdot 3^{1/3} \hbar^2 \pi^{1/3} m^*_{d0}/\mathscr{E}_{gi}$. The experimental points fit well such a straight line and the values of the effective density-of-states mass near the conduction band edge and of the effective width of the forbidden interaction band, found from the intercept and the slope of this line, are $m^*_{d0} = 0.12 m_0$ and $\mathscr{E}_{gi} = 0.19$ eV. The density of states corresponding to these values is shown in Fig. 4.8 only in that region where it does not coincide completely with the experimental curve obtained using no models.

The value obtained for m^*_{d0} is in agreement with the results of the Shubnikov – de Haas and cyclotron resonance measurements, while the value of $\mathscr{E}_{gi}$ is close to the optical width of the forbidden band.

Comparison of the results with the formulas for the Cohen model can be carried out conveniently by considering by the quan-

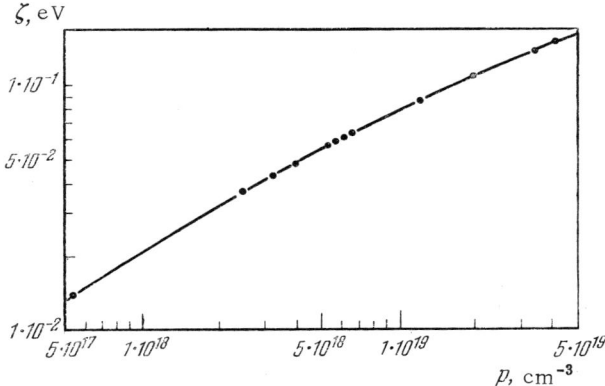

Fig. 4.9. Dependence of the Fermi level on the hole density in p-type PbTe at T = 77°K [336].

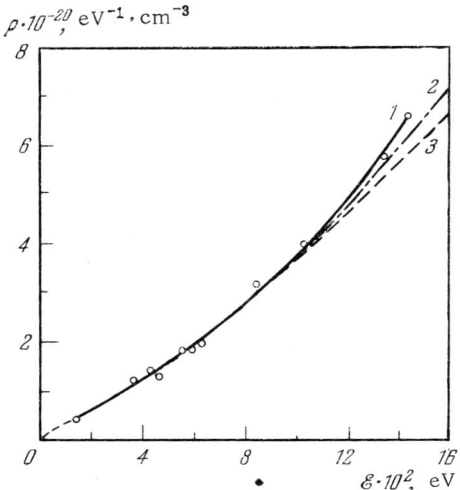

Fig. 4.10. Dependence of the density of states in the valence band of PbTe on the energy [336]: 1) curve obtained from the experimental values of the thermoelectric power in a strong magnetic field; 2) theoretical curve for the Kane model and $\mathscr{E}_{gi}$ = 0.15 eV; 3) theoretical curve for the Cohen model and $\mathscr{E}_{gi}$ = 0.08 eV.

tity $\rho/\mathscr{E}^{1/2}$ as a function of $\mathscr{E}$. According to the formula

$$\frac{\rho}{\mathscr{E}^{1/2}} = \frac{2^{1/2} m_{d0}^{*3/2}}{\pi^2 \hbar^3} \left(1 + \frac{2\mathscr{E}}{\mathscr{E}_{gi}}\right), \qquad (4.18)$$

this dependence should be linear, in agreement with the observations for $\mathscr{E} \leq 0.18$ eV ($n \leq 5 \cdot 10^{19}$ cm^{-3}). The parameters determined from the slope of the line and from the intercept of the ordinate axis are $m_{d0}^* = 0.12 m_0$ and $\mathscr{E}_{gi} = 0.11$ eV.

The Kane and Cohen models give the same value of the density-of-states mass at the bottom of the conduction band and this value is in agreement with the results found using other methods. The values of the effective width of the forbidden interaction band $\mathscr{E}_{gi}$, determined from these simple models, can be compared with the optical value of the forbidden band, which is $\mathscr{E}_g = 0.19$ eV (cf. §5.4C). Since the effective masses in PbTe are governed by the interaction of six bands which are relatively close to one another (cf. §6.2A), the effective forbidden interaction band may, in general, be either larger or smaller than the true forbidden band.

Measurements of the thermoelectric power in strong magnetic fields have shown that the Kane and Cohen nonparabolic models are capable of describing the density-of-states function with suitably selected parameters. The models begin to differ when the ratio of the longitudinal and transverse effective masses is found as a function of the carrier density, as is the case in investigations of the Shubnikov – de Haas effect (cf. §5.2C).

Measurements of the thermoelectric power in strong magnetic fields have been carried out also on samples of p-type PbTe with hole densities from $5 \cdot 10^{17}$ cm^{-3} to $4 \cdot 10^{19}$ cm^{-3}. The chemical potential of p-type PbTe is plotted as a function of the hole density in Fig. 4.9 and the dependence of the density of states on the energy is shown in Fig. 4.10; these dependences have been obtained from an analysis of the experimental data. The effective density-of-states mass near the top of the valence band is found to be $m_{d0}^* = 0.13 m_0$ and the effective value of the forbidden band width in the Kane model is $\mathscr{E}_{gi} = 0.15$ eV. Similar investigations yield $m_{d0}^* = 0.12 m_0$, $\mathscr{E}_{gi} = 0.17$ eV for n-type PbSe, $m_{d0}^* = 0.13 m_0$, $\mathscr{E}_{gi} = 0.16$ eV for p-type PbSe, assuming that the optical value of the forbidden band width of PbSe is $\mathscr{E}_g = 0.16$ eV. The corresponding values for n-type PbS are $m_{d0}^* = 0.22 m_0$, $\mathscr{E}_{gi} = 0.29$ eV, $\mathscr{E}_g = 0.28$ eV.

Thus, the two-band Kane model with an effective width of the interaction forbidden band $\mathscr{E}_{gi}$ equal to the true forbidden band width, can be used to describe the nonparabolicity of n- and p-type samples of all three chalcogenides.

D. Determination of the Effective Mass from Nernst—Ettingshausen Effect Measurements

The need to satisfy the condition $(\mu H/c)^2 \gg 1$ in measurements of the thermoelectric power in strong magnetic fields makes it difficult to determine the effective mass by this method at high temperatures because at these temperatures the mobility is low. A method based on the measurements of the transverse weak-field Nernst – Ettingshausen effect, the thermoelectric power, and the Hall mobility is more suitable for the case of low mobilities. This method was employed by Kaidanov et al. to investigate the nonparabolicity and temperature dependence of the effective mass of n-type PbTe [334], p-type PbTe [531], n-type PbSe [532], p-type PbSe [533], and n-type PbS [534].

The results obtained by this method have been analyzed using the many-ellipsoid model of the conduction band in which the dependence of the energy on the crystal momentum near each extremum is described by the formula

$$\sum_i \frac{\hbar^2 k_i^2}{2m_{i0}^*} = \gamma(\mathscr{E}). \tag{4.19}$$

The nonparabolicity of the band is governed by the function $\gamma(\mathscr{E})$. For this model, we can find a combination of the Nernst – Ettingshausen coefficient Q, the Hall coefficient R, the thermoelectric power α_0, and the conductivity σ of degenerate samples, which is independent of the scattering mechanism; by measuring the quantities which occur in such a combination we can obtain the effective density-of-states mass m_d^* at the Fermi level, which is defined by the formula

$$m_d^* = N^{2/3}(m_1^* m_2^* m_3^*)^{1/3}, \tag{4.20}$$

$$m_i^* = \frac{1}{\hbar^2} \frac{1}{k_i} \frac{\partial \mathscr{E}}{\partial k_i}. \tag{4.21}$$

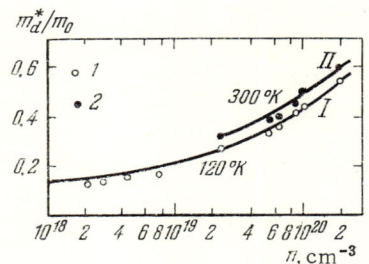

Fig. 4.11. Dependence of the effective density-of-states mass m_d^* on the carrier density in n-type PbTe [334]: 1), 2) experimental points at 120 and 300°K, respectively; I), II) theoretical curves calculated for the Kane model (I: $\mathscr{E}_{gi} = 0.21$ eV, $m_{d0}^* = 0.12 m_0$; II: $\mathscr{E}_{gi} = 0.27$ eV, $m_{d0}^* = 0.17 m_0$).

The relationship for the determination of the value of m_d^* is

$$m_d^* = \frac{\hbar^2 e\, (3\pi^2 n)^{2/3}}{\pi^2 k_0^2 T}\left(\alpha_0 - \frac{Q}{R\sigma}\right).$$

(4.22)

The quantities which occur in (4.22) have been measured for PbTe samples with electron densities from $2 \cdot 10^{18}$ cm^{-3} to $2 \cdot 10^{20}$ cm^{-3} in the temperature range 100–500°K. The values obtained for m_d^* are shown in Fig. 4.11 as a function of the carrier density. The experimental points fit well the theoretical curves plotted using the Kane model, for which the function $\gamma\,(\mathscr{E})$ has the form

$$\gamma = \mathscr{E}\left(1 + \frac{\mathscr{E}}{\mathscr{E}_{gi}}\right).$$

(4.23)

The effective width of the forbidden interaction band $\mathscr{E}_{gi}$ has been assumed to be equal to the optical value of the forbidden band width (0.21 eV at 120°K and 0.27 eV at 300°K). The value of the effective density-of-states mass at the bottom of the conduction band m_{d0}^* has been determined from the condition of the best agreement between the experimental and calculated curves; at 120°K it has been found that $m_{d0}^* = 0.12 m_0$, in agreement with the values obtained from the measurements of the thermoelectric power in a strong magnetic field, the Shubnikov–de Haas effect, and cyclotron resonance. This value and the room-temperature value $m_{d0}^* = 0.17 m_0$ are somewhat lower than those found from the measurements of the thermoelectric power in the absence of a magnetic field. The quantity $\Delta = \partial \ln m_{d0}^* / \partial \ln T$ is 0.35–0.45 in the temperature range 120–300°K and is close to the value found from measurements of the thermoelectric power.

Similar investigations of other lead chalcogenides have yielded the following values at 120°K: $m_{d0}^* = 0.13 m_0$, $\mathscr{E}_{gi} = 0.23$ eV for p-

Sec. 4.1] THERMOELECTRIC POWER AND THERMOMAGNETIC EFFECTS 171

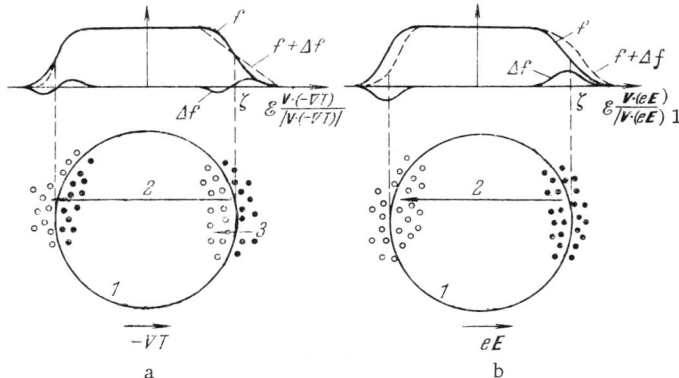

Fig. 4.12. Carrier distribution function and relaxation processes: a) in the presence of a temperature gradient; b) in the presence of an electric field (1 represents the Fermi surface; 2 is in elastic process; 3 is in inelastic process with a small change in crystal momentum.

type PbTe, $m_{d0}^* = 0.11 m_0$, $\mathscr{E}_{gl} = 0.23$ eV for n-type PbSe, $m_{d0}^* = 0.11 m_0$, $\mathscr{E}_{gl} = 0.21$ eV for p-type PbSe. Kaidanov and Chernik [532, 533] have reported that, at high carrier densities (of the order of 10^{20} cm^{-3}), the nonparabolicity of the conduction and valence band of PbSe is considerably stronger than that predicted by the Kane model.

E. Investigation of Scattering Mechanisms by Thermoelectric Methods. Electron—Electron Collisions

Knowledge of the band structure allows us to use the values of the thermoelectric power and other properties to determine the scattering parameter r. This value has been determined in many investigations, for example, in [228, 249, 331, 332, 334, 339]. As a rule, the values obtained are fairly close to $r = -0.5$. In some cases, the reported values of r are closer to zero. In such cases the authors have usually concluded that the obtained values of r indicate that the acoustical scattering predominates. However, it must be pointed out that the effective scattering parameter r for the polar scattering in the nonparabolic region is approximately the same as the parameters for the acoustical scattering in the parabolic range.

The scattering parameter $r = \partial \ln \tau / \partial \ln \mathscr{E}$ of n-type PbTe has been determined at 77°K [249], taking into account the nonparabolicity. For a sample with an electron density of $2 \cdot 10^{18}$ cm^{-3} the value of the scattering parameter is -0.3, which cannot be explained simply by the scattering on the acoustical phonons. The scattering paramter deduced from the thermoelectric power, r_α, differs considerably from r_μ found by analysis of the carrier-dependence of the mobility (cf. §3.1C). The difference between the values of r found from the thermoelectric power and the mobility indicates the presence of an inelastic scattering mechanism since a single relaxation time, common to all effects, cannot be introduced in the case of inelastic scattering.

As pointed out already, a temperature gradient acts effectively as a temperature field which disturbs the equilibrium distribution of carriers in a semiconductor. It follows from the transport equation that, in the presence of a temperature gradient and an electric field, the deviation of the distribution function Δf from the Fermi function consists of two terms: one which is proportional to the quantity $(\mathscr{E} - \zeta)\mathbf{v}.(-\nabla T)$, and another which is proportional to $e\mathbf{v} \cdot \mathbf{E}$ (here, e is the charge of the carriers with a suitable sign).

The distribution of carriers in degenerate samples is shown graphically in Fig. 4.12. The effect of a temperature gradient is to broaden the distribution edge on that side on which the quantity $\mathbf{v} \cdot (-\nabla T)$ is positive, while the other edge becomes steeper (Fig. 4.12a). Carriers moving against the temperature gradient behave as if they were hotter than in the equilibrium state. Carriers moving in the opposite direction are colder. An electric field produces a drift of the carrier gas in the direction eE (Fig. 4.12b).

In measurements of the thermoelectric power, thermal conductivity, and Nernst – Ettingshausen effect there is no flow of charge but an energy flux is observed because hot carriers move in one direction and an equal number of cold carriers moves in the opposite direction.

Let us now consider the relaxation processes. A hot carrier may reach the other side of the Fermi surface and reverse its direction of motion, practically without any loss of energy. Such elastic processes result in the relaxation of the distribution in both cases considered here and they affect the electrical conductivity as well as the thermoelectric effects.

Sec. 4.1]   THERMOELECTRIC POWER AND THERMOMAGNETIC EFFECTS   173

In other processes, a hot carrier loses its excess energy, drops below the Fermi level, but retains almost all its crystal momentum. Inelastic processes of this type have practically no effect on the value of the current and, consequently, they do not affect the electrical conductivity. On the other hand, the energy flux is affected considerably by such collisions and, therefore, these processes affect the thermoelectric phenomena. The influence of the elastic scattering can be described formally by introducing different effective relaxation times for different phenomena.

The difference between the values of r_α and r_μ, observed for n-type PbTe, was attributed by Moizhes and Ravich [249] to electron – electron collisions. In view of the law of conseravation of crystal momentum, such electron – electron collisions have no direct effect on the electrical conductivity but they simply redistribute the energy between carriers (cf. §3.1D). In the strongly degenerate case, such a redistribution is unimportant under conditions used in the measurement of the mobility, but it is important in the measurement of the thermoelectric effects since it gives rise to a redistribution of the energy between the fluxes of the hot and cold carriers.

The influence of the electron – electron collisions on the transport effects can be allowed for by one of two methods. In an approximate method credited to Keyes [345], an energy-independent quantity τ_{ee} is used; this quantity has the dimensions of time and represents the rate of approach of the distribution function to the Fermi function in a reference system moving at a velocity equal to the average electron velocity. This allows for the conservation of the total momentum of the electron gas in the electron – electron collisions. The second method, described by Ziman [346], is variational and does not include any arbitrary parameters.

By solving a modified Boltzmann equation, obtained by the Keyes method, Moizhes and Ravich deduced the following formulas for the carrier mobility, the thermoelectric power, the Lorenz number, and the Nernst – Ettingshausen coefficient of a degenerate semiconductor with a Kane conduction band:

$$\mu = \frac{e\tau_{e1}(\zeta)}{m^*_{\chi 0}(1 + 2\zeta/\mathscr{E}_{gi})}, \qquad (4.24)$$

$$\alpha_0 = \frac{\pi^2 k_0^2 T}{3e\zeta}\left[\left(\frac{\partial \ln \tau_{el}}{\partial \ln \mathscr{E}}\right)_{\mathscr{E}=\zeta}\frac{\tau}{\tau_{el}} + \frac{3}{2} - \frac{2\zeta/\mathscr{E}_{gi}}{1+2\zeta/\mathscr{E}_{gi}} + \frac{3}{2}\frac{\zeta/\mathscr{E}_{gi}}{1+\zeta/\mathscr{E}_{gi}}\right],$$

$$L = L_0 \frac{\tau}{\tau_{el}}, \qquad (4.25)$$

$$Q = \frac{\pi^2 k_0^2 T}{3e\zeta}\frac{\mu}{c}\frac{\tau}{\tau_{el}}\left[\left(\frac{\partial \ln \tau_{el}}{\partial \ln \mathscr{E}}\right)_{\mathscr{E}=\zeta}\frac{\tau}{\tau_{el}} - \frac{2\zeta/\mathscr{E}_{gi}}{1+2\zeta/\mathscr{E}_{gi}}\right]. \qquad (4.26)$$

Here, τ_{el} is the relaxation time for the elastic scattering mechanisms; τ is defined by the formula

$$\frac{1}{\tau} = \frac{1}{\tau_{el}} + \frac{1}{\tau_{ee}} \qquad (4.27)$$

and L_0 is the Lorenz number in the Wiedemann–Franz law for degenerate semiconductors:

$$L_0 = \frac{\pi^2}{3}\left(\frac{k_0}{e}\right)^2. \qquad (4.28)$$

It is evident from these formulas that the ratio of the scattering parameters determined from the carrier mobility and thermoelectric power is equal to the ratio of the actual Lorenz number and L_0:

$$\frac{r_\alpha}{r_\mu} = \frac{L}{L_0} = \frac{\tau}{\tau_{el}}. \qquad (4.29)$$

We shall show later (cf. §4.2A) that the Lorenz number for degenerate lead chalcogenides at temperatures of the order of 50-100°K is 40-60% smaller than the universal constant L_0. The ratio L/L_0 is a convenient measure of the inelasticity of scattering. Theoretical estimates show that collisions between carriers are primarily responsible for the deviation of L from L_0. Further calculations, carried out by the authors of the present book and their colleagues, have shown that the observed inelasticity is partly due to the polar scattering but collisions between carriers produce a much stronger deviation of L/L_0 from unity than does the polar scattering.

Analysis of the data on the Nernst–Ettingshausen effect carried out by Kaidanov, Chernik, et al. [334, 532], has also yielded a value of the scattering parameter which does not agree with the assumption that the scattering is purely of the acoustical origin and that the matrix element of the carrier – phonon interaction is independent of the energy. Our calculations have shown that the theoretical curves can be made to agree with the room-temperature experimental data by assuming that carriers are scattered on the acoustical and optical phonons and by taking into account the energy dependence of the matrix elements for the acoustical scattering due to nonparabolicity. As in the mobility calculations, no adjustable parameters have been used except for the deformation potential constant. At liquid-nitrogen temperature we must take into account collisions between carriers but this does not require introduction of any new adjustable parameters. Thus, the values of the thermoelectric power and of the Nernst – Ettingshausen coefficient of lead chalcogenides are governed by the scattering of carriers on the acoustical and optical phonons as well as by scattering between carriers themselves.

4.2. Thermal Properties

In this section, we shall consider mainly the thermal conductivity of lead chalcogenides. We shall also give some information on their specific heat and thermal expansion.

The total thermal conductivity $\varkappa$ of a semiconductor represents the transport of heat by phonons (the lattice thermal conductivity $\varkappa_l$), free electrons or holes (the electronic thermal conductivity $\varkappa_e$), photons, i.e., electromagnetic radiation ($\varkappa_{em}$), electron – hole pairs in the intrinsic conduction region (the ambipolar thermal conductivity $\varkappa_a$), excitons ($\varkappa_{ex}$), and spin waves ($\varkappa_{sp}$):

$$\varkappa = \varkappa_l + \varkappa_e + \varkappa_a + \varkappa_{em} + \varkappa_{ex} + \varkappa_{sp} \qquad (4.30)$$

We shall ignore the transport of heat by excitons and spin waves because these mechanisms have not been observed in lead chalcogenides. The reader interested in these mechanisms is referred to papers on this subject [347, 348].

The relative importance of a given heat conduction mechanism depends on the temperature range, the band structure of a semiconductor, its degree of doping, etc.

The thermal conductivity yields information on several properties of solids. By investigating the lattice thermal conductivity, we can obtain information on the presence of various defects in the

TABLE 4.1. Information on Solids Which Can Be Obtained from Thermal Conductivity Investigations

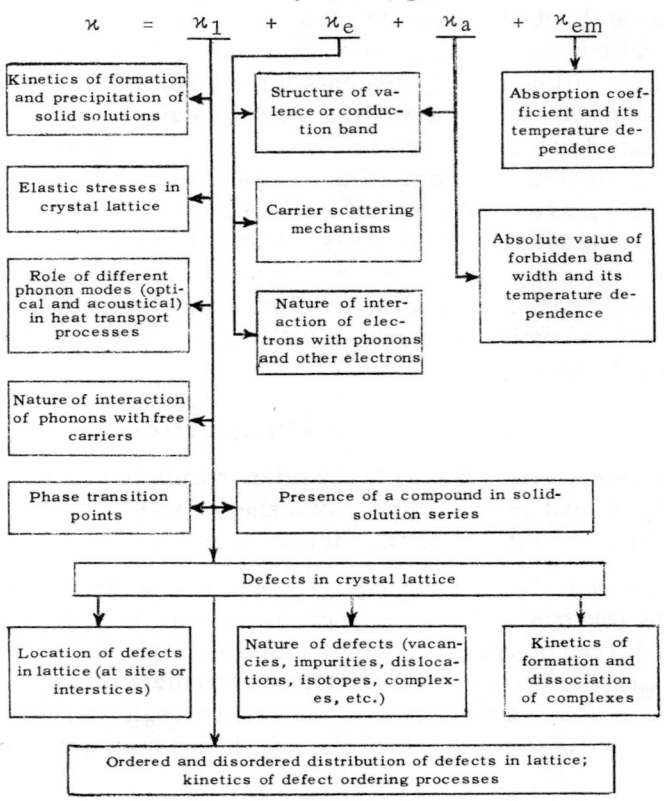

lattice: charged and neutral impurities, vacancies, complexes, dislocations, elastic stresses, etc. The electronic and ambipolar conductivities can be used to deduce information on the carrier scattering mechanism, forbidden band width, temperature dependence of this width, etc.

Table 4.1 shows schematically what information can be obtained about the properties of a solid by investigating various components of its total thermal conductivity.

Knowledge of the total thermal conductivity is also of practical importance. The thermal conductivity governs directly the efficiency of thermoelectric cooling units and generators (cf. §7.1A) and it is necessary to know its value in the design of diodes, transistors, and lasers.

We shall consider briefly certain conclusions of the heat conduction theory (mainly the qualitative conclusions) which are essential to a discussion of the experimental data obtained for PbTe, PbSe, and PbS. Readers who wish to obtain more detailed information on the theory of heat conduction are recommended to consult the various reviews and monographs on this subject [346, 349-354].

The transport of heat by any elementary excitations (phonons, electrons, photons, etc.) can be described by the formula

$$\varkappa = \frac{1}{3} C_v v^2 \tau. \tag{4.31}$$

where C_v is the specific heat at constant volume; v is the velocity of propagation of an elementary excitation and τ is its average relaxation time (the time taken to travel a mean free path).

A. Electronic Thermal Conductivity

We shall consider first the electric component of the thermal conductivity. In lead chalcogenides with carrier densities of about 10^{17} cm^{-3} this component is small, but in heavily doped samples (n≈10^{19}-10^{20} cm^{-3}) it is comparable with the lattice thermal conductivity. Comparison of the electrical [Eqs. (3.1) and (3.2)] and thermal [Eq. (4.31)] conductivities shows that they are both proportional to the carrier relaxation time τ. Bearing in mind that the following formula applies to a nondegenerate electron gas

$$C_v = \frac{3}{2} n k_0 \tag{4.32}$$

and that the average kinetic energy of carriers is $3k_0T/2$, we obtain the Wiedemann–Franz law:

$$\varkappa_e = L_0 \sigma T, \tag{4.33}$$

where L_0 is a constant known as the Lorenz number. This simple approach does not give the exact value of this constant but only its order of magnitude: $3(k_0/e)^2/2$. A more rigorous calculation requires an allowance for the energy distribution of electrons, energy dependence of the relaxation time, and thermoelectric field. For a parabolic band and elastic carrier scattering the Lorenz number is [355]:

$$L_0 = \left[\frac{(r+7/2)F_{r+5/2}(\zeta^*)}{(r+3/2)F_{r+1/2}(\zeta^*)} - \frac{(r+5/2)^2 F_{r+3/2}^2(\zeta^*)}{(r+3/2)^2 F_{r+1/2}^2(\zeta^*)}\right]\left(\frac{k_0}{e}\right)^2. \quad (4.34)$$

Because this formula includes the reduced Fermi level ζ^* and the scattering parameter r, the Lorenz number given by this formula depends weakly on the temperature and parameters of a material (Fig. 4.13). The Lorenz number for nondegenerate samples ($\zeta^* < 0$, $|\zeta^*| \gg 1$) is given by

$$L_0 = (r+5/2)\left(\frac{k_0}{e}\right)^2 \quad (4.35)$$

and depends only on the scattering parameter. For strongly degenerate samples ($\zeta^* \gg 1$), the Lorenz number is a universal constant

$$L_0 = \frac{\pi^2}{3}\left(\frac{k_0}{e}\right)^2. \quad (4.36)$$

The band nonparabolicity introduces changes into the expressions for the Lorenz number, with the exception of the formula which applies in the strongly degenerate case (Appendix A and [356]) When the effective mass increases with the energy, the value of the Lorenz number becomes somewhat smaller than that for a parabolic band because the nonparabolicity reduces the scattering parameter r. However, the nonparabolicity does not affect the Lorenz number of degenerate semiconductors. If the temperature is not too high, the Lorenz number is

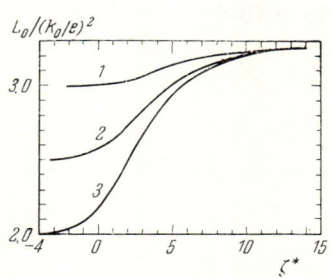

Fig. 4.13. Dependence of $L_0/(k_0/e)^2$ on ζ^* for various mechanisms of carrier scattering: 1) $r = 1/2$; 2) $r = 0$; 3) $r = -1/2$.

Sec. 4.2] THERMAL PROPERTIES 179

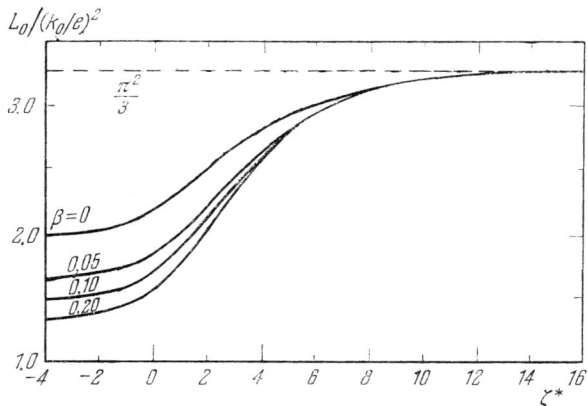

Fig. 4.14. Dependence of $L_0/(k_0/e)^2$ on ζ^*, taking into account the band nonparabolicity for various values of $\beta = k_0 T / \mathscr{E}_g$. The curve for $\beta = 0$ corresponds to curve 3 in Fig. 4.13. The calculation assumes that carriers are scattered by the acoustical phonons.

again little affected by the nonparabolicity because samples with high carrier densities are degenerate and those with low carrier densities have practically all the electrons in the parabolic region. The influence of the nonparabolicity must be allowed for only at high temperatures when the degeneracy is lifted and there are high-energy electrons.

This is illustrated well by the curves plotted in Fig. 4.14, where the dependence $L_0/(k_0/e)^2$ on ζ^*, calculated using Eq. (A.8a) (derived in Appendix A), is given for the case of the acoustical scattering of carriers.

A calculation of the Lorenz number, carried out by Smirnov and Ravich [357] using formulas which allow for the band nonparabolicity [Eqs. (A.6a) and (A.8a)], yielded results in good agreement with the experimental values of the electronic thermal conductivity of heavily doped n-type PbTe samples in the temperature range where the electron-gas degeneracy is lifted.

Equations (4.34)-(4.36) are valid only in the case of elastic carrier scattering. If the scattering is inelastic, we cannot use the same relaxation time for all the processes and the ratio of the effective relaxation times for the electrical and thermal conductivity appears in the expression for the Lorenz number. As pointed out in §4.1E, these relaxation times may be different.

It is known that due to the inelastic scattering of electrons in metals the Lorenz number below the Debye temperature may vary from 0 to $(\pi^2/3)(k_0/e)^2$, depending on the purity of a metal.*

The first to point out the difference between the Lorenz number and L_0 for degenerate n-type PbTe samples were Shalyt and Muzhdaba [358]. They determined the electronic thermal conductivity by measuring the total thermal conductivity in a strong magnetic field. This method is based on the fact that the electronic conductivity decreases in strong magnetic fields as $\varkappa_e \propto 1/H^2$, due to the bending of carrier trajectories.

In a sufficiently strong magnetic field the electronic thermal conductivity is negligibly small and the total conductivity is simply equal to the lattice conductivity. The difference between the values of the thermal conductivity in the presence of a magnetic field and in a strong field is equal to the electronic thermal conductivity. This method is the simplest and most reliable way of determining the Lorenz number. But it has the disadvantage of being limited as far as the range of carrier densities and temperatures is concerned. In samples with a low carrier density, the electronic thermal conductivity is negligible compared with the lattice component. In semiconductors with a high carrier density or in experiments at high temperatures, the mobility is relatively low and it is difficult to reach the conditions which satisfy the criterion of strong magnetic fields $(\mu Hc/)^2 \gg 1$. Strong pulsed magnetic fields, which are easier to generate than static fields, cannot be used in this method because of the slowness with which a suitable temperature is established.

The thermal conductivity of degenerate samples of n-type PbTe, p-type PbTe, p-type PbSe, and n-type PbS with carrier densities of the order of 10^{18}-10^{19} cm^{-3} has been measured in a magnetic field at 20-110°K [358, 361]. At liquid-nitrogen temperature the values of the Lorenz number are 40-60% lower than L_0. Such a deviation of the Lorenz number from L_0 indicates a considerable inelasticity of the scattering processes. When the temperature is reduced below 77°K the Lorenz number approaches L_0 (Fig. 4.15).

---

*Henceforth we shall use L to denote the Lorenz number in the case of inelastic scattering of carriers.

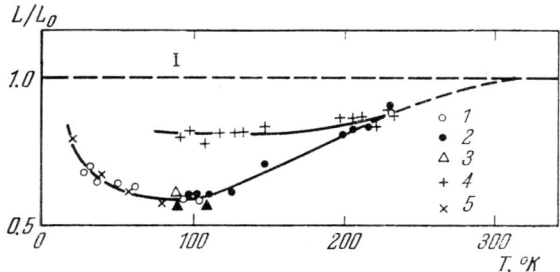

Fig. 4.15. Temperature dependence of L/L_0 for n-type PbTe: 1) $n = 5 \cdot 10^{18}$ cm^{-3}; 2) $n = 2 \cdot 10^{18}$ cm^{-3} (1 and 2 represent the results taken from [358]); 3) $n = 6.3 \cdot 10^{18}$ cm^{-3}; 4) $n = 8.9 \cdot 10^{18}$ cm^{-3} (3 and 4 represent results taken from [360]); 5) values calculated taking into account electron – electron collisions in a sample with $n = 5 \cdot 10^{18}$ cm^{-3}.

Moizhes and Ravich [249] analyzed the carrier-density dependences of the mobility and thermoelectric power (cf. §4.1E) and they also concluded that scattering of carriers in lead chalcogenides at 77°K is inelastic.

Smirnov et al. [360, 535] analyzed the temperature and carrier-density dependences of the thermal conductivity of n-type PbTe and n-type PbSe ($n = 10^{18}$–10^{20} cm^{-3}) in the temperature range 80–300°K and obtained information on the ratio L/L_0. The ratio L/L_0 for samples with electron densities of 10^{18}–10^{19} cm^{-3} increases approximately from 0.6 to 1 when the temperature and carrier density increase (Fig. 4.15). The value of L/L_0 for samples with carrier densities higher than $3 \cdot 10^{19}$ cm^{-3} is close to 1, i.e., scattering is elastic throughout the investigated temperature range.

Information on the Lorenz number can be obtained also by analysis of the dependences of various thermomagnetic effects on the magnetic field [535]. Such an analysis has been carried out for the thermoelectric power of all three investigated materials at 77°K. The obtained values of L/L_0 are quite close to those given in other papers.

We shall now consider the mechanisms of inelastic scattering in lead chalcogenides. One of the inelastic scattering mechanisms is the interaction with the optical lattice vibrations. If this mechanism predominates in a degenerate semiconductor, the Lorenz number is [359]:

$$L = \frac{L_0}{1 + (3/2\pi^2)(\hbar\omega_l/k_0T)^2 [\ln(4\zeta/\hbar\omega_l) - 1/3]}. \quad (4.37)$$

If this formula were applicable to lead chalcogenides, we could have explained the difference between the Lorenz number L and L_0 by the polar scattering. However, screening by free carriers alters considerably the formula for the Lorenz number. Moreover, scattering by the acoustical phonons and impurities is important at the carrier densities considered here. Therefore, using more exact formulas, we have concluded that the polar scattering cannot explain the observed deviation of L/L_0 from 1. For example, in the case of n-type PbTe with an electron density of $2 \cdot 10^{18}$ cm^{-3}, allowance for the polar and elastic scattering gives $L/L_0 \approx 0.8$ [535], while the experiment gives $L/L_0 \approx 0.6$ [358]. We have already mentioned that collisions between carriers affect appreciably the thermoelectric power and the Nernst − Ettingshausen effect of lead chalcogenides. We shall now consider the effect of the electron − electron collisions on the Lorenz number.

In the strongly degenerate case, the electron − electron collisions do not affect the electrical conductivity but give rise to an additional thermal resistivity W_{ee} so that the total thermal resistivity is $W = 1/\varkappa_e = W_0 + W_{ee}$, where $W_0 = 1/L_0\sigma T$ is the thermal resistivity due to the elastic scattering processes. When the electron − electron collisions are taken into account, the Lorenz number becomes

$$\frac{L}{L_0} = \frac{1}{1 + W_{ee}/W_0}. \quad (4.38)$$

The ratio W_{ee}/W_0 can be estimated from the following formula:

$$\frac{W_{ee}}{W_0} = \frac{2\pi^4 e^3 (k_0T)^2 (k_F r_{sc})^3 \mu n}{\varepsilon_\infty^2 \hbar^3 v_F^4 k_F^3}, \quad (4.39)$$

where k_F and v_F are, respectively, the crystal momentum and the velocity of electrons at the Fermi level; r_{sc} is the screening radius; ε_∞ is the high-frequency permittivity. Estimates for samples with carrier densities of about $2 \cdot 10^{18}$ cm^{-3} give a value of $L/L_0 \approx 0.5$-0.6 at 77°K for all three chalcogenides. This value is very close to the experimental data. Thus, the difference between the Lorenz number and L_0 is mainly due to collisions between carriers; allowance for the polar scattering reduces somewhat the theoretical values of L/L_0.

We shall now use the conclusion about the importance of collisions between carriers to reconsider the dependence of L/L_0 on the carrier density and temperature. It follows from Eq. (4.39) that the value of W_{ee}/W_0 should decrease with increasing carrier density because of an increase of the crystal momentum and carrier velocity at the Fermi level. The ratio L/L_0 should approach unity, in agreement with the experimental data.

The temperature dependence of the Lorenz number at $T < 77°K$ can be described by a formula which follows from Eqs. (4.38) and (4.39):

$$\frac{L}{L_0} = (1 + \text{const} \cdot \mu T^2)^{-1}. \qquad (4.40)$$

When the carrier mobility increases rapidly due to reduction in temperature, the Lorenz number varies quite slowly. At sufficiently low temperatures the mobility varies slowly and the ratio L/L_0 increases. Experimental points taken from [358] fit a curve calculated using Eq. (4.40) (Fig. 4.15). Allowance for the polar scattering does not alter the good agreement between the theory and experiment.

When the temperature is increased above 77°K the degeneracy is gradually lifted and Eq. (4.40) ceases to apply. The strength of the electron – phonon scattering increases rapidly, the scattering becomes more and more elastic, and $L/L_0 \to 1$.

Thus, the results obtained for the Lorenz number of lead chalcogenides can be explained by the influence of collisions between carriers.

An estimate of L/L_0 for PbTe – PbSe solid solutions with carrier densities of the order of 10^{18} cm^{-3} has been reported in [360]. The nature of changes in the ratio L/L_0 for the composition 95% PbTe + 5% PbSe is the same as that for pure PbTe samples with the

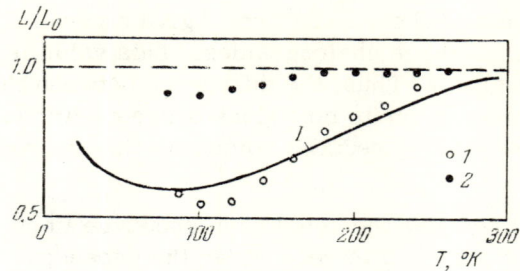

Fig. 4.16. Temperature dependence of L/L_0 for PbTe – PbSe solid solutions: 1) 5% PbSe + 95% PbTe; 2) 40% PbSe + 60% PbTe; I) L/L_0 for PbTe samples with $n \approx 10^{18}$ cm^{-3}.

same carrier density. The behavior of L/L_0 is different for the composition 40% PbSe + 60% PbTe (Fig. 4.16). When the concentration of the second component increases, the mobility of the carriers decreases rapidly and this reduces the relative importance of the electron – electron scattering so that the ratio L/L_0 tends to unity.

B. Ambipolar Diffusion

In the intrinsic conduction region, a considerable contribution to the thermal conductivity is made by the component due to the ambipolar diffusion of electrons and holes. This effect can be represented qualitatively in the following manner [354]. In the intrinsic conduction case, the number of electrons and holes at the hot end of a sample is higher than that at the cold end. This causes diffusion of electron – hole pairs. Such a diffusion gives rise to the absorption of some energy $\mathscr{E}$ at the hot end, which is necessary for the formation of a pair and this energy is evolved at the cold end because of the recombination of electrons and holes. The energy $\mathscr{E}$ consists of the kinetic energy of an electron and a hole and of the energy $\mathscr{E}_g$, which is equal to the forbidden band width; the latter energy is required in order to transfer an electron from a filled band to the conduction band. For the majority of semiconductors, $\mathscr{E}_g \gg k_0 T$ and, therefore, the energy carried by an electron – hole pair is approximately equal to $\mathscr{E}_g$ and much greater than the energy carried by electrons and holes traveling separately. In lead chalcogenides, the contribution to the thermal conductivity from such ambipolar diffusion should be fairly large because the electron transfer to the conduction is easy since the forbidden band of these compounds is relatively narrow.

The formula for the ambipolar component of the thermal conductivity, $\varkappa_a$, deduced by Davydov and Shmushkevich [362] is

$$\varkappa_a = L_0 T \frac{2\sigma_p \sigma_n}{\sigma_p + \sigma_n} \left(\frac{\mathscr{E}_g}{2k_0 T} + r_n + r_p + 3 \right)^2. \tag{4.41}$$

In lead chalcogenides with carrier densities of $(2\text{-}5) \cdot 10^{17}$ cm^{-3}, the value of $\varkappa_a$ is appreciable even at 300-350°K. The value of the forbidden band width $\mathscr{E}_g$ can be calculated from Eq. (4.41), if the other parameters are known.

Determination of the forbidden band width of lead chalcogenides at high temperatures meets with considerable experimental difficulties. The available electrical methods (Hall effect and electrical conductivity) allow us to determine only the value of $\mathscr{E}_{g0}$ which is the forbidden band width at 0°K. The temperature dependence of $\mathscr{E}_g$ is usually determined by an optical method based on an analysis of the behavior of the fundamental absorption edge. However, in substances with a fairly narrow forbidden band, particularly at high temperatures, the determination of the fundamental absorption involves the subtraction of a strong background due to the absorption by free carriers (impurity or intrinsic carriers). An accurate determination of the forbidden band width from the fundamental absorption edge is difficult if the law which the absorption coefficient obeys near the absorption edge is not known. Therefore, it is interesting to mention a method for the determination of $\mathscr{E}_g$ and its temperature dependence from measurements of the electrical conductivity, thermoelectric power, and thermal conductivity of p- and n-type samples in the extrinsic (impurity) and mixed conduction regions, which has been described in [363]. This method has been used [274, 364] to determine $\mathscr{E}_g$ of PbTe and PbSe in the temperature range 400-800°K and has been found to be fairly simple and accurate.

In this method, the value of $\mathscr{E}_g$ is calculated from the formula

$$\varkappa_1^{1/2} \varkappa_2^{1/2} \sigma_1 \sigma_2 = M_1^0 M_2^0 (\mathscr{E}_g^* + 5 + r_1 + r_2)^2 \exp(\mathscr{E}_g^*) \tag{4.42}$$

The subscript "1" refers to a p-type sample and the subscript "2" to an n-type sample. We shall now define the symbols used in Eq. (4.42) for p-type samples (the definition is similar for n-type samples):

$$\varkappa_1 = \varkappa_a/[(k_0/e)^2 T \sigma_1(T)], \quad \sigma_1 = \sigma_1(T)/\sigma_0(T);$$

$\sigma_0(T)$ is the value of the electrical conductivity $\sigma_1(T)$, determined experimentally in the extrinsic conduction region and extrapolated to high temperatures; $M_1^0 = \frac{B}{n_{10}}\left(\frac{m_1}{m_0}\right)^{3/2}$ (n_{10} is the impurity concentration, $B = 4.82 \cdot 10^{15} T^{3/2}$ is the density of states for free electrons, m_1 is the density-of-states effective mass of holes); $\mathscr{E}_g^* = \mathscr{E}_g/k_0 T$. The accuracy of the determination of $\mathscr{E}_g$ by this method is, according to calculations reported in [363], not less than ±0.01 eV.

Figures 4.17a and 4.17b show the dependences of the thermal resistivity of the crystal lattice W_l ($W_l = 1/\varkappa_l$) for PbTe and PbSe samples with carrier densities of about 10^{18} cm^{-3} [364]. The lattice thermal conductivity is found from the expression (the photon thermal conductivity will be discussed later):

$$\varkappa_1 = \varkappa - \varkappa_e - \varkappa_{em}. \tag{4.43}$$

The electronic thermal conductivity is calculated using Eqs. (4.33) and (4.34). We shall show later that, according to the theory, the lattice thermal resistivity at temperatures above the Debye temperature can be represented in the form of a straight line: $W_l = $ const $\cdot$ T. The departure from the linear dependence in Figs. 4.17a and 4.17b at high temperatures is caused by the appearance of an additional thermal conductivity due to electron−hole pairs, which is defined as the difference between $\varkappa_l$, calculated using Eq. (4.43), and the value of the same conductivity extrapolated toward high temperatures in accordance with the law $\varkappa_l = $ const $\cdot$ T.

Figures 4.18 and 4.19 give the temperature dependences of the electrical conductivity σ and the parameter $M_{1,2}^0/B$ for PbTe and PbSe p- and n-type samples, which are necessary in the calculation of $\mathscr{E}_g$ using Eq. (4.42). The values of $M_{1,2}^0/B$ have been determined from the experimental data on the thermoelectric power in the extrinsic conduction region and extrapolated to high temperatures. $M_{1,2}^0/B$ increases as the temperature rises because of an increase in the effective mass. The strong temperature dependence of this parameter for p-type PbTe samples is due to the presence of two valence bands in this compound: the light-hole and heavy-hole bands.

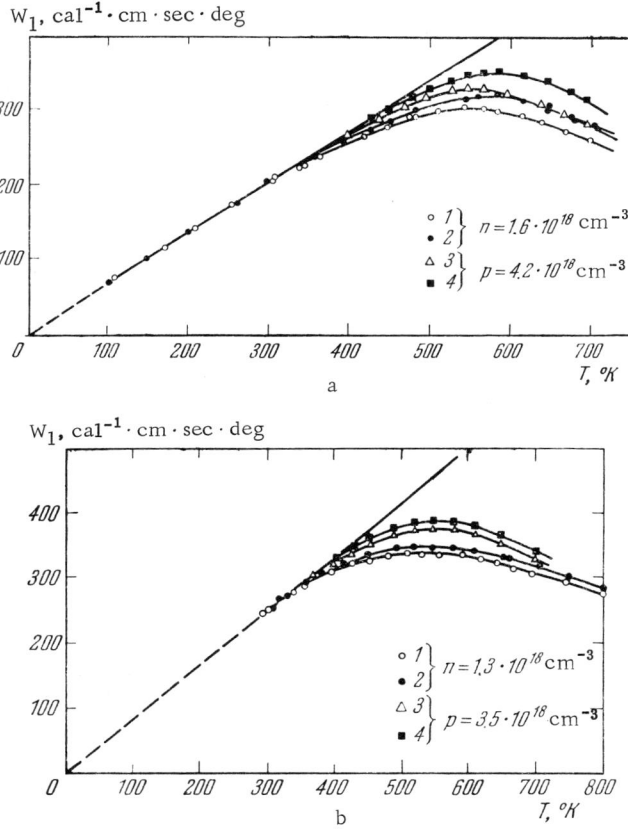

Fig. 4.17. Temperature dependences of the lattice thermal resistivity of PbTe (a) and PbSe (b) [364]: 1), 3) experimental data ($\varkappa - \varkappa_e$); 2), 4) experimental data after allowance for the photon (electromagnetic) component of the conductivity ($\varkappa - \varkappa_e - \varkappa_{em}$). The dashed line represents extrapolation of the high-temperatured data reported in [355].

The results of a calculation of $\mathscr{E}_g$ by means of Eq. (4.42) are presented in Fig. 4.20.* In the temperature range 400–500°K we find that $d\mathscr{E}_g/dT$ is close to $+4 \cdot 10^{-4}$ eV/deg.

---

*The absolute value of $\mathscr{E}_g$ for PbTe, calculated using Eq. (4.42), is still not sufficiently accurate since we have not allowed for the complex nature of the valence band of this compound. However, an estimate of $\mathscr{E}_g$ taking into account the influence of the second valence band shows that it affects the results only at high temperatures and the effect is small.

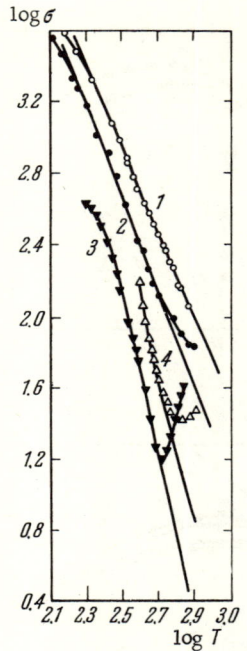

Fig. 4.18. Temperature dependence of the electrical conductivity of n- and p-type samples of PbTe and PbSe: 1) n-type PbTe; 2) n-type PbSe; 3) p-type PbTe; 4) p-type PbSe.

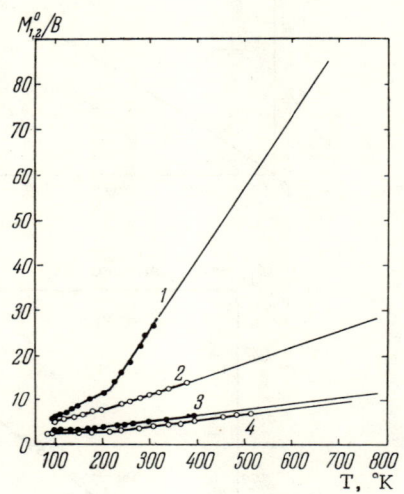

Fig. 4.19. Temperature dependences of the parameter $M^0_{1,2}/B$ for n- and p-type PbTe and PbSe samples: 1) p-type PbTe; 2) p-type PbSe; 3) n-type PbSe; 4) n-type PbTe.

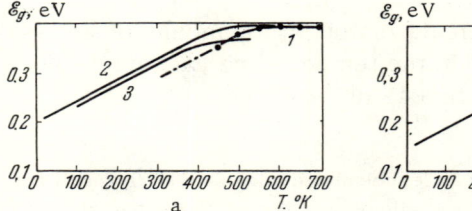
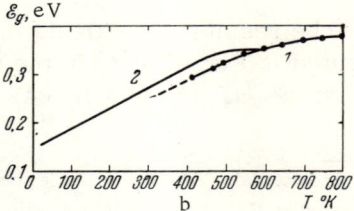

Fig. 4.20. Temperature dependences of the forbidden band width of PbTe (a) and PbSe (b): 1) calculations using Eq. (4.42) [364]; 2) results reported in [64]; 3) results reported in [123].

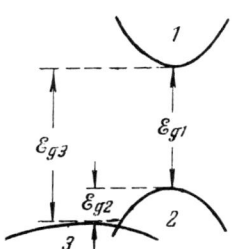

Fig. 4.21. Schematic representation of the band structure of p-type PbTe.

The forbidden band width of PbTe remains constant, beginning from about 550-600°K. Such behavior of $\mathscr{E}_g$ has been explained in [267], where it is assumed that the edge of the heavy-hole band does not vary with temperature, while the gap between the light-hole and conduction bands increases in accordance with the law $d\mathscr{E}_g/dT = +4 \cdot 10^{-4}$ eV/deg. A graphical plot of the energy band structure for parameters selected in this way (Fig. 3.10) shows that the gap between the conduction band and the nearest valence band (i.e., the forbidden band width $\mathscr{E}_g$, which is determined experimentally) should remain constant from about 500°K. In contrast to PbTe, $\mathscr{E}_g$ of PbSe does not become constant at high temperatures and, therefore, the results reported by Gibson [64] remain unconfirmed.

C. Influence of a Complex Valence Band on the Thermal Conductivity of p-type PbTe

It follows from the Hall effect measurements (cf. §3.2B) and from other effects that the valence band of p-type PbTe consists of two subbands containing heavy and light holes. According to the theory given in [365], semiconductors with the band structure shown in Fig. 4.21 have two additional terms in the total thermal conductivity, apart from those included in Eq. (4.30). These terms are the thermal conductivity due to additional ambipolar diffusion from band 3 to band 1 ($\varkappa'_a$) and the conductivity due to hole transitions from band 2 to band 3 ($\varkappa_{h-l}$). The latter conductivity is due to the fact that the equilibrium densities of holes in the light- and heavy-hole bands 2 and 3 depend on temperatures and, therefore, at the hot end of the sample the number of holes in band 2 is greater than that at the cold end. Carriers diffuse toward the cold end, where they are transferred from band 2 to band 3, giving up an energy $\mathscr{E}_{g2}$ (cf. Fig. 4.21), i.e., we have here a mechanism similar to the ambipolar diffusion case described earlier. The components of the thermal conductivity $\varkappa'_a$ and $\varkappa_{h-l}$ can be written in the form

$$\varkappa'_a = \frac{T\sigma_3\sigma_1}{\sigma_1 + \sigma_3}(\alpha_{01} - \alpha_{03})^2, \qquad (4.44)$$

and

$$\varkappa_{h-l} = \frac{T\sigma_2\sigma_3}{\sigma_2 + \sigma_3}(\alpha_{03} - \alpha_{02})^2, \qquad (4.45)$$

where σ_1, σ_2, σ_3, α_{01}, α_{02}, α_{03} are, respectively, the electrical conductivity and thermoelectric power for bands 1, 2, and 3. The dependences given by Eqs. (4.44) and (4.45) are valid for any degree of degeneracy of the electron gas and any mechanism of carrier scattering. For a parabolic band, classical statistics, and scattering by the acoustical vibrations of the crystal lattice, Eqs. (4.44) and (4.45) become

$$\varkappa'_a = \left(\frac{k_0}{e}\right)^2 \frac{T\sigma_1\sigma_3}{\sigma_1 + \sigma_3}\left(\frac{\mathscr{E}_{g3}}{k_0T} + 4\right)^2, \qquad (4.46)$$

$$\varkappa_{h-l} = \left(\frac{k_0}{e}\right)^2 \frac{T\sigma_2\sigma_3}{\sigma_2 + \sigma_3}\left(\frac{\mathscr{E}_{g2}}{k_0T}\right)^2. \qquad (4.47)$$

Saakyan and Smirnov [366] used Eqs. (4.46) and (4.47) to estimate qualitatively the contribution of $\varkappa'_a$ and $\varkappa_{h-l}$ to the total thermal conductivity of p-type PbTe with hole densities of $1.7 \cdot 10^{18}$ cm^{-3} and $9 \cdot 10^{19}$ cm^{-3}.

Their calculations for samples with $p = 1.7 \cdot 10^{18}$ cm^{-3} were carried out for 600-700°K, i.e., in the region where an appreciable additional thermal conductivity due to the ambipolar diffusion has been observed. Saakyan and Smirnov used the band model of p-type PbTe suggested in [267]. According to this model, the light-hole band drops below the maximum of the heavy-hole band because of a temperature-induced change in the forbidden band width.

The calculated value of $\varkappa_{h-l}$ is found to be smaller, about $3 \cdot 10^{-5}$ cal·cm^{-1}·sec^{-1}·deg^{-1}, than the experimentally determined additional thermal conductivity due to the ambipolar diffusion at these temperatures (of the order of 10^{-3} cal·cm^{-1}·sec^{-1}·deg^{-1}).

On the other hand, the contribution of the "additional" ambipolar diffusion ($\varkappa'_a$) is found to be considerable and amounts to about 0.7 of the thermal conductivity due to transitions from the heavy-hole band to the conduction band. Obviously, this value depends on the selected energy band model and contains a large number of parameters so that the calculated value can be strongly overestimated because of the inaccuracy of the calculations.

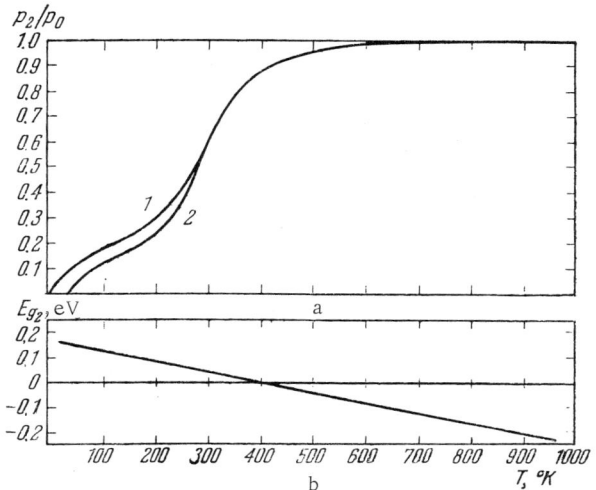

Fig. 4.22. Temperature dependences of the ratio of the heavy (p_2) and total (p_0) hole densities (a) and of the energy gap between the heavy- and light-hole bands (b) for p-type PbTe [267]: 1) $p_0 = 9 \cdot 10^{19}$ cm^{-3}; 2) $p_0 = 5 \cdot 10^{19}$ cm^{-3}.

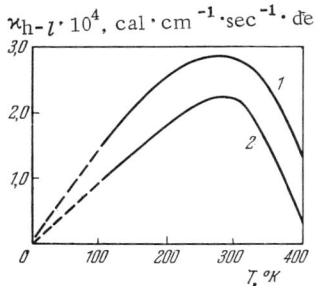

Fig. 4.23. Temperature dependences of the additional thermal conductivity $\varkappa_{h-l}$ for p-type PbTe samples [367]: 1) $p = 1.58 \cdot 10^{20}$ cm^{-3}; 2) $p = 2.93 \cdot 10^{20}$ cm^{-3}.

In the case of p-type PbTe samples with a hole density of $9 \cdot 10^{19}$ cm^{-3}, we can expect a contribution to the thermal conductivity only due to transitions between the heavy- and light-hole bands ($\varkappa_{h-l}$). As before, we can use the energy band model of p-type PbTe proposed in [267]. We then find [cf. Fig. 4.22 and Eq. (4.47)] that at 0°K the heavy-hole band is empty, the electrical conductivity due to this band is zero, and, according to Eq. (4.47), $\varkappa_{h-l} = 0$.

At high temperatures (T > 800°K), all the carriers are in the heavy-hole band (Fig. 4.22). In this case, the electrical conductivity due to the light holes is equal to zero and again we have $\varkappa_{h-l} = 0$. At ~400°K, $\mathscr{E}_{g2} = 0$ and $\varkappa_{h-l}$ is again equal to zero [cf. Eq. (4.47)]. Since these calculations are

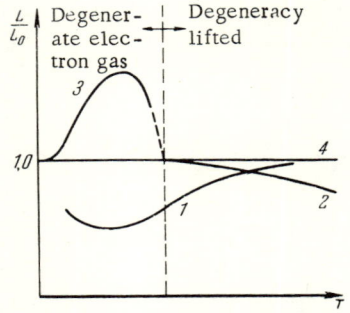

Fig. 4.24. Schematic representation of the temperature dependence of L/L_0 for n- and p-type PbTe. 1) Carrier density of the order of 10^{18} cm^{-3}. The Lorenz number decreases due to the electron–electron interaction. When the carrier density is increased, $L/L_0 \to 1$. 2) Band nonparabolicity at fairly high carrier densities reduces the Lorenz number. 3) Interband scattering of the light and heavy holes in p-type PbTe increases the Lorenz number [367]. 4) Value of the Lorenz number calculated from Eq. (4.34).

only qualitative, we can use Eq. (4.47) to estimate the thermal conductivity component $\varkappa_{h-1}$, in spite of the fact that this equation is valid only for a nondegenerate electron gas and the electron gas in a sample with $p = 9 \cdot 10^{19}$ cm^{-3} is degenerate (but not very strongly) in the temperature range 300–500°K.

Calculations show that the contribution of $\varkappa_{h-1}$ to the total thermal conductivity of doped p-type PbTe samples is not large. It should be allowed for only near room temperature [in this case, $\varkappa_{h-1} \approx 3 \cdot 10^{-4}$ cal·cm^{-1}·sec^{-1}·deg^{-1}, while $\varkappa_{tot} \approx 1 \cdot 10^{-2}$ cal·cm^{-1}·sec^{-1}·deg^{-1}]. The value of $\varkappa_{h-1}$ is negligible at T > 400°K.

Smirnov et al. [367] used the standard method and the values of σ_2, σ_3, α_{02}, and α_{03} found from the thermoelectric power, electrical conductivity, and Hall effect of samples with hole densities of $1.58 \cdot 10^{20}$ cm^{-3} and $2.93 \cdot 10^{20}$ cm^{-3}, and they determined $\varkappa_{h-1}$ from the general formula (4.45). The results of their calculations are presented in Fig. 4.23.

Finally, the presence of two valence bands in p-type PbTe under the conditions of a strong interband interaction between the light and heavy holes may, as demonstrated in [367], increase considerably the Lorenz number. The strongest effect is observed when the value of $(\zeta - \mathscr{E}_0)/k_0 T$ lies between 0 and +1 ($\mathscr{E}_0$ is the energy of the edge of the heavy-hole band).

The conclusions about the Lorenz number, reached in subsections A and C of the present section, are summarized schematically in Fig. 4.24.

D. Photon Thermal Conductivity

Heat may be transported at high temperatures in lead chalcogenides by electromagnetic radiation (photons).

In an investigation of the thermal conductivity of glasses, Genzel [368] demonstrated that, in the case of small temperature drops across a sample and when the reciprocal of the absorption coefficient $1/\alpha$ is small compared with the linear dimensions d of a sample, the transport of heat by electromagnetic radiation can be allowed for by introducing an additional thermal conductivity

$$\varkappa_{em} = \frac{16}{3} \frac{\mathcal{N}^2 \sigma_0 T^3}{\alpha}. \qquad (4.48)$$

If $1/\alpha > d$, we find that Eq. (4.48) becomes

$$\varkappa_{em} = \frac{16}{3} \mathcal{N}^2 \sigma_0 d T^3 \qquad (4.49)$$

($\mathcal{N}$ is the refractive index and σ_0 is the constant in the Stefan–Boltzmann law). In the derivation of Eq. (4.48), it is assumed that the substance is optically isotropic and the absorption coefficient is independent of wavelength. Equation (4.48) can be deduced easily from Eq. (4.31) by replacing in the latter the specific heat C_V by the specific heat of the photon gas $q_{em} = 16\ \mathcal{N}^3 T^3 \sigma_0/c$, and by assuming that $v = c/\mathcal{N}$ (where c is the velocity of light) and $\tau v = 1/\alpha$ [369]. The formulas for $\varkappa_{em}$ in the case of complex lattices with an absorption coefficient depending on the wavelength were deduced by Moizhes [369].

The absorption coefficient of lead chalcogenides is fairly high, particularly in the case of doped samples (cf., for example, [64]) and, therefore, according to Eq. (4.48), $\varkappa_{em}$ should be small. A large contribution of this component to the total thermal conductivity can be expected only for undoped single-crystal samples at temperatures above 400-500°K. The value of $\varkappa_{em}$ for polycrystalline samples should be considerably smaller than that for single crystals with the same carrier density because of the scattering of photons by grain boundaries.

A rough (and somewhat overestimated) value of $\varkappa_{em}$ was obtained in [364] for PbTe samples with a carrier density of the order

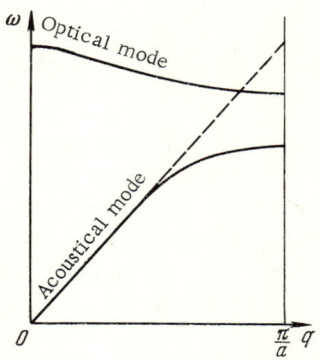

Fig. 4.25. Schematic representation of the dependence of the frequency of the lattice vibrations on the wave vector (a is the distance between atoms in the lattice).

of 10^{18} cm^{-3}. According to this estimate, $\varkappa_{em}$ at 570°K should be about 4-5% of the total thermal conductivity (cf. Figs. 4.17a and 4.17b).

E. Lattice Thermal Conductivity and Specific Heat

We shall now show that the behavior of the lattice thermal conductivity below and above a certain characteristic temperature θ (known as the Debye temperature) is different. Therefore, we shall consider briefly the specific heat and a method for the determination of the Debye temperature from the specific heat.

Figure 4.25 shows schematically the dependence of the frequencies of the crystal lattice vibrations on the wave vector. In fact, a real vibration spectrum of a crystal has three acoustical modes (two transverse and one longitudinal) and (3s − 3) optical models (s is the number of atoms in the unit cell).

The short-wavelength vibrations (corresponding to large values of q) have wavelengths equal to double the distance between neighboring atoms, while the long wavelengths (corresponding to low values of q) are equal to double the length of the whole crystal.

We shall introduce the concepts of the phase and group velocities, which will be required later. The phase velocity v_p represents the velocity of the phase of a monochromatic wave, while the group velocity v_g is the velocity of a wave packet and, consequently, the velocity at which the energy is transported by waves.

The formulas for the phase and group velocities can be written in the form [270]:

$$v_p = \frac{\omega}{|q|}, \quad v_g = \left(\frac{d\omega}{dq}\right). \tag{4.50}$$

For the long acoustical waves, $v_p = v_g = v_0$, where v_0 is the velocity of sound.

According to the theory of Einstein, who assumed that all the atoms in a solid vibrate at the same frequency ν, the specific heat per gram-atom at constant volume can be written in the form

$$C_v = 3R \left(\frac{h\nu}{k_0 T}\right)^2 \exp\left(\frac{h\nu}{k_0 T}\right) \bigg/ \left[\exp\left(\frac{h\nu}{k_0 T}\right) - 1\right]^2. \tag{4.51}$$

In 1912, Debye considered the acoustical vibrations of the crystal lattice and introduced, instead of a single Einstein frequency, a frequency distribution in an isotropic elastic continuum

$$Z(\nu) = 4\pi V \left(\frac{1}{v_l^3} + \frac{2}{v_t^3}\right) \nu^2 = a\nu^2 \tag{4.52}$$

(V is the volume of a solid; v_l and v_t are the velocities of the longitudinal and transverse acoustic waves). Debye also assumed that all three acoustical modes satisfied, at all values of q, the dispersion law $\omega = v_0 q$ (shown by the dashed straight line in Fig. 4.25). In fact, this dispersion law is obeyed rigorously only at low values of q.

Bearing in mind that the total number of normal vibrations of N particles in a solid is 3N, Debye defined the limiting frequency ν_{max}, corresponding to the minimum wavelength:

$$\int_0^{\nu_{max}} Z(\nu) \, d\nu = 3N. \tag{4.53}$$

The specific heat per mole of a substance at constant volume can therefore be written as an integral of the Einstein function

$$C_v = D(\theta/T) = 9R(T/\theta)^3 \int_0^{\theta/T} \frac{(\theta/T)^4 \exp(\theta/T)}{[\exp(\theta/T) - 1]^2} \, d(\theta/T), \tag{4.54}$$

where $D(\theta/T)$ is the Debye function which is tabulated in [370],* and

---

*Detailed information on the specific heat of solids can be found in a monograph by Ansel'm [270].

$\theta = h\nu_{max}/k_0 = \hbar\omega_{max}/k_0$ is known as the Debye temperature for the acoustical modes. Similarly, we can introduce the concept of the Debye temperature for the optical modes:

$$\theta_{l,t} = \frac{\hbar\omega_{l,t}^0}{k_0}, \qquad (4.55)$$

where $\omega_{l,t}^0$ is the limiting frequency of the longitudinal or transverse optical modes.

In the extreme cases of very low temperature ($T \ll \theta$) we find that $C_V \propto T^3$, while at high temperatures ($T \gg \theta$) the specific heat C_V tends to a constant value which, according to the Dulong – Petit law, should be 6 cal·(g-atom)$^{-1}$·deg^{-1}.

The experimental values of the specific heat are usually found at constant pressure. The specific heat at constant volume is frequently found from the formula

$$C_p - C_v = AC_p^2 T, \qquad (4.56)$$

where $A = 2.14 \cdot 10^{-2}/T_{mp}$ mole/cal (T_{mp} is the melting point).

The Debye temperature θ can be determined by comparing the experimentally measured value of C_V with the theoretical Debye formula (4.54).

There are also other methods of determining θ and $\theta_{l,t}$: from the melting point (Lindemann's formula), expansion coefficients, compressibility, neutron spectroscopy results, x-ray data, measurements of the infrared absorption, Mossbauer effect data, etc.*

The specific heat of the lead chalcogenides has been reported in [372-374]. The most accurate values of C_p (Fig. 4.26) and the Debye temperature [calculated from C_V using Eq. (4.54)] at 20-260°K are given in [372]. The average Debye temperatures θ for PbTe, PbSe, and PbS are, respectively, 130, 160, and 220°K.†

---

*Details of the available methods for the determination of the Debye temperature can be found in Blackman's article [371].
†The dependence $\theta(T)$ for PbTe is given also in Fig. 4.43.

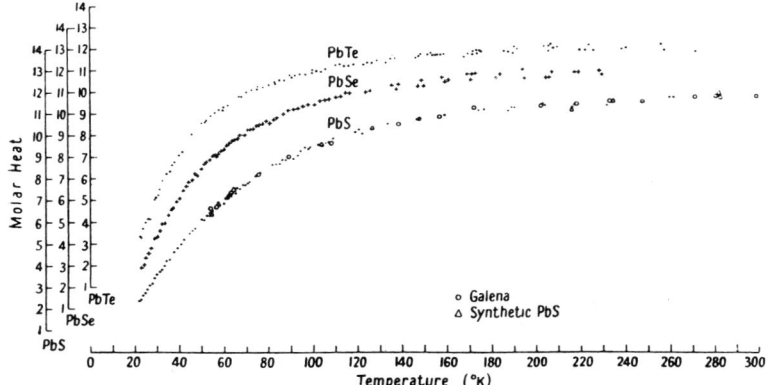

Fig. 4.26. Temperature dependences of C_p for PbTe, PbSe, and PbS.

We shall now consider briefly some conclusions which follow from the theory of the lattice thermal conductivity and which are necessary for the explanation of the experimental results obtained for lead chalcogenides.

In 1914, Debye showed that the thermal resistance of a solid was due to the anharmonicity of atomic vibrations. When vibrations are considered in the harmonic approximation, the thermal resistance is found to be zero since harmonic waves satisfy the principle of linear superposition according to which waves are propagated independently of the crystal and are not scattered by each other.

Debye deduced a formula for the thermal conductivity of the crystal lattice of the same type as our Eq. (4.31).

The Debye theory was modified by Peierls [346, 350] in accordance with the quantum-mechanical ideas; Peierls introduced the interaction of phonons, regarded as quanta of the vibrational spectrum of a crystal.

The scattering in the Peierls' theory is described as the annihilation of one (or two) phonons and the creation of two (or one) phonons. Several types of collision are possible. In one type, known as the N (normal) process:

$$\omega_1 + \omega_2 = \omega_3, \tag{4.57}$$

$$\mathbf{q}_1 + \mathbf{q}_2 = \mathbf{q}_3, \tag{4.58}$$

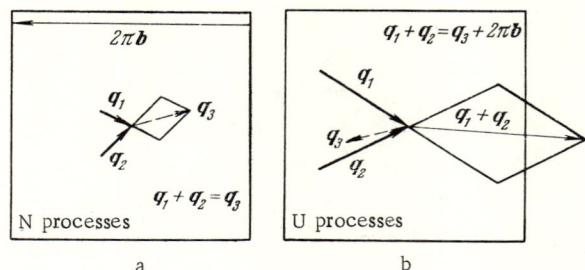

Fig. 4.27. Schematic representation of collision processes described by Eqs. (4.58) and (4.59) [375].

the total momentum is conserved and the direction of the energy flux remains constant (ω and q are, respectively, the angular frequency and the wave vector of a phonon). Thus, the simplest type of phonon interaction gives rise to infinite thermal conductivity.

Peierls showed that only the processes resulting in a change of the total momentum and, consequently, a change in the flux energy during collision of the type

$$\mathbf{q}_1 + \mathbf{q}_2 = \mathbf{q}_3 + 2\pi\mathbf{b} \tag{4.59}$$

($\mathbf{b}$ is the reciprocal lattice vector), may give rise to a finite thermal resistivity. Peierls called this type of scattering the umklapp or U processes.

The two processes are shown diagramatically in Fig. 4.27. Each of the squares shown in that figure represents a unit cell in the reciprocal lattice multiplied by a factor of 2π. In the description of the state of the lattice by means of the vector $\mathbf{q}$, we need consider only one unit cell of the reciprocal lattice, because a wave with a wave vector $\mathbf{q}$ is identical to every wave for which $\mathbf{q} + 2\pi\mathbf{b}$ [375]. In the U processes (Fig. 4.27b), the vector $\mathbf{q}_1 + \mathbf{q}_2$ projects outside the square. A vibration $\mathbf{q}_1 + \mathbf{q}_2$ excited in this way is identical with a vibration $\mathbf{q}_3$ because $\mathbf{q}_3$ differs from $\mathbf{q}_1 + \mathbf{q}_2$ only by $2\pi\mathbf{b}$. In the N processes (Fig. 4.27a), the vector $\mathbf{q}_3$ lies within the square.

The N processes give rise to considerable difficulties in the theory of the lattice thermal conductivity, particularly at low temperatures, when the phonons in a crystal have long wavelengths (small values of $\mathbf{q}$) and the N processes represent the main kind

Sec. 4.2] THERMAL PROPERTIES 199

of phonon – phonon interaction. Although the N processes do not give rise to a thermal resistance, they may affect the lattice thermal conductivity indirectly by redistributing phonons between various modes. These phonons are then scattered by other processes.

At higher temperatures, short-wavelength phonons (with large values of **q**) are excited and the probability is higher for the processes in which the sum of the wave vectors of two interacting phonons lies outside the first Brillouin zone.

The N and U processes are the main types of collisions associated with the effect of anharmonic forces. However, higher-order processes are also possible when four phonons interact simultaneously.

Equation (4.31) can be generalized to the transport of heat by phonons [376]:

$$\varkappa_l = \frac{3}{(2\pi)^3} \int v_g^2 \cos\vartheta \tau(\mathbf{q}) C_p(\mathbf{q}) d^3\mathbf{q}, \qquad (4.60)$$

where C_{ph} is the phonon specific heat [cf. Eq. (4.51)]; $\tau(\mathbf{q})$ is the total phonon relaxation time (the factor 3 occurs because of the inclusion of three modes of the vibrational spectrum). In principle, we can calculate $\varkappa_l$ exactly but the absence of data on the actual vibrational spectra of crystals and on anharmonic forces and the complex procedures necessary to obtain the exact solutions of the Boltzmann equation make such a calculation practically impossible. It is usual to employ a simple model which can be used in calculations. It is assumed that all the phonon scattering processes can be represented by a relaxation time which depends on the frequency and temperature. It is also assumed that a crystal is elastically isotropic and that there is no dispersion in the vibrational spectrum ($\omega = v_0 q$). The latter condition (known as the Debye approximation) is, as already pointed out, valid only at low temperatures.

If several scattering processes take place in a solid and these processes satisfy Eq. (4.59) the total relaxation time can be represented in the form

$$\tau^{-1} = \sum_j \tau_j^{-1}. \qquad (4.61)$$

The situation is more complicated when the reciprocals of the relaxation times of the N and U processes have to be added (in this case, the U processes are understood to be all the processes in which the momentum of a crystal is not conserved: they include the umklapp processes, as well as the scattering by crystal boundaries, impurities, isotopes, dislocations, etc.). As already pointed out, the N processes conserve the total momentum in a crystal. Consequently, strictly speaking, it is not permissible to add the relaxation time of the normal processes to the relaxation time of those processes in which the momentum of a crystal is not conserved.

However, Callaway [376] introduced a combined relaxation time

$$\tau_c^{-1} = \tau_N^{-1} + \tau_U^{-1} \tag{4.62}$$

(the subscripts N and U refer to the N and U scattering processes), solved the Boltzmann equation using this relaxation time, and obtained a total relaxation time τ of the type

$$\tau = \tau_c \left(1 + \frac{\beta'}{\tau_N}\right). \tag{4.63}$$

The thermal conductivity of an isotropic crystal lattice can thus be written in the following form, including the N processes:

$$\varkappa_l = \frac{v_{\text{rp}}^2}{2\pi^2} \int \tau_c \left(1 + \frac{\beta'}{\tau_N}\right) C_{\text{ph}}\, q^2 dq. \tag{4.64}$$

Callaway calculated also the quantity β':

$$\beta' = \int_0^{\theta/T} \frac{\tau_c}{\tau_N} \frac{e^x x^4 dx}{(e^x - 1)^2} \bigg/ \int_0^{\theta/T} \frac{1}{\tau_N}\left(1 - \frac{\tau_c}{\tau_N}\right) \frac{e^x x^4 dx}{(e^x - 1)^2}, \tag{4.65}$$

where $x = \hbar\omega/k_0 T$. When $\tau_U \to \infty$ (the U processes are absent) $\tau_c \to \tau_N$, $\beta' \to \infty$, and, according to Eq. (4.64), $\varkappa_l \to \infty$, in agreement with the definition of the N processes.

Calculations of $\varkappa_l$ using Eq. (4.60), carried out by several workers for low temperatures (not exceeding 80°K) by means of

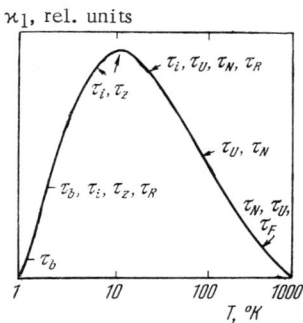

Fig. 4.28. Schematic representation of the temperature dependence of the thermal conductivity. Temperature intervals, in which different scattering processes are dominant, are indicated.

computers, are in good agreement with the experimental data for a large number of materials (cf., for example, [376, 377] and other papers).

One of the difficulties in such calculations is the selection of the constant coefficients which occur in the formulas for τ_j. This makes the calculated values of the thermal conductivity somewhat indeterminate.

The most important phonon scattering processes are: the scattering by crystal boundaries (τ_b) [378, 379]; the scattering by impurities and isotopes (τ_i) [353, 380]; three-phonon collisions of the N (τ_N) [353, 381] or of the U (τ_U) [353, 377] type; four-phonon collisions (τ_F) [382]; electron–phonon interactions (τ_z) [383]; and resonance scattering (τ_R) [384, 385]. The values of the relaxation times for these processes are given in the cited papers. Figure 4.28 shows schematically at which temperatures the various scattering processes predominate.

We must stress once more that Eq. (4.60) can be used only at low temperatures (T < θ) at which the Debye approximation for the vibrational spectrum is valid. When the temperature is increased the number of short-wavelength phonons increases rapidly and the Debye approximation becomes less acceptable. This is because the Debye approximation is based on the assumption that the frequency is a linear function of the wave vector, i.e., that all the phonons of a given mode have the same group velocity. This assumption is satisfied well only by the acoustical phonons of long wavelengths, which are important at low temperatures. In fact, the group velocity of the acoustical phonons decreases with decreasing wavelength and vanishes when the wave vector reaches the boundary of the first Brillouin zone. However, Eq. (4.60) can nevertheless be used, with some qualifications, to calculate the thermal conductivity of the lattice of solid solutions at high temperature because, in this case, the importance of the short-wavelength phonons decreases because of their scattering by crystal lattice imperfections resulting from the introduction of a second component.

According to Peierls' theory, the high-temperature (T > θ) lattice thermal resistivity in the case of three-phonon processes varies in accordance with the law

$$W_1 = \frac{1}{\varkappa_1} = \text{const} \cdot T, \qquad (4.66)$$

and, because of the excitation of the total phonon spectrum, any defect gives rise to an additional thermal resistivity which is independent of temperature [353]. Experiments carried out on a large number of compounds have confirmed these conclusions and, moreover, the dependence of Eq. (4.66) has been found to be valid at relatively low temperatures, of the order of θ/3. Holland [377] gave a theoretical explanation of this observation by considering separately the contributions of the longitudinal and transverse lattice vibrations to the thermal conductivity.

Four-phonon collisions [382] at high temperatures yield the dependence $W_l \propto \text{const} \cdot T^2$.

The absolute thermal conductivity at high temperatures can be calculated using various semiempirical formulas suggested by Ioffe [349], Keyes [386], Dugdale and MacDonald [387], Tavernier [388], or Leibfried and Schlomann (cf. [346]). The Leibfried–Schlomann formula is most widely used. In the form given by Ziman [346], it is

$$\varkappa = \varkappa_0 \frac{\theta}{T}, \qquad (4.67)$$

where

$$\varkappa_0 \approx 5 \cdot 10^{-8} \frac{\overline{A} a \theta^2}{\gamma_G^2} \text{ W} \cdot \text{cm}^{-1} \cdot \text{deg}^{-1}$$

($\overline{A}$ is the average atomic weight; a is the lattice constant in Å; θ is the Debye temperature in °K; γ_G is the Gruneisen parameter).

All the formulas and theoretical treatments given so far apply solely to the acoustical modes. However, at higher temperatures, the optical phonons may contribute to the thermal conductivity.

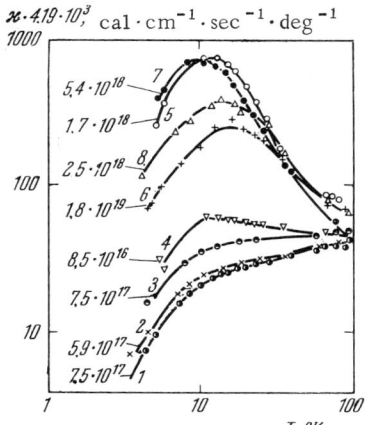

Fig. 4.29. Temperature dependences of the thermal conductivity obtained by Greig [23]: 1) PbS; 2), 3), 4) n-type PbS; 5) p-type PbS; 6) p-type PbS + Ag; 7) p-type PbSe; 8) PbTe.

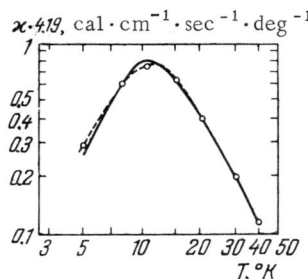

Fig. 4.30. Temperature dependence of the thermal conductivity of a PbS single crystal (curve 5 of Fig. 4.29). The continuous curve represents Greig's experimental data [23]; the points and the dashed curve represent the calculation reported in [404].

We shall now consider the experimental data on the lattice thermal conductivity of lead chalcogenides.

Many reports of measurements or calculations of the lattice thermal conductivities of PbTe [23, 389-399], PbSe [23, 209, 355, 391, 400, 401] and PbS [23, 237, 391, 402, 403] have been published.

Low Temperatures. The low-temperature (4-100°K) thermal conductivities of natural and synthetic PbTe, PbSe, and PbS single crystals were measured by Greig [23] (Fig. 4.29). Agrawal and Verma [404] calculated the absolute value of the lattice thermal conductivity for one of the single crystals of PbS investigated by Greig [23]; they used Eqs. (4.60), (4.61), (4.63), and the relaxation time values considered earlier in the present section. The electronic thermal conductivity was found to be negligibly small. Selecting reasonable values of the constants occurring in the expressions for τ_j^{-1} of Eq. (4.61) and allowing for the phonon – phonon scattering (N and U processes), as well as for the scattering by impurities, crystal boundaries, dislocations, and conduction electrons, Agrawal and Verma were able to obtain an absolute value and a temperature dependence of the thermal conductivity of PbS (Fig. 4.30) in good agreement with the experimental data.

TABLE 4.2. Values of Thermal
Conductivities $\varkappa_e$ and $\varkappa_l$ for PbS [237]
(in $cal \cdot cm^{-1} \cdot sec^{-1} \cdot deg^{-1}$)

T, °K	$\varkappa_l \cdot 10^3$	$\varkappa_e \cdot 10^3$	
		$r = -1/2$	$r = 1/2$
100	13.9	0.036	0.053
200	7.52	0.018	0.027
300	6.02	0.013	0.020
400	4.49	0.012	0.018

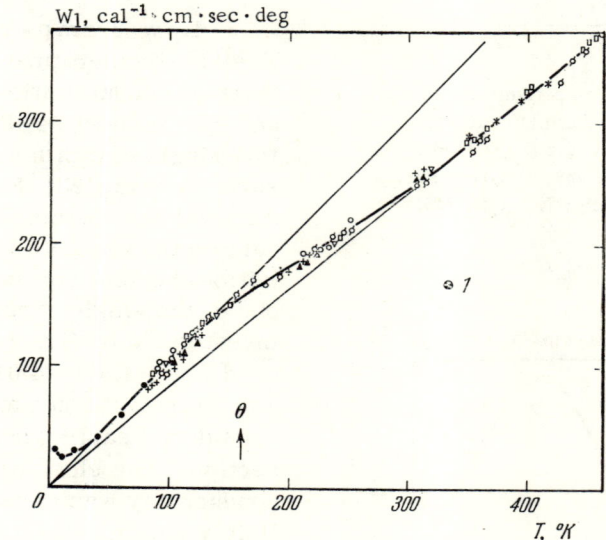

Fig. 4.31. Temperature dependence of the lattice thermal resistivity of PbSe [355]: 1) results taken from Greig [23].

Moderate Temperatures. The fullest data on the thermal conductivities of undoped ($n \approx 10^{17}$ cm^{-3}) PbTe, PbSe, and PbS samples at moderate temperatures (100-400°K) are given in the papers of Devyatkova, Smirnov, et al. [237, 355, 389]. In this range of temperatures, samples with a low carrier density exhibit no thermal conductivity components due to photons or ambipolar diffusion and the electronic thermal conductivity of such samples is negligibly small compared with the lattice thermal conductivity. By way of example, Table 4.2 gives the electronic and lattice thermal conductivity components of PbS [237].

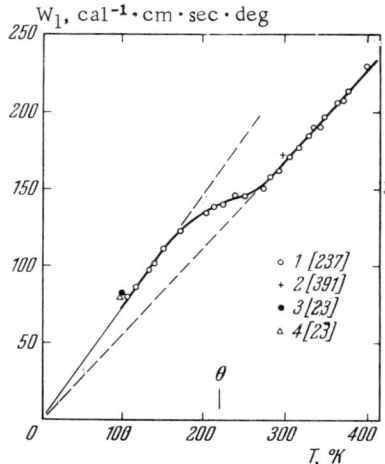

Fig. 4.32. Temperature dependence of the lattice thermal resistivity of PbS: 1) [237]; 2) [391]; 3), 4), [23] (2 and 3 represent synthetic single crystals; 1 and 4 represent natural PbS).

Figures 4.31, 4.32, and 4.36 (line 1) give the values of the lattice thermal resistivities (W_l) of PbSe, PbS, and PbTe taken from [355, 237, 389]. For all three compounds, the values of W_l at 100°K are in agreement with the values reported by Greig [23].

For PbTe, $W_l = \text{const} \cdot T$ throughout the investigated range of temperatures and it vanishes when it is extrapolated to 0°K. On the other hand, the lattice thermal resistivities of PbSe and PbS can be represented in the form of two linear segments: a low-temperature line $W_l = AT$ and a high-temperature line $W_l = BT$, where $A \neq B$. Both lines pass through zero when extrapolated to 0°K. Between these two linear segments, there is a transition region near the Debye temperature. The results obtained can be explained by the participation of the optical phonons in the thermal conductivity.

The optical phonons are not excited below the Debye temperature and in that temperature range the transport of heat is solely due to the acoustical phonons.

The optical phonons are gradually excited on approach to the Debye temperature. These phonons can alter the thermal conductivity in various ways:

1) the thermal conductivity may increase when the additional transport of heat is due to the optical phonons of high dispersion, when there is no scattering of the acoustical on the optical phonons;

2) the thermal conductivity may decrease when the acoustical and optical phonons interact strongly and are scattered by one another (in this case the optical phonons should have a low dispersion and make no contribution to the thermal conductivity);

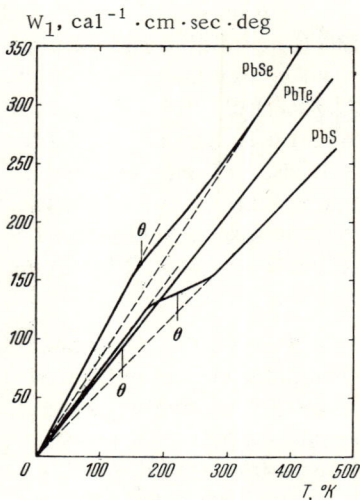

Fig. 4.33. Temperature dependences of the lattice thermal resistivities of PbTe, PbSe, and PbS.

3) the thermal conductivity may be unaltered if the optical phonons have a low dispersion and the scattering of the acoustical on the optical phonons is slight;

4) the first and second mechanisms may combine to give: a) no change in the thermal conductivity; b) predominance of the first mechanism over the second (an increase of the thermal conductivity), or c) predominance of the second mechanism over the first (a reduction of the thermal conductivity).

In the case of PbTe, mechanism 4a is possible, while in the case of PbSe and PbS mechanisms 1 or 4b are likely. Steigmeier and Kudman [405] investigated the thermal conductivity of $A^{III}B^{V}$ compounds and found a considerable scattering of the acoustical on the optical phonons above the Debye temperature. A similar effect has been reported earlier [406] for some alkali halide compounds.

Figure 4.33 shows the temperature dependences of W_1 for all three lead chalcogenides. We can see that the thermal conductivities of PbS and PbTe are similar at low temperatures. Efros [407] has calculated very approximately the vibrational spectra of PbS and PbTe. His results can be used for a qualitative comparison of the results obtained on the thermal conductivity. According to Efros [407], the acoustical mode spectra of PbTe and PbS are identical but the optical spectra differ very considerably (this result was confirmed by experimental studies of the phonon spectra of PbTe and PbS (422, 536]). This is why their thermal conductivities differ at high temperatures.

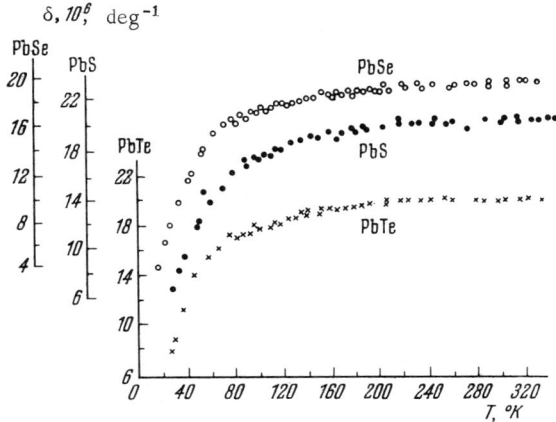

Fig. 4.34. Temperature dependences of the linear expansion coefficients of PbTe, PbSe, and PbS.

F. Thermal Expansion of the Lattice and Its Relationship to the Thermal Conductivity

In real crystals, the average distance between atoms increases with increasing temperature. The linear expansion coefficient δ is proportional to the anharmonicity coefficient β:

$$\delta \propto \beta \qquad (4.68)$$

(β is the coefficient in the expression

$$F = c' \frac{\Delta x}{x_0} + \beta \left(\frac{\Delta x}{x_0}\right)^2 + \ldots,$$

where F is the force acting between atoms; x_0 and Δx are, respectively, the distance between atoms at rest and the change in that distance due to an increase in temperature).

Frenkel' demonstrated [408] that the lattice thermal conductivity is usually inversely proportional to the square of β:

$$\varkappa_l \propto \beta^2. \qquad (4.69)$$

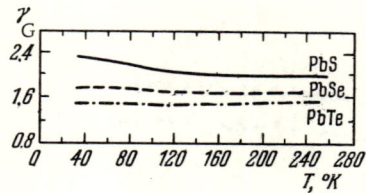

Fig. 4.35. Temperature dependences of the Grüneisen constants of PbTe, PbSe, and PbS [411].

Thus, we obtain the following relationship between $\varkappa_1$ and δ:

$$\frac{1}{\varkappa_1} \propto \delta^2. \qquad (4.70)$$

The relationship (4.70) was confirmed experimentally and theoretically by Zhuze [409] and Kontorova [410].

The linear expansion coefficients of PbTe, PbSe, and PbS were measured by Novikova and Abrikosov [411] and by other workers [412, 413]. The average values of δ are given in Fig. 4.34.

According to the theory, the volume thermal expansion coefficient δ' (for cubic crystals $\delta' = 3\delta$) increases as T^3 at $T \ll \theta$, but at $T \gg \theta$ it tends to a constant value. Grüneisen established a relationship between δ' and other thermodynamic quantities [414, 415]:

$$\delta' = \gamma_G \frac{C_v \xi_T}{V}, \qquad (4.71)$$

where ξ_T is the isothermal compressibility and

$$\gamma_G = -\frac{d \log \theta}{d \log V} \qquad (4.72)$$

is the Grüneisen constant. The value of γ_G, which describes the deviation of the vibration spectrum from the harmonic approximation, can be regarded as a measure of the anharmonicity of atomic vibrations in a solid. The value of γ_G for PbTe, PbSe, and PbS are given in Fig. 4.35 [411].

Using the data for the linear expansion coefficient [411] and the lattice thermal conductivity [237, 355, 389], we checked the relationship (4.70) for PbTe, PbSe, and PbS. We found that it was not satisfied by lead chalcogenides. This was probably because of the complex temperature dependences of the lattice thermal conductivities of PbSe and PbS (cf. Fig. 4.33).

Knowing the value of the Grüneisen constant and the Debye temperature, we can calculate the absolute value of the lattice ther-

TABLE 4.3. Lattice Thermal Conductivities
(in cal·cm^{-1}·sec^{-1}·deg^{-1}) at T = 300°K
and Debye Temperatures
of PbS, PbSe, and PbTe

Compound	$\varkappa_{l.\,calc} \cdot 10^3$ $T=300\,°K$	$T=\theta$	$\varkappa_{l.\,exp} \cdot 10^3$ $T=\theta$	$\varkappa_{l.\,calc}/\varkappa_{l.\,exp}$ $T=\theta$	$\overline{A}$ (average atomic weight)
PbS	6.02	7.15	106	15	119.64
PbSe	3.90	6.10	98.5	16	143.08
PbTe	4.80	10.20	103	10	167.41

mal conductivity using the Leibfried – Schlomann formula (4.67). The results of such calculations are given in Table 4.3.

As for many other compounds, the Leibfried – Schlomann formula overestimates the values of the lattice thermal conductivities of lead chalcogenides.

Comparing the second, third, and sixth columns of Table 4.3, we can see that the usual reduction in the lattice thermal conductivity with increasing average atomic weight does not apply to lead chalcogenides.

G. Influence of Impurities on the Thermal Conductivity of Lead Chalcogenides at Temperatures T ≥ Θ. Solid Solutions

In the present state of the theory of heat conduction, it is fairly difficult to obtain a quantitative estimate of the influence of a given impurity on the lattice thermal conductivity above the Debye temperature. There are, however, several empirical formulas, among which the most widely used is the Ioffe formula [349, 391, 416].

Let us assume that N impurity atoms are distributed at random between N_0 host lattice atoms in 1 cm^3 of a crystal. We shall introduce the concept of the cross section of an impurity atom

$$s = \phi a^2 \qquad (4.73)$$

(a is the lattice constant). The number of scattering events, due to the presence of N impurity atoms, is governed by the product of the number of such atoms and their cross section $N\phi a^2$. For a simple cubic lattice $N_0 a^3 = 1$. Thus, the number of collisions with impurity atoms can be assumed to be $\frac{N}{N_0}\frac{\phi}{a}$, and the total number of collisions per 1 cm path can be taken as $\frac{1}{l_0} + \frac{N}{N_0}\frac{\phi}{a}$ (instead of $1/l_0$ in the same crystal when no impurities are present; here, l_0 is the mean free path). Then, $\varkappa_0/\varkappa_{\text{imp}}$ becomes

$$\frac{\varkappa_0}{\varkappa_{\text{imp}}} = 1 + \frac{N}{N_0}\phi\frac{l_0}{a}. \qquad (4.74)$$

The formula given by Eq. (4.74) is valid at low impurity concentrations. A. V. and A. F. Ioffe [391] show that, when $\phi \leq 1$, the impurities are located at the lattice sites, when $\phi > 1$ the impurities are interstitial, and when $\phi < 1$, the impurity atoms are located at defects.

An attempt to describe the influence of point defects on the lattice thermal conductivity at temperatures $T \geq \theta$ has been made in three theoretical papers [417, 418, 419]. All three papers are based on the Debye spectrum and the Rayleigh formula ($\tau^{-1} \propto \omega^4$) although it is known that the Debye approximation and the Rayleigh formula do not describe sufficiently accurately the behavior of short-wavelength phonons. However, when the Debye spectrum and the Rayleigh formula are used simultaneously the errors tend to cancel out. The total relaxation time used in [417, 418] is

$$\frac{1}{\tau} = \frac{1}{\tau_i} + \frac{1}{\tau_U}, \qquad (4.75)$$

where $1/\tau_i = A\omega^4$ and $1/\tau_U = B\omega^2$ (τ_i and τ_U are the relaxation times for the scattering by defects and in the U processes). The total value of τ is substituted into Eq. (4.60). After integration, we find that

$$\varkappa_{\text{imp}} = \frac{k_0}{2\pi^2 v_0}\frac{\omega_0}{B}\tan^{-1}\left(\frac{\omega_{\max}}{\omega_0}\right), \qquad (4.76)$$

where $\omega_0^2 = B/A$ and has the dimensions of the frequency; ω_{max} is maximum frequency in the Debye model; $\bar{v}_0$ is the average velocity of sound. The expression for the thermal conductivity of a defect-free crystal ($\varkappa_0$) is calculated in the same way but it is assumed that $A = 0$. Then

$$\varkappa_0 = \frac{k_0 \omega_{max}}{2\pi^2 \bar{v}_0 B}. \tag{4.77}$$

Combining Eqs. (4.76) and (4.77), we obtain

$$\varkappa_{imp} = \varkappa_0 \frac{\omega_0}{\omega_{max}} \tan^{-1}\left(\frac{\omega_{max}}{\omega_0}\right). \tag{4.78}$$

In order to determine $\varkappa_{imp}$ we have to know ω_0/ω_{max}. We find from Eq. (4.77) that

$$\frac{\omega_0^2}{\omega_{max}^2} = \frac{B}{A\omega_{max}^2} = \frac{k_0}{2\pi^2 \bar{v}_0 \varkappa_0 \omega_{max} A}. \tag{4.79}$$

If we assume that $\varkappa_0$ is known, we need to know only A in order to determine ω_0/ω_{max}. It is difficult to calculate A when the influence of defects is manifested by a local change in the elastic properties of a crystal but it can be calculated quite simply if phonons are scattered by defects resulting from a local change in the density (for example, in the case of isotopes). In this case, we have

$$A = \frac{\sum_i x_i (M_i - \bar{M})^2}{4\pi \bar{v}_0^3 N \bar{M}^2}, \tag{4.80}$$

where x_i is the relative concentration of atoms of mass M_i; N is the number of atoms per unit volume; $\bar{M}$ is the average mass of atoms which is given by the expression $\bar{M} = \sum_i x_i M_i$.

When the scattering by isotopes is small compared with the U processes, then — because A is small — we find that $\omega_{max} \ll \omega_0$ and Eq. (4.78) assumes the form

$$\varkappa_{imp} = \varkappa_0 \left(1 - \frac{1}{3}\frac{\omega_{max}^2}{\omega_0^2}\right). \tag{4.81}$$

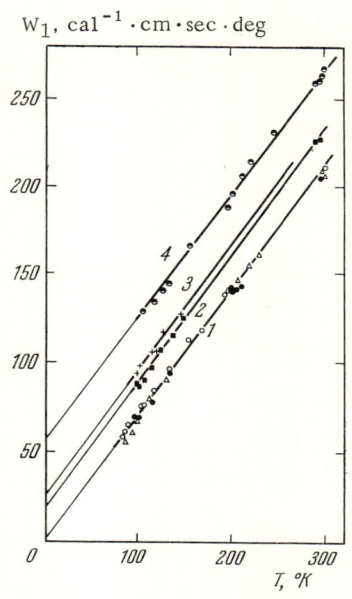

Fig. 4.36. Temperature dependences of the lattice thermal resistivity of PbTe: 1) $n = (4-5) \cdot 10^{17}$ cm^{-3}; 2) $n = 3.1 \cdot 10^{19}$ cm^{-3}; 3) $n = 5.5 \cdot 10^{19}$ cm^{-3}; 4) $n = 1.2 \cdot 10^{20}$ cm^{-3} (1 represents a stoichiometric sample; 2-4 are iodine-doped samples).

A calculation of the temperature dependence of the additional thermal resistivity due to the presence of impurities can be found in [419]. In the case of low impurity concentrations, ΔW is independent of temperature, which follows from the general theory of heat conduction [353]. In the presence of a large number of impurities (for example, in solid solutions), $\varkappa_{imp} \propto T^{-1/2}$ and it depends on the concentration of impurities (defects) as $A^{-1/2}$.

The influence of Cl, Br, and I impurities on the thermal conductivity of n-type single-crystal and polycrystalline samples of PbTe has been investigated in detail in the temperature range 80-300°K [389]. In this investigation, samples had electron densities ranging from 10^{17} to 10^{20} cm^{-3}. The scattering of electrons in heavily doped n-type PbTe is mainly elastic, as demonstrated in §4.2A. Therefore, the Lorenz number for electrons can be calculated using Eq. (4.34). Figure 4.36 gives the temperature dependence of the lattice thermal conductivity of iodine-doped samples. (Similar results have been obtained also for n-type PbTe doped with Br and Cl.) It is evident from this figure that, in full agreement with the theory developed for temperatures $T \geq \theta$, the additional thermal resistivity due to impurities, $\Delta W = W_{imp} - W_l$, is constant at each impurity concentration and does not vary with temperature. ΔW is proportional to the concentration of impurities ($\Delta W = \text{const} \cdot n$) and is independent of the nature of the halogen (Fig. 4.37). Moreover, ΔW is in no way related to the difference between the masses of the foreign (substituent) and host atoms or to the difference between their atomic radii, which indicates indirectly a change in the elastic

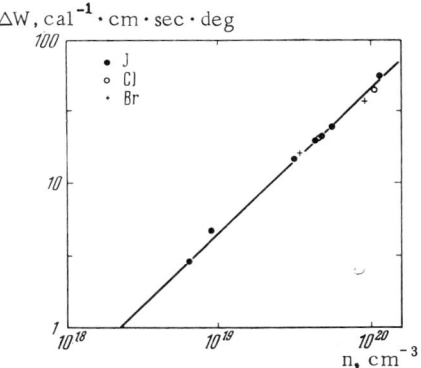

Fig. 4.37. Dependence of the additional thermal resistivity on the carrier density in PbTe samples doped with iodine, chlorine, and bromine.

TABLE 4.4. Changes in Lattice Thermal Conductivity of PbTe Due to Introduction of Halogen Impurities

Halogen concentration, cm^{-3}	Reduction of $\varkappa_l$ (in %) compared with $\varkappa_l$ of stoichiometric sample	
	T = 100°K	T = 300°K
$\sim 10^{20}$	45	21
$3 \cdot 10^{19}$	16	6

binding forces in a crystal. Therefore, Eqs. (4.78) or (4.81) cannot be used.

Halogen impurities scatter phonons strongly (cf. Table 4.4). The formula (4.74) is used in [389] to find the factor ϕ, which occurs in the expression for the phonon scattering cross section; l_0, which occurs in Eq. (4.74), is estimated from the expression $\varkappa_0 = C_V l_0 v_g/3$. The value of l_0 is frequently estimated experimentally at moderate and high temperatures using either the longitudinal velocity of sound v_l or the average velocity $\bar{v}_0 = (v_l + 2v_t)/3$. Moizhes et al. [420] used Ge and NaI as examples to show that the group velocity can be 2-5 times smaller than $\bar{v}_0$ or v_l. According to the calculations of Efros [407], $v_g = \bar{v}_0/3$ for PbTe. The parameter ϕ, calculated from Eq. (4.76), has approximately the same value for Cl, Br, and I impurities and is fairly large (about 6-3.7). On the other hand, an estimate of ϕ for neutral Sn and Se impur-

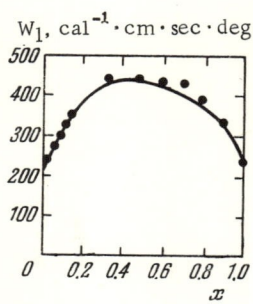

Fig. 4.38. Dependence of the lattice thermal resistivity on the composition of $PbSe_xTe_{1-x}$ solid solutions at $T = 300°K$. The points represent the results reported in [400]; the continuous curve is taken from [343].

ities in PbTe gives a value of the order of 0.7. The strong scattering of phonons by charged impurities can be explained by the high permittivity of PbTe (cf. §3.4B). A charged impurity placed in a medium with a high value of ε_0 distorts the crystal lattice around it quite strongly because of the polarization and this distortion is independent of the mass or atomic radius of the impurity, but is governed solely by its charge.

The thermal conductivity of PbTe — PbSe solid solutions has been investigated on several occasions [391, 400, 416]. Fuller information is given in [343] on polycrystalline p- and n-type samples with carrier densities of 10^{17}–10^{18} cm^{-3}. For samples with $n \approx 10^{18}$ cm^{-3}, the electronic component of the thermal conductivity has been calculated using the Lorenz number given in §4.2A. The lattice thermal resistivities of the PbTe — PbSe solid solutions of various compositions are shown in Figs. 4.38 and 4.39. It is found that the lattice thermal resistivities of solid solutions (with the exception of the composition 20% PbSe — 80% PbTe) are parallel to the thermal resistivity of the PbTe lattice on the side of the compositions close to PbTe and to the thermal resistivity of the PbSe lattice on the PbSe side, i.e., they are the same kind as those in the case of lightly doped PbTe (ΔW is independent of temperature). No tendency of solid solutions to become amorphous at high concentrations of the second component has been observed. The temperature dependence $\varkappa_{imp} \propto T^{-1/2}$, reported in [418, 419], is not obeyed by PbTe — PbSe solid solutions.

The absolute value of $\varkappa_{imp}/\varkappa_0$ of PbTe — PbSe solid solutions can be calculated using Eq. (4.78). Drabble and Goldsmid [352] used the experimental data reported in [400] to calculate $\varkappa_{imp}/\varkappa_0$. The experimental data were found to disagree with this theory. This disagreement can be attributed either to the use of the Debye approximation or to the fact that Eq. (4.78) makes allowance only for a change in the mass but ignores a change in the interatomic forces.

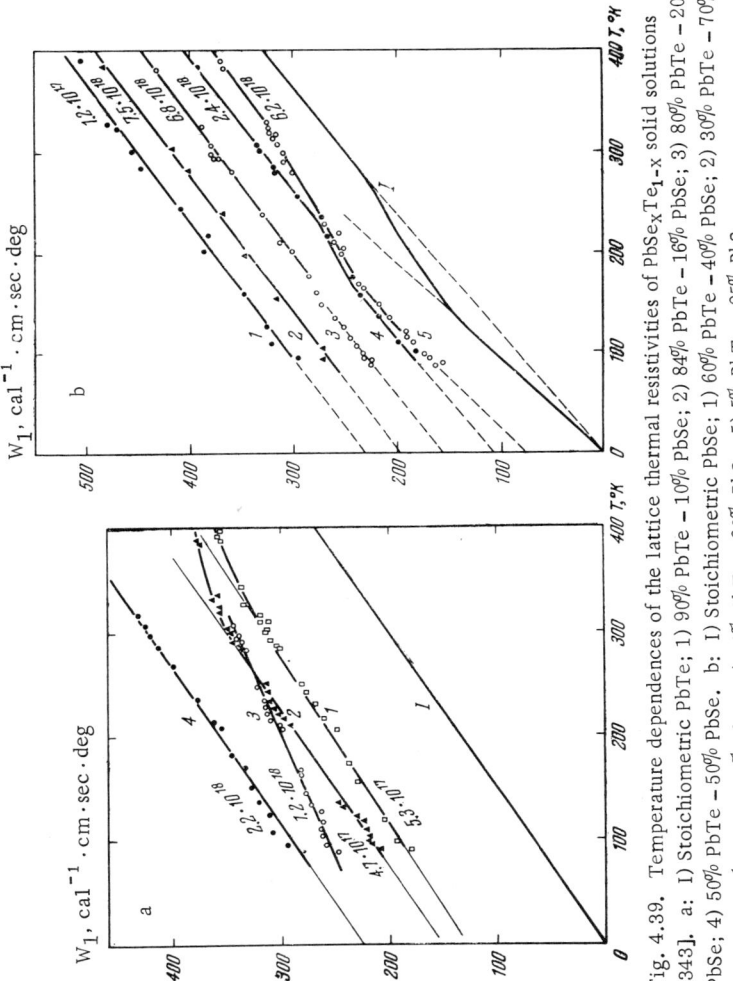

Fig. 4.39. Temperature dependences of the lattice thermal resistivities of $PbSe_xTe_{1-x}$ solid solutions [343]. a: I) Stoichiometric PbTe; 1) 90% PbTe − 10% PbSe; 2) 84% PbTe − 16% PbSe; 3) 80% PbTe − 20% PbSe; 4) 50% PbTe − 50% PbSe. b: I) Stoichiometric PbSe; 1) 60% PbTe − 40% PbSe; 2) 30% PbTe − 70% PbSe; 3) 20% PbTe − 80% PbSe; 4) 10% PbTe − 90% PbSe; 5) 5% PbTe − 95% PbSe.

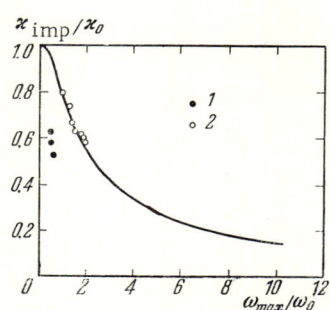

Fig. 4.40. Dependence of $\varkappa_{imp}/\varkappa_0$ on ω_{max}/ω_0 for PbTe – PbSe solid solutions: 1) calculated using $\bar{v}_0$ and Eq. (4.78); 2) calculated using v_g and Eq. (4.78).

This may lead to underestimation of the parameter A and, therefore, overestimation of the value of ω_0/ω_{max} [cf. Eq. (4.79)] and, consequently, of $\varkappa_{imp}/\varkappa_0$ [cf. Eq. (4.78)].

Using his own experimental data (similar to those reported in [343, 400]), Saakyan [421] also calculated $\varkappa_{imp}/\varkappa_0$ from Eq. (4.78), but he took into account that $\bar{v}_0 = 3v_g$ for PbTe and PbTe – PbSe solutions [407] and obtained good agreement between the theory and experiment (Fig. 4.40).

H. Phonon Spectrum of PbTe

Recently, Cochran et al. [422] used the neutron-spectroscopy method to determine the phonon spectrum of PbTe.* The measurements were carried out in a (110) plane and the frequencies of the vibration modes were determined for wave vectors directed along the three symmetry axes and along the zone boundary (Fig. 4.41).

In their analysis of the experimental data, Cochran et al., used models similar to those described earlier for alkali halide crystals. Comparisons between the experimental and theoretical results were carried out chiefly for two models:

I) a rigid ion model (in this case, the polarizability is ignored);
II) a shell model, which allows for the polarizability of both ions of lead telluride.

Figure 4.41 shows the experimental and calculated data for the dependence $\nu(\mathbf{k})$. We can see that the agreement between the theory and experiments is better for model II.

However, it must be mentioned that the shell model does not agree with the experimental data for lead telluride as satisfactorily as in the case of alkali halide crystals NaI and KBr [423], whose

---

*Similar measurements were carried out later on PbTe and PbS [536].

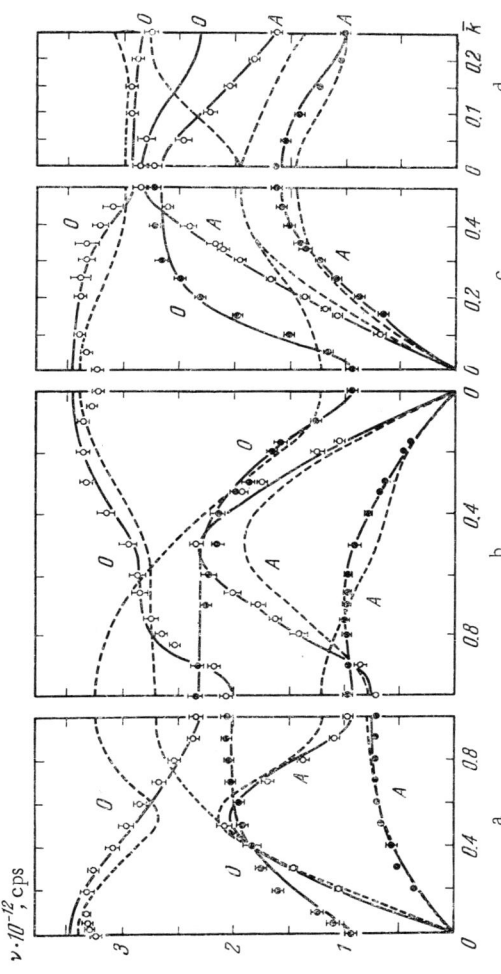

Fig. 4.41. Phonon spectrum of PbTe at T = 296°K [422]. a) $\mathbf{k} = (0, 0, \bar{k})$; b) $\mathbf{k} = (\bar{k}, \bar{k}, 0)$; c) $\mathbf{k} = (\bar{k}, \bar{k}, \bar{k})$; d) $\mathbf{k} = (0.5 - \bar{k}; 0.5 - \bar{k}; 0.5 + \bar{k})$. The points are the experimental values; the dashed curves were calculated using model I, and the continuous curves calculations using model II.

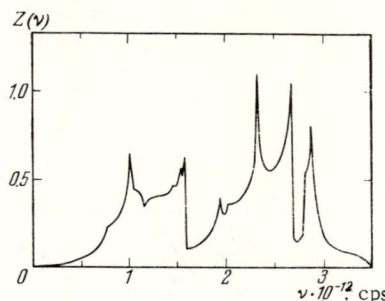

Fig. 4.42. Frequency distribution of the vibrations in PbTe [422] (calculated using model II).

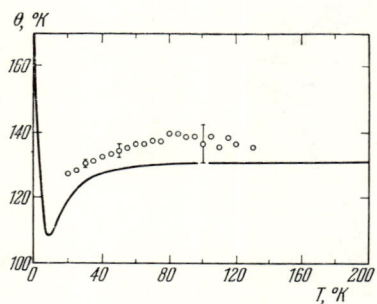

Fig. 4.43. Temperature dependence, $\theta(T)$, of the Debye temperature of PbTe. The continuous curve was calculated using model II [422]; the points are the experimental results of Parkinson and Quarrington [372].

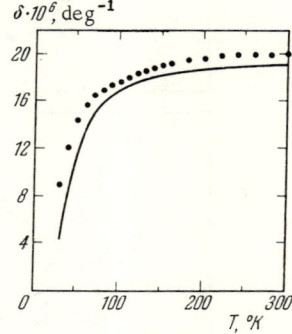

Fig. 4.44. Temperature dependence of the linear expansion coefficient of PbTe. The continuous curve was calculated using model II [422]; the points are the experimental results of Novikova and Abrikosov [411].

high-frequency permittivity is $\varepsilon_\infty \approx 2.5$. This model is less satisfactory for germanium and silicon [424], whose permittivities are $\varepsilon_\infty \approx 15$, than for alkali halides. Therefore, it is not surprising that its disagreement with the experimental data for PbTe ($\varepsilon_\infty \approx 33$) is even greater.

This comparison shows that the shell model describes the experimental data satisfactorily if the polarizability of the ions is low.

Sec. 4.2] THERMAL PROPERTIES 219

Using a method described by Gilat and Dolling [425] and model II, Cochran et al. [422] obtained the frequency distribution of vibrations for PbTe (Fig. 4.42). The form of the dependence $Z(\nu)$ differs from the distribution given by Eq. (4.52), assumed in the Debye model of the specific heat. The results given in Fig. 4.42 were used to determine the specific heat and to calculate the Debye temperature θ. The results of such a calculation are compared with the experimental data of Parkinson and Quarrington [372] in Fig. 4.43. In spite of the fact that $Z(\nu)$ obtained in [422] differs from the dependence given by Eq. (4.52), the theory and experiment* are in satisfactory agreement (there is a systematic deviation of about 5%).

The data obtained for $\nu(\mathbf{k})$ are used in [422] to calculate the linear expansion coefficient, employing a method developed in [426, 427]. The anharmonicity in model II is allowed for by the introduction of a suitable parameter [426, 427].

The results for δ obtained by a calculation reported in [422] and the experimental results of Novikov and Abrikosov [411] are given in Fig. 4.44. The agreement between experiment and theory is again farily satisfactory.

---

*The Debye temperature in the experiments of Parkinson and Quarrington [372] was determined using the Debye model and the frequency distribution given by Eq. (4.52).

CHAPTER V

MAGNETIC PROPERTIES

The title of this chapter is somewhat imprecise. In addition to the intrinsic magnetic properties of lead chalcogenides, we shall also consider the magneto-optical effects, magnetoplasma phenomena, cyclotron resonance, tunnel effect in magnetic fields, and oscillations of the magnetoresistance and ultrasound absorption in magnetic fields. These effects are associated with the quantization of electron states in a magnetic field or, when the quasi-classical approach is used, with the bending of the electron orbits. The galvanomagnetic and thermomagnetic effects in nonquantizing magnetic fields have already been discussed (cf. §§3.2 and 4.1).

In a magnetic field, a three-dimensional allowed band, which is degenerate only in respect of the spin, splits into a series of one-dimensional Landau bands, each of which splits again into two sub-bands, corresponding to the opposite directions of the spin. The carrier energy is then given by

$$\mathscr{E} = \left(l + \frac{1}{2}\right)\hbar\omega_c + \frac{\hbar^2 k_H^2}{2 m_H^*} + m_s \mu_B g H. \tag{5.1}$$

Here, ω_c is the cyclotron frequency

$$\omega_c = \frac{eH}{m_c^* c}, \tag{5.2}$$

where m_c^* is the cyclotron effective mass, whose value for an ellipsoidal band is given by

$$m_c^* = m_\perp^* \left(\frac{K}{K\cos^2\vartheta + \sin^2\vartheta}\right)^{1/2}, \tag{5.3}$$

where ϑ is the angle between the axis of revolution of the ellipsoid and the magnetic field; l is the oscillatory (Landau) quantum number, which assumes intergral values $l = 0, 1, 2, 3, \ldots$; k_H and m_H^* are the crystal momentum and the effective mass along the magnetic field direction; $\mu_B = e\hbar/2m_0c$ is the Bohr magneton; m_s is the quantum number associated with the angular momentum and equal to $\pm 1/2$; g is the spin splitting factor, whose value for an ellipsoidal band is given by

$$g = \pm (g_\parallel^2 \cos^2 \vartheta + g_\perp^2 \sin^2 \vartheta)^{1/2}. \tag{5.4}$$

The quantization of the energy levels in a magnetic field governs the contribution of free carriers to the diamagnetic and paramagnetic susceptibility of crystals. It also gives rise to a considerable change in the energy dependence of the density of states. The density of states can be obtained from Eq. (5.1) [270] and is given by

$$\rho(\mathscr{E}) = \frac{(2\,m_H^*)^{1/2} m_c^* \omega_c}{(2\pi\hbar)^2} \sum_{lm_s} \left[\mathscr{E} - \left(l + \frac{1}{2} \right) \hbar\omega_c - m_s \mu_B g H \right]^{-1/2}. \tag{5.5}$$

The above formula shows that the energy dependence of the density of states in a magnetic field is strongly nonmonotonic. At points

$$\mathscr{E} = \left(l + \frac{1}{2} \right) \hbar\omega_c + m_s \mu_B g H$$
$$(l = 0, 1, 2, \ldots;\ m_s = \pm 1/2)$$

the density of states becomes infinite. These singular points correspond to the lower edges of the Landau bands ($k_H = 0$). In the intervals between these singularities, the density of states, (considered as a function of the energy) decreases. When the magnetic field is varied, the Landau bands shift relative to the Fermi level and the density of states near this level oscillates. Since transport and other effects in degenerate samples are mainly due to carriers with energies close to the Fermi level, oscillations of the density of states near this level give rise to oscillations of the effects themselves with the magnetic field intensity. Oscillations of the density of states may also be manifested indirectly in transport phenomena through the energy dependence of the relaxation time, because this dependence is governed, to a considerable extent, by the nature of the density of states.

It also follows from Eq. (5.1) that the minimum carrier energy (measured from the band edge in the absence of a magnetic field) is not equal to zero but is given by $(\hbar\omega_c - |g|\mu_B H)/2$. Therefore, the forbidden band width depends on the magnetic field intensity and this gives rise to a shift of the fundamental absorption edge and of the recombination radiation edge in the magnetic field. Since the probability of optical transitions in the fundamental absorption region depends strongly on the density of states, oscillations of this density give rise to oscillations of the absorption coefficient near the fundamental edge.

5.1. Diamagnetism and Paramagnetism

A. Magnetic Susceptibility

The magnetic susceptibility of a solid consists of the contributions from electrons in filled bands (the lattice susceptibility), electrons at lattice defects, and free carriers. The first two components are very difficult to estimate. The temperature dependence of the magnetic susceptibility is governed primarily by the free-carrier contribution, which is given by the following formula for the case of a parabolic band:

$$\chi_{fc} = \frac{\mu_B^2}{\eta} \frac{\partial n}{\partial \zeta} \left[\frac{g^2}{4} - \frac{1}{3}\left(\frac{m_0}{m^*}\right)^2 \right], \qquad (5.6)$$

where η is the density of the solid.

For nondegenerate samples, we have $\partial n/\partial \zeta = n/k_0 T$. The first term in square brackets (the paramagnetic contribution) is due to the fact that free carriers have an intrinsic magnetic moment and its relative importance is governed by the value of the g factor. The second term in the square brackets (the diamagnetic contribution) depends strongly on the effective mass. Lead chalcogenides have large g factors and small effective masses (the relationship between the g factor and the effective mass is considered in §6.2).

The magnetic susceptibility of PbTe [428] and PbSe [333, 429, 430] in weak magnetic fields was determined by measuring the force acting on a sample in an inhomogeneous magnetic field. These measurements of the susceptibility were carried out between 100°K and 500-600°K. The specific magnetic susceptibility, defined as the

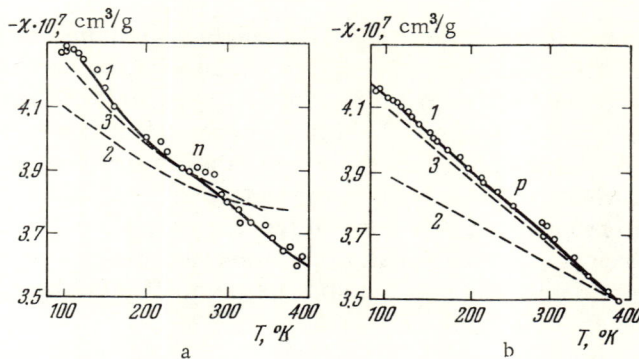

Fig. 5.1. Temperature dependences of the magnetic susceptibility of heacily doped n-type (a) and p-type (b) PbSe samples [333]: 1) experimental curve; 2) theoretical curve for a parabolic band; 3) theoretical curve for the Kane model.

magnetic moment per unit mass in a unit magnetic field was found to be negative (i.e., lead chalcogenides are diamagnetic) and approximately equal to $-0.4 \cdot 10^{-6}$ cm^3/g at low temperatures; its absolute value was found to decrease with increasing temperature (Fig. 5.1). Chlorine and silver impurities affect considerably the temperature dependence of the magnetic susceptibility of lead chalcogenides [428, 429].

Detailed calculations of the magnetic susceptibility of lead chalcogenides have not been carried out but attempts have been made to explain the temperature dependence of the susceptibility by the free-carrier contribution [333, 428-430]. The temperature dependence of the magnetic susceptibility in the parabolic band case is too weak to explain the experimental data obtained for heavily doped PbSe (Fig. 5.1) [333, 430]. The agreement between the theory and experiment is considerably better for the nonparabolic Kane band, whose shape is governed by the interaction of two bands, which are degenerate only in respect of the spin and which lie at the same point in the Brillouin zone. Therefore, in this case the conduction and valence bands are of the same shape and the magnetic susceptibility of n- and p-type samples does not differ greatly.

A more rigorous calculation of the magnetic susceptibility of free carriers in lead chalcogenides should include an allowance for the anisotropy of the effective masses and of the g factor (corresponding to the positions of the extrema at the L points of the Brillouin zone), for departures of the value of the g factor from 2, and for the energy dependences of the effective mass and g factor. A considerable scatter of the available experimental and theoretical values of the g factor (cf. Table 6.9) makes it difficult to reliably calculate the magnetic susceptibility. Approximate estimates, in which we have used Eq. (5.6) and values obtained from measurements of the Shubnikov – de Haas effect (cf. Table 5.1), have indicated that the paramagnetic and diamagnetic terms for lead chalcogenides are of the same order of magnitude but the absolute value of the paramagnetic term is somewhat smaller than that of the diamagnetic term.

B. de Haas—van Alphen Effect

Oscillations of the magnetic susceptibility in a strong magnetic field (the de Haas – van Alphen effect) are due to oscillations of the density of states and have been observed by Stiles et al. [431] in p-type PbTe and n-type PbS single crystals with carrier densities of the order of 10^{18} cm^{-3}. Stiles et al., were the first to study the de Haas – van Alphen effect in semiconductors.

Stiles et al. used a pulsating magnetic field of up to 125-kOe intensity. A small sample was placed within a coil. An alternating magnetization, produced by the magnetic field, generated an electric signal in the coil, which is proportional to the magnetic susceptibility.

The magnetic susceptibility of degenerate semiconductors oscillates as a function of $1/H$. Measurements of the $1/H$ period and of the oscillation amplitude can be used to find the longitudinal and transverse components of the effective mass and the number of energy ellipsoids. The conditions of applicability of this method for the determination of the band structure parameters, its advantages, and methods for theoretical analysis of the obtained experimental data are the same as those discussed later in connection with the method based on measurements of the Shubnikov – de Haas effect (cf. §5.2). Measurements of the oscillation period as a function of the orientation of the magnetic field relative to the crystallographic axes make it possible to find the effective-mass anisotropy coefficient K and the number of carriers in one ellipsoid. The temperature dependence of the oscillation amplitude yields the value of the cyclotron effective mass m_c^* [cf. Eq. (5.3)]. Comparison of the number of carriers in one ellipsoid with the total carrier density found from measurements of the Hall effect yields the number of ellipsoids.

Stiles et al., analyzed their experimental results [431] and concluded that the valence band of PbTe is described by four strongly elongated ellipsoids with the following parameters: $K = 6.4$, $m_\perp^* = 0.043 m_0$. Apart from oscillations associated with this valence band, Stiles et al., observed oscillations which they attributed to the presence of another (spherical) valence band whose edge is close to the edge of the ellipsoidal band ($\Delta\mathscr{E} \leq 0.004$ eV). However, other investigators have not confirmed the existence of the spherical valence band (cf. §3.2B and §5.5A).

The conduction band of PbS is nearly spherical and consists of four ellipsoids; the effective mass is $0.14 m_0$.

C. Paramagnetism of Free Carriers. Knight Shift

The spin paramagnetism of free carriers not only contributes to the magnetic susceptibility, where it is added to the larger absolute value of the diamagnetic term, but it also gives rise to several other effects. One of them is the wavelength-independent contribution to the Faraday effect, which is proportional to the paramagnetic susceptibility. This phenomenon has been observed in n-type PbS and in other lead chalcogenides (for details see §5.4A).

Another effect associated with the spin paramagnetism is the Knight shift, i.e., a change in the frequency of the nuclear magnetic resonance due to the interaction of nuclei with free carriers. A redistribution of carriers between various spin states produces a finite average magnetic moment per one carrier. The interaction of this moment with the magnetic moment of the nucleus shifts the resonance frequency by an amount proportional to the paramagnetic susceptibility and to the square of the amplitude of the wave function $|\psi(0)|^2$ at the point where the nucleus is located. The latter quantity is a measure of the degree to which the wave functions near an extremum are of the s-type.

The Knight shift in the nuclear resonance of Pb in p-type PbTe was investigated by Weinberg and Callaway [432]. The resonance was observed at a fixed frequency of 7.5173 Mc. The relative shift of the resonance field was $\Delta H/H = 2.34 \cdot 10^{-3}$ at room temperature. When the temperature was increased from 260 to 450°K, the effect increased considerably.

The presence of the Knight shift shows that electrons at the top of the valence band are described by wave functions with a finite density near the Pb nuclei. Among the wave functions which (according to theoretical calculations) may belong to states near the forbidden band at the point L, only the function L_1^6 is of the s-type in the vicinity of the Pb nucleus (the meaning of the symbol L_1^6 is explained in §6.1B). This conclusion, drawn from investigations of the Knight shift, has been used as a starting point in the theoretical calculations of the band structure of PbTe (cf. §6.1C) and has played an important role in the determination of the energy band structure.

The observed increase in the Knight shift with increasing temperature was attributed by Weinberg and Callaway [432] to a rise of the hole density with temperature. This explanation conflicts entirely with the Hall effect data, which indicate that the carrier density in lead chalcogenides is usually independent of temperature in the extrinsic conduction region.

5.2. Shubnikov—de Haas Effect

Oscillations of the resistance of degenerate samples in strong magnetic fields (the Shubnikov – de Haas effect) in lead chalcogenides have been investigated quite thoroughly [240, 339, 433-441]. * These measurements have been carried out on n- and p-type samples of all three chalcogenides in a wide range of carrier densities at 4.2°K or lower temperatures. Oscillations of the longitudinal and transverse magnetoresistance have been observed. Investigations of the Shubnikov — de Haas effect have yielded accurate and detailed information on the band structure of lead chalcogenides. The most comprehensive review of the results obtained from measurements of the Shubnikov – de Haas effect was undertaken by Cuff et al. [439].

A. Conditions for the Observation of Resistance Oscillations in Strong Magnetic Fields

The Shubnikov – de Haas effect has much in common with the de Haas – van Alphen effect, which has already been discussed. The

---
*Oscillations of the thermoelectric power and thermal conductivity of PbTe have also been observed [537].

Shubnikov — de Haas effect is easier to measure but quite stringent requirements must be satisfied by the contacts in measurements of the resistance in strong magnetic fields. The maxima of both effects are exhibited when the bottom of a Landau band coincides with the Fermi level. When these two effects are plotted as a function of 1/H, the maxima are found to be distributed periodically.

An explanation of the Shubnikov — de Haas effect is somewhat more complex than that of the de Haas — van Alphen effect, because the resistance is not only governed by the density of states but also by the relaxation time. Let us consider qualitatively the transverse resistance in a strong magnetic field when carriers are scattered elastically. We shall assume first that there is no scattering. Then, in crossed electric and magnetic fields, all electrons drift along a direction perpendicular to both fields and the drift velocity is cE/H. Assuming that an external voltage is applied along the x axis and a magnetic field along the z axis, we find that the nondiagonal component of the conductivity tensor is

$$\sigma_{xy} = en \frac{c}{H}. \tag{5.7}$$

The motion of an electron is finite along the x direction and, in the absence of scattering, the diagonal component is $\sigma_{xx} = 0$. In the presence of scattering, the centers of oscillators may be shifted along the x direction and σ_{xx} is no longer equal to zero. In calculating this quantity, it is usually sufficient to consider only the contribution of electrons whose energies are close to the Fermi energy (the number of such electrons is proportional to the density of states). Moreover, the value of σ_{xx} is proportional to the intensity of scattering (inversely proportional to the relaxation time) and the intensity of scattering is proportional to the density of final states, which also lie near the Fermi level if the scattering is elastic. For these two reasons, the maxima of σ_{xx} correspond to density-of-states maxima.

Let us now consider the diagonal component of the resistivity tensor ρ_{xx} whose dependence on the magnetic field is measured experimentally. The electron drift along the y direction gives rise to a Hall electric field E_y, which is not limited to any finite value in the ideal case when there is no scattering and E_x is constant. The density of the current flowing along the x axis is

$$j_x = \sigma_{xx} E_x + \sigma_{xy} E_y. \tag{5.8}$$

In the absence of scattering the first term is equal to zero because $\sigma_{xx} = 0$, but the second term is infinite because the Hall field can be infinite. An infinitely high current density along the direction of the applied electric field E_x corresponds to a zero value of the resistivity $\rho_{xx} = 0$. When scattering in a strong magnetic field ($\mu H/c \gg 1$) is allowed for, the second term in Eq. (5.8) is still much larger than the first, i.e., the current along the x axis is mainly due to the Hall field E_y but now this field is finite. Having found this value when there is no current along the y direction and bearing in mind that in an isotropic material $\sigma_{xx} = \sigma_{yy}$ and $\sigma_{xy} = -\sigma_{yx}$, we obtain the following formula:

$$\rho_{xx} = \frac{E_x}{j_x} = \frac{\sigma_{xx}}{\sigma_{xy}^2}.$$

Since σ_{xy} does not oscillate, the value of ρ_{xx} assumes maximum values only when σ_{xx} has maxima, i.e., when the Fermi level coincides with the energy at which the density of states has a maximum.

Formally, this explanation of the Shubnikov – de Haas effect can be presented as follows. In a strong transverse magnetic field the diagonal component of the resistivity tensor ρ_{xx} is proportional to $<1/\tau>$ (cf., for example, [270]). The angular brackets denote integration over the energy range [cf. Eq. (3.5)]; the integrand includes the density of states and the derivative of the Fermi function ($-\partial f/\partial \mathscr{E}$), which – in the case considered – can be replaced by the δ function and, instead of integration, one can use the value of the integrand in which the substitution $\mathscr{E} = \zeta$ is made. Since $1/\tau$ is proportional to the density of states, ρ_{xx} is proportional to the square of the density of states. Thus, the magnetoresistance has maximum values in magnetic fields at which the Fermi level crosses the bottom of a Landau band. This conclusion was reached first by Shalyt and Efros [442].

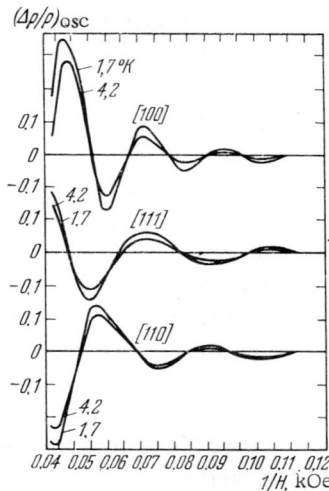

Fig. 5.2. Oscillatory component of the transverse magnetoresistance of n-type PbSe as a function of 1/H at 4.2 and 1.7°K [434]. The current was directed along [110]; the magnetic field, along [100], [111], and [110].

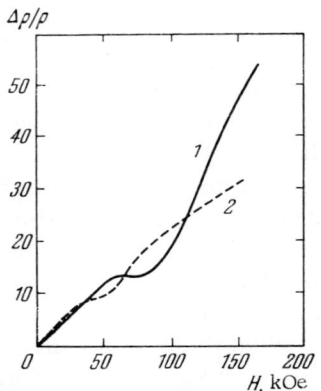

Fig. 5.3. Dependences of the transverse (1) and longitudinal (2) magnetoresistance of n-type PbTe on strong magnetic fields at 4.2°K [240]. The magnetic field was applied along [100].

In order to observe appreciable oscillations of the magnetoresistance, we must satisfy the condition

$$\hbar\omega_c > k_0 T, \tag{5.9}$$

i.e., the separation between the Landau levels should exceed the thermal broadening of the Fermi level. Moreover, the separation between the Landau levels should be greater than the broadening of these energy levels due to scattering. This condition is identical with the requirement that carriers in a magnetic field must make several revolutions during the period represented by the relaxation time:

$$\omega_c \tau \gg 1, \quad \text{or} \quad \mu \frac{H}{c} \gg 1. \tag{5.10}$$

Finally, a semiconductor should be degenerate:

$$\zeta \gg k_0 T. \tag{5.11}$$

Lead chalcogenides are excellent materials for the observation of quantum oscillations in magnetic fields. Their small effective masses give rise to large separations between the Landau levels [cf. Eq. (5.9)]. The higher carrier mobility at low temperatures ensures that the condition (5.10) is satisfied. The absence of impurity levels in the forbidden band means that a high carrier density is retained even at temperatures below 4°K so that strong degeneracy is observed right down to the very lowest temperatures [cf. Eq. (5.11)]. The disadvantage of lead chalcogenides is the inhomogeneity of the samples, which effectively broadens the Fermi level.

Cuff et al. [434] investigated the Shubnikov – de Haas effect in lead chalcogenides using magnetic fields up to 26 kOe. In such magnetic fields, the Fermi energy coincides with the Landau levels corresponding to $l > 1$ and several oscillations of the magnetoresistance are observed (Fig. 5.2). In sufficiently strong magnetic fields, all carriers are in the first Landau band, corresponding to $l = 0$ (the quantum limit). Using magnetic field pulses up to 150 kOe intensity, Kanai et al. [240] observed a nonmonotonic increase of the resistance with increasing magnetic field when the Fermi level passes through the bottom of the first Landau band (Fig. 5.3).

B. Determination of the Anisotropy Coefficient and of the Number of Ellipsoids

Since the magnetoresistance maxima are observed when the Fermi level coincides with the bottom of a Landau band, we can show that the magnetoresistance oscillation period (expressed in terms of 1/H) is $\Delta(H^{-1}) = \mathcal{A}^{-1}$, in the case of ellipsoidal constant-energy surfaces; the quantity $\mathcal{A}$ is given by

$$\mathcal{A} = \frac{\pi^2 c \hbar}{2e} \left(\frac{3n}{\pi N}\right)^{2/3} \frac{K^{1/6}}{(K \cos^2 \vartheta + \sin^2 \vartheta)^{1/2}} \tag{5.12}$$

(ϑ is the angle between the direction of the magnetic field and the angle of revolution of the ellipsoid). When a magnetic field is ap-

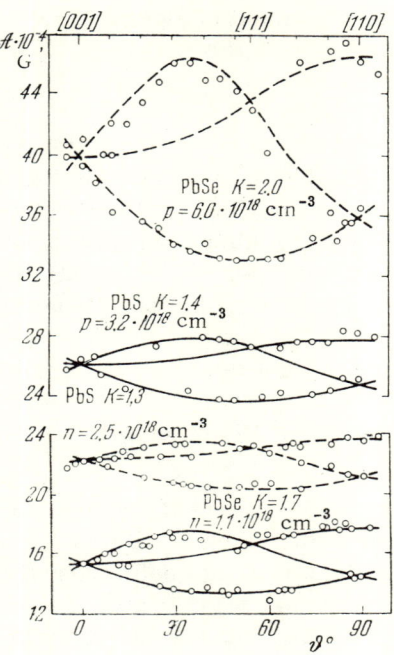

Fig. 5.4. Reciprocal of the period of the Shubnikov – de Haas oscillations in PbSe and PbS as a function of the magnetic field orientation rotating in a (110) plane [439]. The experimental points and theoretical curves are given for the <111> model and the values of the coefficient K cited in this figure.

plied along a direction other than [100], the ellipsoids are oriented in different ways with respect to the magnetic field and two or more systems of oscillations are observed.

For a Fermi surface of arbitrary shape, the oscillation period is given by

$$\mathcal{A} = \frac{\hbar c}{2\pi e} A, \qquad (5.13)$$

where A is the area of the extremal cross section (perpendicular to the magnetic field direction) of the Fermi surface.

By measuring the oscillation period for various directions of the magnetic field, we can determine the number of equivalent ellipsoids N and the anisotropy coefficient K. Both these quantities are expressed in terms of $\mathcal{A}_\|$ and $\mathcal{A}_\perp$, which are proportional to cross sections, which are parallel ($\vartheta = 90°$) and perpendicular ($\vartheta = 0°$) to the ellipsoid axis:

$$N = 3\pi^2 n \left(\frac{\hbar c}{2e}\right)^{3/2} (\mathcal{A}_\|^2 \mathcal{A}_\perp)^{-1/2}. \qquad (5.14)$$

The total carrier density n is found from the Hall effect in a strong magnetic field. This method gives N=4 for all three lead chalcogenides of n- and p-type [439]. It follows that in PbSe and PbS, as well as in PbTe, the extrema of the conduction and valence bands are located at the Brillouin zone boundary along the <111> directions.

The angular dependence of the reciprocal of the oscillation period $\mathcal{A}$ is governed by the anistropy coefficient K. Figure 5.4 shows the dependence of the value of $\mathcal{A}$ on the angle between the magnetic field direction and one of the cubic axes of n- and p-type PbSe and PbS crystals. The magnetic field is rotated in a (110) plane. The value of K can be determined either from curves such as those given in Fig. 5.4 or from the formula

$$K = \left(\frac{\mathcal{A}_\|}{\mathcal{A}_\perp}\right)^2. \qquad (5.15)$$

The values of the anisotropy coefficient K obtained in this way are listed in Table 5.1. The extrema of PbSe and PbS are characterized by prolate ellipsoids with a weak anisotropy (particularly in the case of PbS).

C. Determination of Effective Mass and Band Nonparabolicity

It is evident from Eq. (5.13) that the oscillation period is independent of the effective carrier mass. Therefore, the cyclotron effective mass can be determined from the temperature dependence of the amplitudes of the Shubnikov – de Haas oscillations. In particular, the amplitude of the transverse magnetoresistance os-

TABLE 5.1. Quantities Determined from
the Shubnikov—de Haas Effect [439]

Compound	Type of conduction	$m^*_{\perp 0}/m_0$	$m^*_{\parallel 0}/m_0$
PbTe	n	0.024±0.003	0.24±0.05
	p	0.022±0.003	0.31±0.05
PbSe	n	0.040±0.008	0.070±0.015
	p	0.034±0.007	0.068±0.015
PbS	n	0.080±0.010	0.105±0.015
	p	0.075±0.010	0.105±0.015

Compound	Type of conduction	K	$\mathscr{E}_{gi}$, eV	$\mathscr{E}_{P\perp}$, eV	$|g_\parallel|$
PbTe	n	10±1.5	0.16±0.04	6.3±0.4	45±8
	p	14±2	0.14±0.04	6.4±0.4	51±8
PbSe	n	1.75±0.2	0.22±0.04	5.2±0.5	27±7
	p	2.0±0.2	0.18±0.03	5.5±0.4	32±7
PbS	n	1.3±0.1	0.29±0.04	3.4±0.4	12±3
	p	1.4±0.1	0.33±0.04	4.9±0.4	13±3

cillations is proportional to [434]:

$$\left(\frac{\Delta\rho}{\rho}\right)_{\text{osc}} \propto T \sinh^{-1} \frac{2\pi^2 k_0 T}{\hbar\omega_c}. \qquad (5.16)$$

Since the measurements reported in [434] were carried out in the temperature range 1.2-4.2°K, all the band structure parameters obtained in this investigation apply in practice at 0°K. The values of the transverse effective mass $m^*_\perp = [(1/\hbar^2)(1/k_\perp)(\partial\mathscr{E}/\partial k_\perp)]^{-1}$ were determined in a range of carrier densities sufficiently high to find, for each band, the transverse effective mass at the band edge as well as the band nonparabolicity parameters. The energy dependence of the transverse effective mass is of the same form for the two simple energy band models considered in §6.2B. From

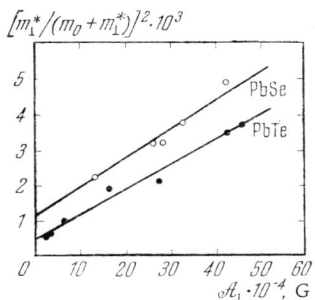

Fig. 5.5. Dependences of the transverse effective mass on the extremal cross section of the Fermi surface for p-type samples of PbTe and PbSe [439]. The experimental points obtained from the Shubnikov–de Haas effect fit well the straight lines whose parameters correspond to the values of $m^*_{\perp 0}$ and $\mathscr{E}_{gi}$, listed in Table 5.1.

the formulas obtained for these models we can deduce the relationship*

$$\left(\frac{m^*_\perp}{m_0 \pm m^*_\perp}\right)^2 =$$

$$\left(\frac{m^*_{\perp 0}}{m_0 \pm m^*_{\perp 0}}\right)^2 + 4.64 \cdot 10^{-8} \frac{\mathscr{A}_\perp}{\mathscr{E}_{P\perp}},$$

(5.17)

where the quantity $\mathscr{A}_\perp = (\hbar c/2e)/k^2_\perp$ is expressed in gauss, while the quantity $\mathscr{E}_{P\perp} = 2m_0(P/\hbar)^2$ is expressed in electron volts ($P_\perp$ is the effective matrix element of the transverse component of the momentum operator). The "+" and "−" signs are used to denote the valence and conduction bands, respectively.

An analysis of the experimental results by means of Eq. (5.17) yields the values of $m^*_{\perp 0}$ and $\mathscr{E}_{P\perp}$. Figure 5.5 shows the dependence of the quantity $[m^*_\perp(m_0 + m^*_\perp)]^2$ on $\mathscr{A}_\perp$ for p-type samples of PbTe and PbSe. The experimental points obtained by measuring the Shubnikov–de Haas effect fit well the straight lines whose parameters correspond to the values of $m^*_{\perp 0}$ and $\mathscr{E}_{gi}$ listed in Table 5.1. The effective matrix element of the momentum is found directly from the slope of the straight lines with an error of about 10%. The values of the transverse effective mass at the band edge found from the intercept on the ordinate axis are less accurate. The effective width of the forbidden interaction band is found from the formula

$$\mathscr{E}_{gi} = \mathscr{E}_{P\perp} \frac{m^*_{\perp 0}}{m_0 \pm m^*_{\perp 0}}.$$

(5.18)

---

*In contrast to the formulas in §6.2B, Eq. (5.17) allows for the contribution of the free-electron mass to the transverse effective mass, which results in a small correction of the effective mass.

The parameters deduced from the Shubnikov – de Haas effect are listed in Table 5.1. The effective width of the forbidden interaction band agrees reasonably well with the optical value of the forbidden band width, obtained from magneto-optical measurements (cf. §5.4C). However, in contrast to the optical forbidden band width, the interaction band width increases monotonically along the chalcogenide series from PbTe to PbS.

The anisotropy coefficients of p-type samples of a given lead chalcogenide are somewhat larger than the coefficients of n-type samples of the same chalcogenide and they decrease by an order of magnitude along the chalcogenide series from PbTe to PbS. The anisotropy coefficients of the conduction and valence bands of PbSe and PbS depend weakly on the carrier density. Thus, the longitudinal effective masses, like the transverse masses, of these semiconductors are almost as independent of the energy. The energy dependence of the anisotropy coefficient of PbTe is stronger than that of PbSe and PbS but it is still not sufficiently large to assume that the longitudinal effective mass is not affected by the band occupancy. The quantity $(m_{\parallel}^* - m_{\parallel 0}^*)/m_{\parallel 0}^*$ is only half the quantity $(m_{\perp}^* - m_{\perp 0}^*)/m_{\perp 0}^*$.

In contrast to compounds belonging to the InSb group, the value of the matrix element of the transverse component of the momentum operator decreases appreciably along the chalcogenide series from PbTe to PbS. Therefore, the transverse effective mass increases along the lead chalcogenide series much more rapidly than the nonomonotonic variation of the forbidden band width. For example, the transverse masses at the valence band edges of PbTe, PbSe, and PbS are, respectively, $0.022m_0$, $0.034m_0$, and $0.075m_0$, while the optical values of the forbidden band width are, respectively, 0.19, 0.16, and 0.28 eV. The longitudinal effective masses decrease nonmonotonically along the chalcogenide series from PbTe to PbS (for the valence bands of PbTe, PbSe, and PbS, these masses are, respectively, $0.31m_0$, $0.068m_0$, and $0.105m_0$). We shall consider these relationships again in Chap. VI.

D. Spin Splitting and the Determination of the Effective g Factors

All n- and p-type lead chalcogenides exhibit either a second harmonic or a doublet structure in the magnetoresistance oscillations. The cause of this effect is the spin splitting of the Landau levels, whose value is governed by the g factor. The spin splitting is within the limits $(0.5 \pm 0.1)\hbar\omega_c$. The values of the longitudinal g factors near the band edges obtained from measurements of the Shubnikov – de Haas effect are given in Table 5.1.

5.3. Tunnel Effect in a Magnetic Field

Rediker and Callawa [322] observed a strong dependence of the tunnel current through a p – n junction in PbTe at 4.2°K on a magnetic field applied perpendicularly to the current. In a magnetic field of 66.5 kOe, the tunnel effect decreased appreciably so that the negative resistance region disappeared (Fig. 5.6). When the magnetic field was rotated in the p – n junction plane, which (in different diodes) was perpendicular to the crystallographic axes [100], [110], and [111], the tunnel current in strong magnetic fields was found to exhibit a considerable anisotropy. The nature of the angular dependence of this effect indicates a complex band structure in PbTe with prolate ellipsoids along the <111> directions.

A theory of the tunnel effect in PbTe-type semiconductors subjected to a transverse magnetic field was constructed by Aronov and Pikus [323]. Considering the motion of an electron in crossed electric and magnetic fields (the electric field is assumed to be strong), Aronov and Pikus showed that the energy corresponding to the drift velocity, as well as the energy acquired by an electron in a Larmor orbit, are comparable with the forbidden band width $\mathscr{E}_g$, and, therefore, the nonparabolicity of the electron spectrum plays an essential role (cf. §5.2A). Aronov and Pikus assumed that the structure of the conduction and valence band edges, degenerate only in respect of the spin, is determined fully by the interaction of these two bands; we shall call this the "Kane" model.* Then, the motion of an electron is described by a two-band equation which differs, in the simplest case, from the Dirac equation only in one respect: instead of the velocity of light, the maximum velocity is $s = (\mathscr{E}_g/2\, m_{d1}^*)^{1/2}$, where m_{d1}^* is the density-of-states effective mass in one ellipsoid.

---

*The structure of the valence band in this model differs considerably from that in the model developed by Kane [205] for InSb.

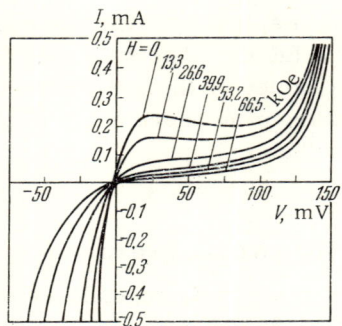

Fig. 5.6. Current — voltage characteristics of PbTe p–n junctions at 4.2°K in transverse magnetic fields up to 66.5 kOe [322]. The current was directed along [110].

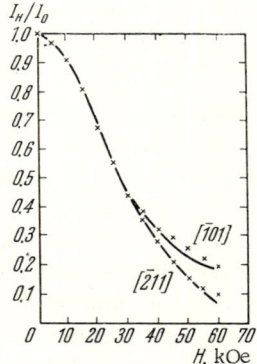

Fig. 5.7. Dependences of the tunnel current on the magnetic field intensity for two field directions corresponding to the maximum and minimum current through a PbTe tunnel diode (theoretical curves [323] and experimental points [322]). The current was directed along [111]. The following values were used in calculations: $K = 11$, $E = 4.8 \cdot 10^4$ V/cm, $\mathscr{E}_g = 0.19$ eV, $m^*_{\perp 0} = 0.027\ m_0$.

In the case of ellipsoidal bands, it is convenient to introduce quantities which are proportional to the magnetic and electric fields:

$$H' = \frac{s}{c}\left(\sum_{i=1}^{3} H_i^2 \frac{m_i^*}{m_{d1}^*}\right)^{1/2}, \quad (5.19)$$

$$E' = \left(\sum_{i=1}^{3} E_i^2 \frac{m_{d1}^*}{m_i^*}\right)^{1/2}, \quad (5.20)$$

where H_i, E_i, and m_i^* are the components of the electric and magnetic fields and of the effective mass along the principal axes of a given ellipsoid. The effective masses in the above formulas are the values near the band extrema (in the model considered, they are equal for electrons and holes).

When $H' < E'$, the motion of an electron is unrestricted and the current through the p–n junction is due to direct transitions; the value of the current is found from formulas deduced for the case when there is no magnetic field but E in these formulas must be replaced with $[(E')^2 - (H')^2]^{1/2}$, i.e., the applied magnetic field reduces the effective electric field in the p–n junction and, therefore, it weakens the tunnel effect. The theoretical and experimental

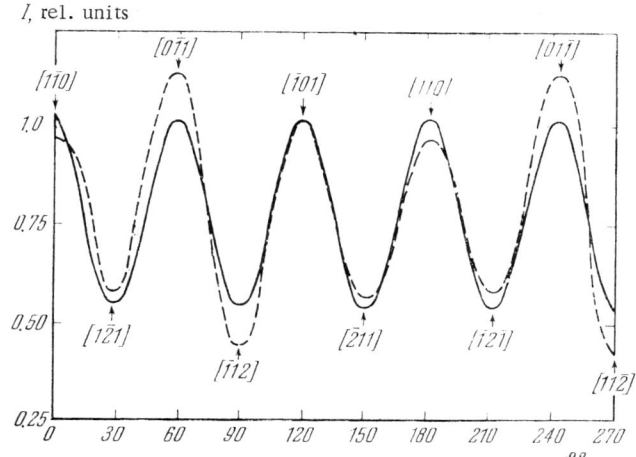

Fig. 5.8. Dependence of the tunnel current through a PbTe diode on the magnetic field direction rotating in a (111) plane for H = 53.2 kOe, E = 4.8 · 10⁴ V/cm (experimental data from [322]; theoretical curve from [323]).

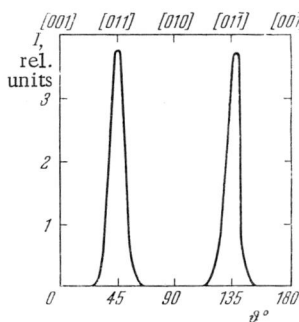

Fig. 5.9. Theoretical dependence of the tunnel current through a PbTe diode on the direction of a magnetic field rotating in a (100) plane for H = 60 kOe, E = 5.05 · 10⁴ V/cm [323].

dependences of the tunnel current on the magnetic field intensity are shown in Fig. 5.7 for two directions of the magnetic field at which the tunnel current assumes maximum and minimum values in its dependence on the magnetic field orientation. The corresponding angular dependences of the tunnel current at a fixed magnetic field intensity are shown in Fig. 5.8. Both Figs. 5.7 and 5.8 apply to a p−n junction oriented in a (111) plane.

The following parameters were used in plotting the theoretical curves: $\mathscr{E}_g = 0.19$ eV, $m_{\perp}^* = 0.027 m_0$, $K = 11$. These values are in agreement with the results obtained from measurements of the optical absorption in a magnetic field (§5.4C), from the Shubnikov—de Haas effect (§5.2B and §5.2C), and from cyclotron resonance (§5.5A). The value of the electric field in the p—n junction was selected so that the ratio of the maximum to minimum current agreed with the experimental value. These values of the electric field are in order-of-magnitude agreement with the results of theoretical estimates and measurements of the p—n junction capacitance.

The good agreement between the theoretical and experimental curves shows that measurements of the dependence of the tunnel current on the magnetic field intensity can provide a reliable method for the determination of the parameters of a semiconductor and the most accurate method for measuring the electric field in a p—n junction. This agreement is also a proof of the validity of the model examined later (§6.2B) which is used to describe the conduction and valence bands of PbTe. The essence of this model lies in the ellipsoidal shape of the constant-energy surfaces elongated along a <111> direction, the coincidence of the positions of the band extrema in the **k** space, and the nonparabolicity along the longitudinal and transverse directions, whose nature is governed by the interaction between the two bands.

When $H' > E'$, the motion of an electron in crossed electric and magnetic fields is restricted and direct tunnel transitions in a uniform electric field are forbidden by the laws of conservation of energy and momentum. In this case, the tunnel current is possible only due to the scattering by phonons or impurities. Since the value of H' depends on the direction of the magnetic field with respect to the ellipsoid axes, angular dependences may have intervals of angles where $E' > H'$ which alternate with intervals where $E' < H'$ and there is no tunnel current. A theoretical curve for such a case, in the form of isolated peaks, is shown in Fig. 5.9.

5.4. Magneto-Optical Effects

In this and the following section, we shall consider the interaction of lead chalcogenides with electromagnetic radiation in a magnetic field. As in investigations of the optical phenomena, the passage of an electromagnetic wave through matter can be conveniently described by a complex refractive index (cf. §2.1A), which is now a function of the magnetic field intensity. It is then found that, at a given frequency of an electromagnetic wave, we can have two solutions for a complex refractive index, describing the ordinary and extraordinary rays in a medium in which a magnetic field induces birefringence (double refraction).

First, we shall consider the effects observed at wavelengths in the optical range (influence of a magnetic field on the absorption and reflection of light, recombination radiation in a magnetic field, and the Faraday effect). The frequency of electromagnetic waves in the optical range is considerably higher than the cyclotron frequency ($\omega \gg \omega_c$).

A. Faraday Effect

The Faraday effect is the rotation of the plane of polarization of linearly polarized light during its passage through a sample subjected to a magnetic field parallel to the light beam (cf., for example, [130, 443]). We shall consider first the Faraday effect in a weakly absorbing medium ($\mathcal{K}^2 \ll \mathcal{N}^2$), consisting of free atoms.

A linearly polarized wave can be regarded as composed of two circularly polarized waves whose electric vectors rotate at a frequency ω in opposite senses about the direction of propagation of the wave. A magnetic field imposes, on the motion of electrons in the central field of an atom, an additional uniform rotation about the direction of the magnetic field, and the frequency of this rotation is the Larmor value $\omega_L = eH/2m_0c$. The frequencies of rotation of the electric vector of two waves polarized in opposite senses, measured in a rotating system of coordinates coupled to the electron system, differ from the frequency ω by $\pm \omega_L$. Since the refractive index is a function of the frequency, the velocities of propagation of these two waves are different and the phase shift between them, which appears after the passage through the samples, results in the rotation of the plane of polarization of the linearly polarized wave which is composed of two circularly polarized waves; the angle of rotation of the plane of polarization, measured with respect to the initial orientation of the linearly polarized wave, is given by

$$\vartheta_F = \frac{\omega d}{2c} [\mathcal{N}(\omega + \omega_L) - \mathcal{N}(\omega - \omega_L)]. \tag{5.21}$$

In the optical range of frequencies, we usually have $\omega \gg \omega_L$ and the difference between the refractive indices is then $(\partial \mathcal{N}/\partial \omega) 2\omega_L$, which yields the Becquerel formula for the Faraday angle of rotation:

$$\vartheta_F = \frac{e\omega}{2 m_0 c^2} \frac{\partial \mathcal{N}}{\partial \omega} dH. \tag{5.22}$$

Equation (5.22) is more general than might be expected from the assumptions made in its derivation. In particular, at frequencies $\omega \ll \mathcal{E}_g/\hbar$ the optical dispersion in a semiconductor is governed by the interaction of light with free carriers and Eq. (5.22) remains valid for an extrinsic (for example, n-type) semiconductor if m_0 is replaced by the effective electron mass m* (the Larmor frequency should be replaced with $\omega_c/2$). Using Eqs. (5.22), (2.3), and (2.20), we obtain a relationship derived by Stephen and Lidiard [443]:

$$\vartheta_F = \frac{2 \pi n e^3 dH}{m^{*2} c^2 \mathcal{N} \omega^2}. \tag{5.23}$$

The effect has an opposite sign for a p-type sample. It follows from Eq. (5.23) that the product $\mathcal{N}\vartheta_F$ is inversely proportional to the square of the effective mass. This makes the Faraday effect a convenient and accurate method for the determination of the effective mass in the infrared range. The inequalities assumed in the derivation can be easily satisfied under experimental conditions. In contrast to cyclotron resonance, it is not necessary to satisfy the inequality $\omega_c \gg \gamma$, so that measurements can be carried out in a wide range of temperatures and carrier densities. Another advantage of this method of determination of the effective mass is that the scattering mechanism need not be known.

In the case of a many-ellipsoid model with cubic symmetry, the effective mass in Eq. (5.23) should be replaced with the quantity

$$m_F^* = \left(\frac{3K}{K+2}\right)^{1/2} m_\perp^*. \tag{5.24}$$

If an energy band is isotropic but not parabolic, the effective Faraday mass in a degenerate sample is given by

$$m_F^* = \left(\frac{1}{\hbar^2 k} \frac{d\mathcal{E}}{dk}\right)^{-1}. \tag{5.25}$$

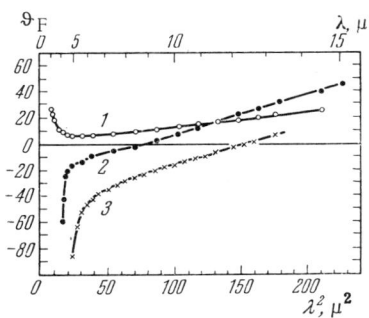

Fig. 5.10. Dependence of the Faraday rotation angle of a PbS sample 0.425 mm thick (n = 3.5·10^{17} cm^{-3}) on the wavelength. Experimental data were taken from [446]. 1) T = 300°K, H = 60.4 kOe; 2) T = 100°K, H = 60.4 kOe; 3) T = 4.2°K, H = 38.5 kOe.

The expression for the general case is given in [443] and an equation for the generalized Kane band is given in Appendix A.

Measurements of the Faraday effect due to the presence of free carriers (the interband Faraday effect) in lead chalcogenides have been carried out by many workers [118, 122, 123, 127, 444-446]. Usually, the dependence $\mathcal{N}\vartheta_F$ on λ^2 is a straight line, in agreement with Eq. (5.23) (Fig. 5.10). Some deviation from the linear dependence has been reported for p-type PbTe [118] and p-type PbS [127]. However, in those cases when $\mathcal{N}\vartheta_F$ depends linearly on λ^2, the straight lines do not pass through the origin of coordinates. If we subtract the component proportional to λ^2 from the measured value of the Faraday effect, we obtain a quantity which is independent of λ but which depends strongly on the carrier density and temperature (it is inversely proportional to temperature). Therefore, in the determination of the effective mass from the Faraday effect by means of Eq. (5.23), it is necessary to subtract the term which is independent of λ from the measured value of $\mathcal{N}\vartheta_\Phi$. In other words, the effective mass should be determined from the slope of the straight line.

Measurements of the Faraday effect in n-type samples yielded the following room-temperature values of the effective electron mass: $0.08m_0$ for PbTe [444], $0.10m_0$ for PbSe [444], and $0.17m_0$ for PbS [122, 127]. The values obtained for the effective mass depend strongly on temperature: at 77°K the effective electron mass of PbS is $0.12m_0$ [127] but at 500°K it rises to $0.28m_0$ [122]. The effective mass of holes is $0.20m_0$ for PbTe at 293°K [122], $0.14m_0$ for PbSe at 300°K [118], and $0.12m_0$ for PbS at 77°K [127].

The interband Faraday effect is observed in the region where $\omega \gtrsim \mathcal{E}_g/\hbar$ [127, 445, 446]. The interband Faraday effect depends in a complex manner on the carrier density and the temperature and its sign changes when these two parameters are varied. In the photon energy range above the fundamental absorption edge, the spectrum of the interband Faraday effect exhibits several maxima and minima, like the absorption spectrum in a magnetic field (this will be discussed in subsection C of the present section). The interband effect may also change its sign when the frequency is increased in agreement with Eq. (5.22), since, in the region where the interband transitions are important, the frequency dependence of the refractive index is nonmonotonic.

Mitchell et al. [446] demonstrated that the appearance of the wavelength-independent term in $\mathcal{N}\vartheta_F$, in the frequency range $\omega < \mathcal{E}_g/\hbar$, is also associated with the interband transitions. In the presence of free carriers, the long-wavelength limit of the interband Faraday effect is not equal to zero and is due to the magnetization of the electron system. A further theoretical investigation of this effect, carried out by Stern [447] showed, in particular, that the wavelength-independent term is proportional to the paramagnetic susceptibility of carriers, which is in agreement with the temperature dependence of this term. Theoretical estimates of the value of this term are in agreement with the experimental data for n-type PbS.

B. Infrared Reflection in a Magnetic Field

The difference between the refractive indices of the two circularly polarized waves also gives rise to a difference between the reflection coefficients of these waves. In a magnetic field, the plasma frequency ω_p is replaced by $\omega_p \pm \omega_c/2$ and the reflection minima for the two waves shift in opposite directions. In the case of linearly polarized light, the reflection spectrum exhibits two minima separated by a frequency ω_c. This has been observed in n-type PbS [445]. The separation between the minima has been used to determine the effective mass of electrons, which increased from $0.12m_0$ to $0.16m_0$ when the temperature was increased from liquid-nitrogen to room temperature; this is in agreement with the results obtained from measurements of the Faraday effect.

C. Fundamental Absorption in a Magnetic Field

Quantization of the electron levels in a magnetic field and associated redistribution of the density of states alter appreciably the

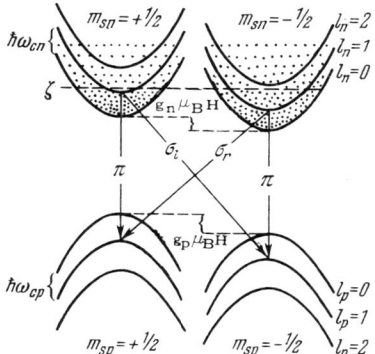

Fig. 5.11. Energy bands in a strong magnetic field, including spin splitting [127]. The arrows indicate σ and π transitions; the occupancy of the conduction band is shown. The energy bands in this figure correspond to positive values of the g factors.

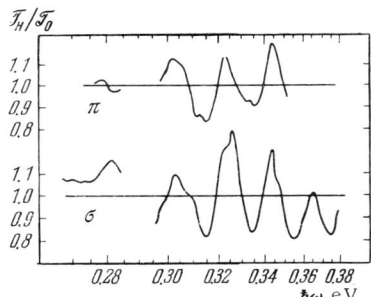

Fig. 5.12. Ratio of the transmission coefficient in a magnetic field to the same coefficient in the absence of a field, plotted as a function of the photon energy for a PbS film at T = 77°K in H = 89 kOe [448]. The light was propagated at right angles to the magnetic field direction; the plane of polarization was either parallel (the π spectrum) or perpendicular (the σ spectrum) to the magnetic field.

form of the optical absorption spectrum near the fundamental edge. The edge shifts in the direction of shorter wavelengths. The strongest absorption is observed when the photon energy is equal to the gap between the Landau levels of electrons and holes, corresponding to $k_H = 0$. Then, Eq. (5.1) yields the expression for the photon energies corresponding to the absorption maxima:

$$\hbar\omega = \mathscr{E}_g + \left(l_p + \frac{1}{2}\right)\hbar\omega_{cp} + \\ + \left(l_n + \frac{1}{2}\right)\hbar\omega_{cn} - \\ - g_p\mu_B H m_{sp} + g_n\mu_B H m_{sn}.$$

(5.26)

TABLE 5.2. Results of Measurements of the Absorption
of Light in a Magnetic Field [450]

Compound	$\mathscr{E}_g$, eV			$\mathscr{E}_{1g}$, eV	$\dfrac{m^*_{np}[100]}{m_0}$ $(T=77\ °K)$	$\dfrac{m^*_\perp}{m_0}$	K	$g_p^{[100]}$	$g_n^{[100]}$
	T=77°K (film on substrate)	T=77°K (unstressed film)	T=4.2°K (unstressed film)						
PbTe	0.203	0.217	0.190	2.1	0.022	0.028	9		
PbSe	0.159	0.176	0.165	2.5	0.030	0.049	1.6	∓22	±22
PbS	0.278	0.307	0.286	4.2	0.055	0.100	1.1	−8.5	+10

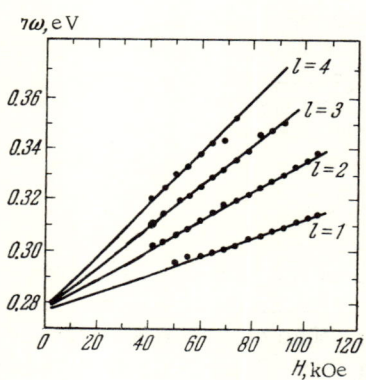

Fig. 5.13. Photon energies corresponding to the minima of the transmission coefficient, plotted as a function of the magnetic field intensity for an n-type PbS film at 77°K (the film was attached to the substrate) [127]. The experimental points fit straight lines which converge on the same point whose position, on the ordinate axis, gives the value of the forbidden band width $\mathscr{E}_g$; the slopes of the lines can be used to determine the reduced effective mass m^*_{np}.

The Landau quantum numbers should satisfy the selection rule $l_n - l_p = 0$. The selection rules for the quantum numbers m_s depend on the directions of propagation and polarization of light with respect to the magnetic field direction. When linearly polarized light, with its electric vector parallel to the magnetic field, is propagated at right angles to the magnetic field, we find that $m_{sn} - m_{sp} = 0$ (the π spectrum, Fig. 5.11). When a circularly polarized wave is propagated along the lines of force of the magnetic field, we find that $m_{sn} - m_{sp} = \pm 1$, where the two signs correspond to the two opposite senses of the circular polarization, so that two spectra are observed: σ_r and σ_l. The superposition of the σ_r and σ_l spectra gives the σ spectrum observed when linearly polarized light, with its electric vector perpendicular to a magnetic field, is propagated at right angles to the magnetic field. The intensities of the σ_r and σ_l lines for a degenerate semiconductor may by unequal because of the different occupancy of states below the Fermi level.

The magnetoabsorption spectra of lead chalcogenides were investigated at 4.2 and 77°K by Mitchell et al. [127, 448-450]. The π and σ spectra exhibit a series of extrema (Fig. 5.12). These extrema are clearly visible when $\omega_{cn} + \omega_{cp} \gg \gamma_n + \gamma_p$. The profile of the lines in the σ spectrum indicates direct allowed transitions.

The absence of the splitting in the σ spectrum demonstrates that the σ_r and σ_l spectra are similar; it follows that $g_n \approx -g_p$ because the σ_r and σ_l lines should be separated by $(g_n + g_p)\mu_B H$. The two π transitions, corresponding to different values of m_s, differ by $(g_p - g_n)\mu_B H$ and the splitting of the π spectrum can be used to determine the magnitudes but not the signs of the g factors of the conduction and valence bands. The signs of the g factors of PbS can be found from the influence of the band occupancy on the ratio of the intensities of the σ_l and σ_r spectra. This yields the values of the g factors listed in Table 5.2.

The absence of the splitting of the σ spectrum in stressed films, whose surfaces coincide with a (100) plane, also indicates that the extrema of the conduction and valence bands are either at the center of the Brillouin zone or along the <111> directions because for other positions the extrema would have shifted in different ways under the influence of stresses in the films [127].

The absorption lines of PbS are at equal distances from one another, which indicates band parabolicity and the absence of band edge degeneracy (apart from the spin degeneracy) [127]. On the other hand, the first lines of PbTe are split unequally, due to the considerable nonparabolicity [449].

Measurements of the optical absorption in a magnetic field provide a very accurate method for the determination of the forbidden band width. By plotting the positions of the absorption maxima as a function of the magnetic field intensity, Mitchell et al. [127, 448-450] obtained straight lines converging on the same point, whose ordinate was equal to $\mathscr{E}_g$ (Fig. 5.13); the slopes of these lines yielded the reduced effective mass m_{np}^*. The accuracy of this method of determination of the forbidden band width is of the order of 1% [127]. Table 5.2 gives the values of the forbidden band width obtained at 77°K for an epitaxial film on a substrate and those obtained at 4.2 and 77°K for a film removed from its substrate. The considerable difference between the values found for a film on a substrate and for a free film is attributed to the influence of a two-dimensional isotropic deformation on the forbidden band width. The forbidden band width of a PbS film removed from its substrate is equal to the width in a natural single crystal [127]. Assuming that, at room temperature, a film attached to a substrate is not deformed and comparing the thermal expansion coefficients of the film and the substrate, Mitchell et al. [127, 450] calculated the deformation potential $\mathscr{E}_{1g}$ (the change in the forbidden band width per unit relative change in volume) whose values (Table 5.2) were in agreement with the results obtained from measurements of the electrical properties under stress (cf. §3.3).

The magneto-optical measurements also yield highly accurate values of the effective mass

$$m^*_{np} = \frac{m^*_n m^*_p}{m^*_n + m^*_p}. \tag{5.27}$$

The reduced effective mass was determined in [127] using two methods: from the separation between neighboring absorption maxima, which is given by

$$\hbar(\omega_{cn} + \omega_{cp}) = \frac{\hbar eH}{m^*_{np}c},$$

and from the slope of the straight lines giving the dependences of the positions of the maxima on the magnetic field intensity.

The values of the electron and hole masses can be determined separately using the Faraday effective data for free carriers or assuming the electron and hole masses to be equal, as was done in [450]. Investigations of the dependence of the reduced effective mass on the magnetic field orientation with respect to the crystal axes can be used to determine the longitudinal and transverse components of the effective masses. This method shows that the effective masses in PbS are almost isotropic and that the absorption spectra are independent of the magnetic field orientation [127]. On the other hand, PbTe exhibits a strong dependence of the spectra on the magnetic field orientation, which indicates an appreciable anisotropy of the effective masses [449]. Values of the reduced effective mass m^*_{np}, the transverse components of the effective electron and hole masses $m^*_\perp$, and the anisotropy coefficients K, are all given in Table 5.2. The effective masses and g factors are in satisfactory agreement with the values obtained from the Shubnikov – de Haas effect (cf. §5.2).

D. Recombination (Laser) Radiation in a Magnetic Field

The magneto-optical absorption of light is related to the influence of a magnetic field on the process which is the opposite of the absorption, i.e., recombination radiation. In particular, Butler et al., observed a shift of the wavelength of the stimulated recombination radiation (laser emission) of PbSe in a magnetic field (cf. §2.2D). They found that the photon energy in the laser emission peak increases proportionally to the magnetic field at a rate of $7.1 \cdot 10^{-8}$ eV/Oe. Such an increase in the photon energy is in agreement with the results deduced from the data on the fundamental absorption in a magnetic field and confirms that the observed laser emission peak is due to direct transitions between the conduction and valence bands.

A magnetic field of about 10-kOe intensity was found to reduce the threshold current density of a semiconductor injection laser from 2000 to 500 A/cm^2. Further increase of the magnetic field intensity produced a slow rise of the threshold current.

Recombination radiation and photoconductivity peaks, associated with multiphoton transitions between the Landau levels, were observed by Button et al. [538] in magnetic fields up to 100 kG at photon energies $\hbar \omega < \mathscr{E}_g$. In contrast to the one-photon transitions, the selection rule for the two-photon transitions was of the form $l_n - l_p = \pm 1$. The forbidden band width and the reduced mass found in this investigation were close to the values obtained from the magneto-optical measurements in the fundamental absorption region [450].

5.5. Low-Frequency Oscillations in a Magnetic Field

In this section we shall consider the interaction of lead chalcogenides with electromagnetic radiation in the radiofrequency range, the appearance of magnetoplasma waves, and the absorption of ultrasound in a magnetic field.

A. Cyclotron Resonance

The resonance absorption of microwaves in the frequency range close to cyclotron resonance has been investigated in n- and p-type PbTe crystals at low temperatures [451-456]. A sample was placed in a resonator located at the end of a waveguide. Electromagnetic waves of $7 \cdot 10^{10}$-cps frequency were used and the applied magnetic field was varied from zero to several kilo-oersteds. As a rule, the magnetic field was directed parallel to the sample surface; measurements were also made with the magnetic field perpendicular to the sample surface [451, 452] and inclined to the sample surface [454]. An electromagnetic wave was propagated normally to the surface, which coincided with a (100) crystallographic surface; a sample with a surface parallel to a (110) plane was also investigated [451]. The electric field vector of the wave was either parallel or perpendicular to the static magnetic field.

Stiles et al. [451] carried out such measurements and obtained a series of the absorbed power maxima as a function of the magnetic field intensity. Applying the theory propounded in [457], Stiles et al. [451] concluded that the absorption maxima were due to the Azbel'–Kaner cyclotron resonance [458], in spite of the fact that, according to their estimates, the cyclotron orbit radius was less than the depth of the skin-effect layer.

It is usually assumed that the Azbel'–Kaner cyclotron resonance is observed under the anomalous skin-effect conditions. If a magnetic field is directed parallel to the surface, carriers move along helical orbits and periodically cross the skin-effect layer where they interact with an electromagnetic wave. If the cyclotron frequency at which carriers are revolving is a multiple of the electromagnetic radiation frequency, the carriers are accelerated by the electric field of the wave and the wave energy is absorbed. Thus, the Azbel'–Kaner cyclotron resonance should be observed at frequencies $\omega = l\omega_c$ ($l = 1$ represents the first harmonic; $l = 2$, 3, ... represents higher harmonics).

Stiles et al., studied the dependence of the positions of the resonance maxima in a p-type PbTe sample with a hole density of $3 \cdot 10^{17}$ cm^{-3} on the angle of rotation of a magnetic field in a (100) plane; they confirmed the <111> model for the valence band of PbTe and obtained values of the effective hole masses ($m^*_{\perp p} = 0.029 m_0$, K = 14.5). Apart from the maxima corresponding to the cyclotron

Sec. 5.5] LOW-FREQUENCY OSCILLATIONS IN A MAGNETIC FIELD   251

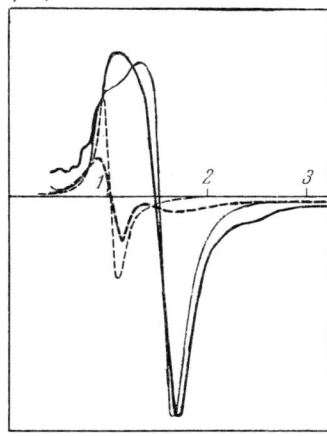

Fig. 5.14. Derivative of the absorbed microwave power, with respect to the magnetic field, plotted as a function of the magnetic field intensity applied to n-type PbTe [459]. The thick curves represent the experimental data; the thin curves represent the theoretical dependences for $\omega\tau = 9$. The continuous curves correspond to an ordinary ray, polarized parallel to the magnetic field; the dashed curves represent an extraordinary ray, polarized at right angles to the magnetic field. The magnetic field was directed along the [100] axis.

frequencies of holes in ellipsoids, Stiles et al., also found maxima which they attributed to almost-isotropic heavy and light holes.

Nii [454] measured the cyclotron absorption in n- and p-type samples of PbTe; he found that the shape of the curve describing the anisotropy of the cyclotron resonance of electrons was identical with the shape of the curves obtained for holes, including isotropic heavy and light holes. The same results was obtained for n-type PbTe by Hansen and Cuff [453]. The existence of such a complex and similar structure in the valence and conduction bands seems unlikely. Moreover, parallel and perpendicular orientations of the electric vector of the wave with respect to the magnetic field yield different dependences of the absorbed power A on the magnetic field (Fig. 5.14), while in Azbel'–Kaner resonance the curves should be similar.

The theory of cyclotron resonance under the classical skin-effect conditions, developed by Nii [454] and by Numata and Uemura [459] for the many-ellipsoid <111> model, makes it possible to explain the observed dependences without assuming the existence of isotropic bands. This theory is based on Maxwell's equations and classical equations of electron motion, which are used to deduce expressions for the complex refractive index in a magnetic field. A special feature of lead chalcogenides, which distinguishes them from materials (such as germanium) investigated earlier by the

252 MAGNETIC PROPERTIES [CHAP. V

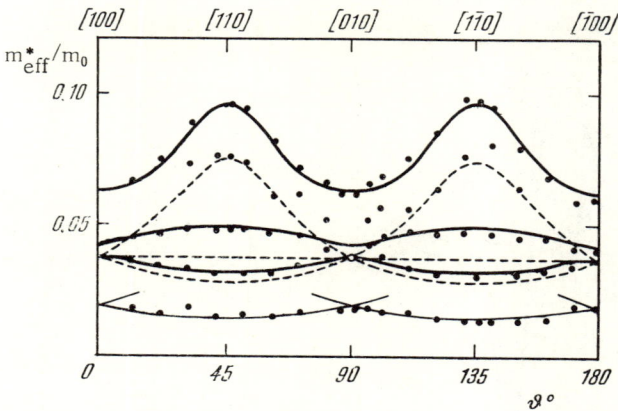

Fig. 5.15. Effective masses determined from the values of the magnetic field in which dielectric anomalies, hybrid resonance, and Azbel' – Kaner resonance were observed in n-type PbTe [455]. The magnetic field was rotated in a (100) plane. The theoretical curves for the dielectric anomalies are continuous and thick; those for the hybrid resonance are dashed; the Azbel' – Kaner resonance curves are continuous and thin; the experimental data are represented by points.

cyclotron resonance method, is the relatively high free-carrier density retained down to the lowest temperatures (1.8°K in Nii's experiments [454]). This effect is called the magnetoplasma cyclotron resonance [455].

The theory predicts that, when the magnetic field is directed along the [100] axis, an electromagnetic wave is propagated in the crystal in the form of an ordinary wave, whose electric vector is aligned along the magnetic field, and an extraordinary wave, whose electric vector is perpendicular to the magnetic field. These two waves have different values of the complex refractive index $\mathcal{N}^*$, which can be used to represent the absorption and reflection coefficients (cf. §2.1A). Figure 5.14 shows the dependence of the absorbed power on the magnetic field for the ordinary and extraordinary waves in the n-type PbTe. This figure includes the experimental curves and the theoretical dependences, obtained by solving the equation for the complex refractive index. The best agreement between the theoretical and experimental curves is ob-

Sec. 5.5] LOW-FREQUENCY OSCILLATIONS IN A MAGNETIC FIELD 253

tained for $\omega\tau = 9$ (τ is the relaxation time). The values of the effective masses used in the calculations were obtained from an analysis of the cyclotron absorption anisotropy.

A theoretical analysis shows that the absorption peaks occur at the same values of the magnetic field at which the square of the complex refractive index vanishes, i.e., we observe what is known as the dielectric anomaly. In the $\tau \to \infty$ case, the absorption is observed only in magnetic fields weaker than that in which $\mathcal{N}^{*2}=0$; in stronger magnetic fields, the electromagnetic wave is not absorbed. When the magnetic field intensity is reduced, the absorption also decreases, vanishing at the value of the magnetic field at which $\mathcal{N}^{*2}=\infty$. At this point, as well as at the point where $\mathcal{N}^{*2}=0$, the derivative of the absorbed power with respect to the magnetic field becomes infinite and peaks of the dA/dH curve should be observed. The positions of the peaks for various orientations of the magnetic field are shown in Figs. 5.15-5.18. The ordinates in these figures give the effective masses defined by the formula

$$m^*_{\text{eff}} = \frac{eH_0}{\omega c},$$

where H_0 are the values of the magnetic field corresponding to the peaks of the absorption derivative.

Figure 5.15 shows the dependence of the effective masses on the angle of rotation of the magnetic field in a (100) plane. The continuous thick curves are obtained by calculation of the dielectric anomalies. These curves do not intersect at any point. The experimental points, representing the strong absorption peaks, are very close to the curves calculated from the dielectric anomalies. In these calculations it is assumed that $\tau = \infty$ and the effective electron masses ($m^*_{\perp n} = 0.024 m_0$, $K = 9.8$) are obtained from the condition of the best agreement between the theoretical and experimental curves at two points. These values of the effective masses are used to calculate the remaining curves and the curves for other orientations of the magnetic field.

The dashed curves represent the angular dependences of the effective masses for which $\mathcal{N}^{*2} = \infty$. In contrast to the curves corresponding to $\mathcal{N}^{*2} = 0$, the dashed curves intersect at several points. When the magnetic field rotates in a (100) plane, two of

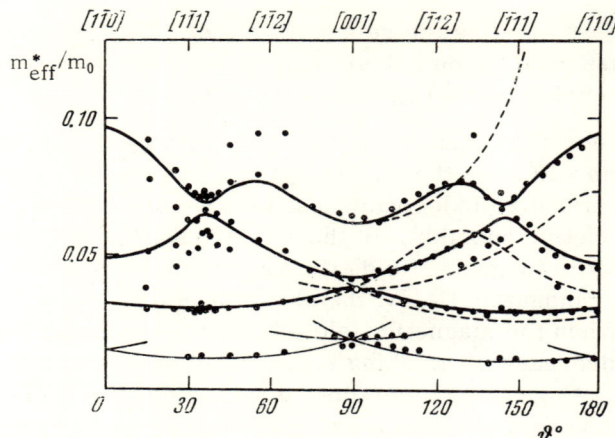

Fig. 5.16. The effective masses determined as in Fig. 5.15 for n-type PbTe in a magnetic field rotated in a (110) plane.

these curves coincide with the curves for the cyclotron masses and the third lies between them. At the points where, according to the theoretical calculations, $\mathcal{N}^{*2} = \infty$, there is a weak structure. This structure is attributed to a hybrid resonance, which is explained as follows.

When a sample contains groups of carriers with two different cyclotron masses, they interact strongly at a frequency intermediate between the cyclotron frequencies of these carriers. When an electromagnetic wave interacts simultaneously with both types of carrier a hybrid resonance is observed. For an arbitrary direction of the magnetic field, the electrons in all four ellipsoids have, in general, different cyclotron masses and three intermediate hybrid frequencies are observed. If two or more ellipsoids are oriented in the same way relative to the magnetic field direction, i.e., if two (or more) cyclotron frequencies coincide, the hybrid resonance is observed at this coincidence frequency. In the case shown in Fig. 5.15, we observe such coincidence of the hybrid and cyclotron frequencies.

There are also three dielectric anomalies $\mathcal{N}^{*2} = 0$ when there are four ellipsoids. At frequencies corresponding to the dielectric anomalies, the orbits of two rotating electrons, belonging to different ellipsoids, have the same eccentricity and they rotate in the

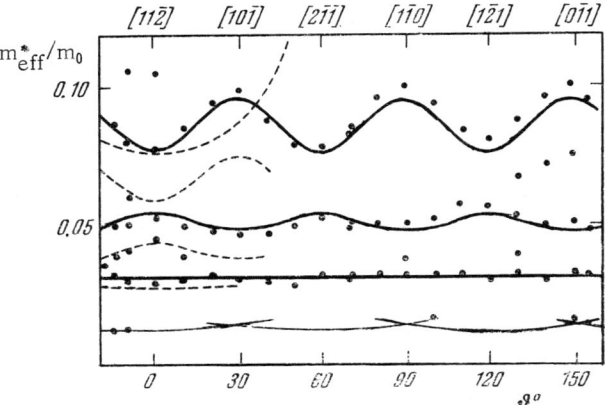

Fig. 5.17. Effective masses determined as in Fig. 5.15 for n-type PbTe in a magnetic field rotated in a (111) plane.

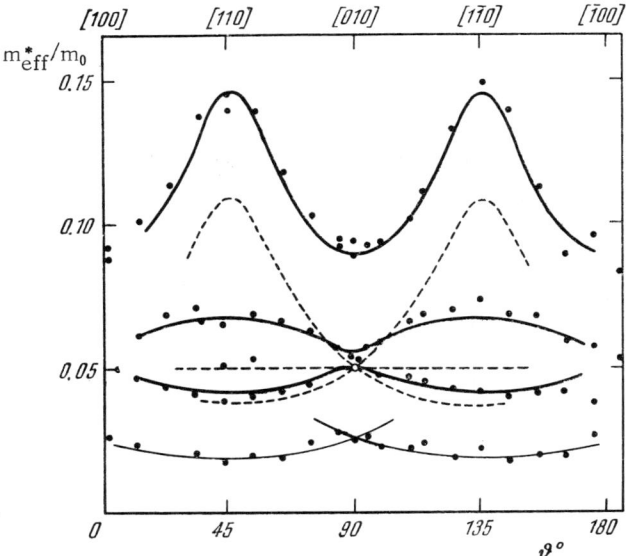

Fig. 5.18. Effective masses determined as in Fig. 5.15 for p-type PbTe in a magnetic field rotated in a (100) plane.

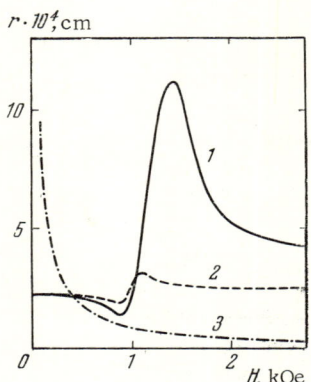

Fig. 5.19. Dependence of the depth of the skin layer and of the radius of the cyclotron orbit on a magnetic field parallel to the [100] direction, observed for PbTe at $\omega\tau = 9$ [459]: 1) depth of the skin layer for an ordinary wave; 2) depth of the skin layer for an extraordinary wave; 3) radius of the cyclotron orbit.

same sense as the electric vector in the medium.

The structure observed in a part of the absorption curve corresponding to weak magnetic fields cannot be explained by the classical skin effect theory and is attributed to the second harmonic of the Azbel'–Kaner cyclotron resonance. The angular dependence of an effective mass which is equal to half the cyclotron mass is shown by a thin continuous line.

Figure 5.16 shows the angular dependence of the effective masses in the case when the magnetic field rotates in a (110) plane. In addition to the three dielectric anomalies and the second harmonic of the Azbel'–Kaner resonance, we observe resonance at one cyclotron frequency and two hybrid frequencies. Another curve, corresponding to $\mathscr{N}^{*2} = \infty$, appears because the magnetic field is not perpendicular to the direction of incidence of the electromagnetic wave. The shape of this curve is governed only by the relative directions of the magnetic field and wave vector of the incident wave and is independent of the orientation of the magnetic field relative to the crystallographic axes.

At values of the magnetic field corresponding to this curve, we observe an absorption which increases monotonically with increasing magnetic field intensity.

When the magnetic field rotates in a (111) plane the curves obtained are similar (Fig. 5.17). In this case, we have three hybrid frequencies, which lie between four cyclotron frequencies.

All these curves were obtained for a sample with an electron density of $2 \cdot 10^{17}$ cm^{-3}. Similar results were obtained for p-type PbTe with the same carrier (hole) density (cf., for example, Fig.

5.18). The values obtained for the effective hole mass and the anisotropy coefficient are: $m_{\perp p}^{*} = 0.031 m_0$, $K = 12.2$.

Fujimoto [456] investigated cyclotron resonance in n-type PbTe and found that the transverse effective mass increased by 17% when the electron density was increased from $1.23 \cdot 10^{17}$ cm^{-3} to $3 \cdot 10^{18}$ cm^{-3}, but the anisotropy coefficient decreased so that the longitudinal effective mass remained constant.

A cylindrical resonator with a high value of Q was used in an investigation of the anisotropy of this effect because the positions of the maxima in the dA/dH curve are important. Comparison of the theoretical shape of the dA/dH curve was made with the experimental curves obtained using a resonator, which ensured that the polarization of the radiation was linear [455].

The depth of penetration of an electromagnetic wave into a sample was calculated as a function of the magnetic field and compared with the radius of the cyclotron orbit. The calculated results, presented in Fig. 5.19, show that, in those intervals of the magnetic fields where the dielectric anomalies and hybrid resonance are observed, the orbit radius is considerably less than the depth of penetration of the magnetic field. The relationship is reversed in weak magnetic fields and this makes it possible to observe the Azbel' – Kaner cyclotron resonance in such fields.

Numata and Uemura [459] considered theoretically cyclotron resonance under uniaxial deformation which produced a relative shift of the equivalent (in the absence of deformation) extrema and resulted in a transfer of carriers from some extrema to others. No experimental investigations of cyclotron resonance under deformation have yet been carried out.

B. Reflection of Microwaves in a Strong Magnetic Field

The reflection of 4- and 8-mm radio waves from PbTe samples was investigated by Sawada et al. [460] at 4.2°K in pulsed magnetic fields up to 150-kOe intensity. A high cyclotron frequency, exceeding considerably the frequency of electromagnetic waves ($\omega_c \gg \omega$), was found to correspond to high values of the magnetic field. The magnetic field was applied either along the direction of propagation

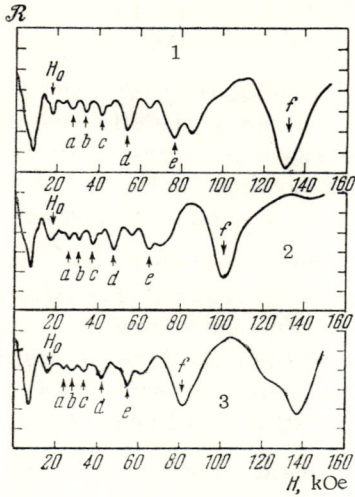

Fig. 5.20. Microwave reflection coefficient as a function of the magnetic field applied to a p-type PbTe sample, 0.94 mm thick, with a hole density of $5 \cdot 10^{17}$ cm^{-3} [460]. The magnetic field was directed parallel to the line of propagation of microwaves. 1) $\omega = 66$ Gc; 2) $\omega = 69.0$ Gc; 3) $\omega = 71.2$ Gc.

of radio waves or at right-angles to this direction, with the electric vector of the wave perpendicular to the magnetic field. One of the aims of this investigation was to determine the static permittivity ε_0.

It is evident from Fig. 5.20 that the reflection coefficient plotted as a function of the magnetic field intensity exhibits several minima. The minima denoted by a-f are attributed to interference effects ("geometric resonance") since the distances between the minima in a given magnetic field are proportional to the thicknesses of the samples. These minima shift in the direction of weaker magnetic fields when the frequency of the radio waves is increased. An analysis of the positions of the interference minima yields $\varepsilon_0 \approx 1 \cdot 10^4$ at a carrier density of $5 \cdot 10^{17}$ cm^{-3} and $\varepsilon_0 = 0.3 \cdot 10^4$ at a carrier density of $1 \cdot 10^{17}$ cm^{-3}.

Using the values obtained for ε_0, Sawada et al. calculated the critical magnetic field in which the real component of the permittivity vanishes at a given frequency and gives rise to a reflection minimum. The value of such a critical magnetic field is

$$H_0 = \frac{4\pi nec}{\omega \varepsilon^0(\omega)}, \qquad (5.28)$$

where $\varepsilon^0(\omega)$ is the permittivity at a given frequency in the absence of free carriers. This quantity is governed by the interband electron transitions and by the lattice polarization. At the frequency used in these experiments, $7 \cdot 10^{10}$ cps, the value of $\varepsilon^0(\omega)$ is close to the static permittivity $\varepsilon^0(\omega) \approx 0.9\varepsilon_0$. The critical magnetic field at a carrier density of $5 \cdot 10^{17}$ cm^{-3} is $H_0 = 20$ kOe. In this field, a reflection minimum is indeed observed; the position of this minimum is independent of the sample thickness and, in agreement with the theory, it is proportional to the carrier density.

The values of the static permittivity obtained by this method depend strongly on the carrier density and differ from the values determined by measuring the p−n junction capacitance (§3.4B), by neutron spectroscopy, and from infrared absorption (cf. §2.1D). The cause of the discrepancies between these values is not known.

C. Helicon Waves

Helicon waves are a variety of plasma waves which may appear in an uncompensated plasma present in a metal or in an extrinsic semiconductor with a high carrier density [461, 462]. They are plane, circularly polarized waves of the electromagnetic type, propagated along the magnetic field direction. Helicon waves are related to the high-frequency Hall current. Their frequency is given by the formula

$$\omega = \frac{c^2}{4\pi} k k_H H R. \qquad (5.29)$$

Here, k and k_H are the absolute value and the projection along the magnetic field direction of the wave vector **k**, which assumes values given by

$$k_i = \frac{n_i \pi}{a_i} \qquad (5.30)$$

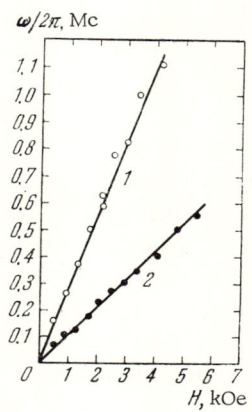

Fig. 5.21. Dependence of the frequency of helicon waves on the intensity of a magnetic field applied to PbTe [463]: 1) magnetic field along the thickness of a plate; 2) magnetic field along the width of the plate.

(n_i are arbitrary integers; the subscript i is used to number the axes parallel to the edges of a plate: i = 1, 2, 3; a_i are the dimensions of the plate); R is the Hall coefficient.

The frequency of helicon waves is $\omega \ll \omega_c$. In order to observe helicon waves, we must satisfy the usual condition $\omega_c \tau \gg 1$. In view of the high mobility in lead chalcogenides at liquid-helium temperature, this condition is quite easy to satisfy.

Helicon waves were observed by Kanai [463] in n- and p-type PbTe at 4.2°K. In these investigations, Kanai used a static magnetic field of the order of several kilo-oersteds. He found that rapid variation of the current in a crystal may excite helicon waves. The investigation was carried out using two coils within which a sample had been placed. In the oscillator method for observing helicons, a voltage pulse was applied to the primary coil. The secondary coil then exhibited damped oscillations which were recorded with an oscillograph. Helicon waves decay rapidly in pure samples and, therefore, it is more convenient to use a resonance method. In this method the voltage is varied sinusoidally and the excited oscillations resonate at the helicon frequency.

Figure 5.21 shows the frequency of helicon waves as a function of the magnetic field intensity. One line represents the case when the field is parallel to the shortest edge of a plate and the other when the magnetic field is parallel to an edge 2.2 times longer. The ratio of the frequencies of helicons for the two directions of the magnetic field is 2.6, which is close to the ratio of the width of the sample and its thickness, in agreement with the theoretical predictions. The proportionality of the helicon frequency to the magnetic field follows from Eq. (5.29). The decay time of helicon waves, which is inversely proportional to the resistivity of a sample, is also in agreement with the theory.

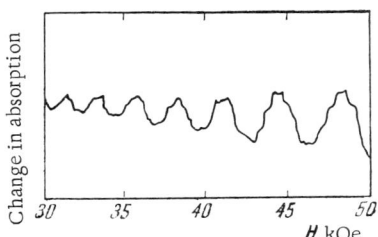

Fig. 5.22. Variation of the absorption of 57 Mc longitudinal ultrasonic waves in p-type PbTe in a magnetic field at 4.2°K [464].

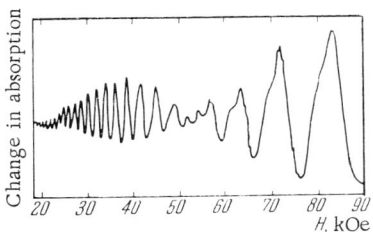

Fig. 5.23. Variation of the absorption of 43 Mc transverse ultrasonic waves in p-type PbTe in a magnetic field at 1.8°K [464].

Measurements of the frequency of helicon waves can be used to determine the Hall coefficient and, hence, the carrier density. This method of determination of the carrier density was employed in [460]. Helicon waves in PbTe were also investigated by others [539-541].

D. Oscillations of Ultrasound Absorption

An acoustical wave in a solid produces electric fields which interact with the carriers basically in the same way as an electromagnetic wave. A characteristic feature of an acoustical wave is its low velocity, which allows us to assume that the generated electric field is static and that it oscillates in space. The decay of an acoustical wave due to the interaction with the carriers is represented, like the decay of an electromagnetic wave, by the conductivity $\sigma(\omega)$. When a quantizing magnetic field is applied, this quantity reaches a maximum for those values of the magnetic field at which the Fermi level coincides with the bottom of one of the Landau bands; when the magnetic field intensity is varied, the absorption of ultrasound oscillates with a definite period when plotted against $1/H$.

De Haas – van Alphen oscillations in the absorption of ultrasound were investigated by Shapira, Lax, and Beckman [464, 465] in PbTe and PbSe at liquid-helium temperatures in magnetic fields up to 130 kOe. Oscillations of the absorption of the longitudinal and transverse ultrasonic vibrations were observed (Figs. 5.22 and 5.23). These oscillations are periodic in $1/H$. In fields $H > 45$ kOe, the oscillations of the transverse ultrasound absorption had a complex structure [464]. The effects of the spin splitting of the electron levels were observed [465]. The oscillations were stronger at high frequencies (of the order of several tens of megacycles) and low temperatures (down to 1.4°K).

The ultrasound was propagated in n- and p-type PbTe samples [464] along a magnetic field parallel to the [100] axis. In p-type PbSe [465], the effect was measured for various directions of the magnetic field. These measurements demonstrated that the valence band of PbSe can be described by four ellipsoids located at the edge of the Brillouin zone and elongated along the <111> directions. The dependence of the oscillation period on the orientation of the magnetic field relative to the crystallographic axes, makes it possible to find the anisotropy coefficient of the effective mass of holes, which is $K = 2.1$ for a sample with a hole density of $7 \cdot 10^{18}$ cm^{-3}, in agreement with the results obtained from the magnetoresistance oscillations [439]. In a PbSe sample with a higher hole density ($3.4 \cdot 10^{19}$ cm^{-3}), some deviation of the Fermi surface from the ellipsoidal shape was observed; this indicates that the longitudinal effective mass depends in a different manner on energy than the transverse mass (cf. §6.2B).

Giant quantum oscillations of the absorption of ultrasound, periodic in $1/H$, were observed in PbTe by Nanney [525].

CHAPTER VI

BAND STRUCTURE AND SCATTERING MECHANISMS (THEORY AND CONCLUSIONS FROM EXPERIMENTAL DATA)

6.1. General Energy Band Structure

The first calculation of the energy band structure of one of the chalcogenides, PbS, was carried out by Bell et al. [466] at a time when the experimental data on the properties of lead chalcogenides were scant and the calculation methods were imperfect (Bell et al. [466] used the method of cells without relativistic corrections). The results of their calculation were contradicted by later experimental data. An energy band structure which is in agreement with experiment was derived theoretically only fairly recently [467-471], after the accumulation of a considerable amount of experimental data on the band structure of all three lead chalcogenides.

We shall start with a list of the principal experimental results which have been used to make the theoretical calculations and to check the conclusions drawn from the theories. The relevant experiments are described in detail in the preceding chapters, where references to the original work are given.

A. Interpretation of the Band Structure from Experimental Data

1. Lead chalcogenides are semiconductors with nonoverlapping conduction and valence bands.

2. Investigations of the optical absorption show that the top of the valence band and the bottom of the conduction band of all three

compounds are are the same point in the **k** space. This conclusion is confirmed by magneto-optical measurements, investigations of recombination spectra, observations of the laser effect, and data on the tunnel effect in p – n junctions.

3. From the same experimental data it follows that the matrix element of the momentum of states corresponding to the conduction and valence band edges is not equal to zero.

4. Measurements of the magnetoresistance, piezoresistance, tunnel effect in a magnetic field, cyclotron resonance, and optical absorption in a magnetic field indicate that the valence and conduction bands of PbTe are described by a many-ellipsoid model with the ellipsoids strongly elongated along the <111> directions.

5. Observations of oscillations of the magneto-resistance (Shubnikov – de Haas effect) in all three n- and p-type compounds and oscillations of the magnetic susceptibility (de Haas – van Alphen effect) in p-type PbTe and n-type PbS show that the carrier density in one extremum is four times smaller than the total density determined by the Hall method. This means that, not only in PbTe but also in PbSe and PbS, the conduction and valence bands are described by a many-ellipsoid model with the extrema located along the Brillouin zone boundary along the <111> directions (point L). This conclusion is confirmed by investigations of the quantum oscillations of the absorption of ultrasound in a strong magnetic field, by measurements of the shift of the fundamental absorption edge with increasing carrier density (due to degeneracy), and by the comparison of the density-of-states effective mass with the conductivity effective mass.

6. The anisotropy coefficient of the effective masses, determined from measurements of the magneto-resistance, magneto-optical absorption, cyclotron resonance, Shubnikov – de Haas and de Haas – van Alphen effects, decreases approximately by an order of magnitude between PbTe and PbS. Electrons and holes in PbS have almost-isotropic effective masses.

7. Almost all the properties of lead chalcogenides are very similar for the three compounds, which indicates that their band structures are also similar.

8. The properties of n- and p-type materials are also similar. Hence, it follows that the edges of the conduction and valence

bands have almost the same structure. In particular, the effective electron and hole masses have similar values.

9. Various measurements have made it possible to find the numerical values of the band structure parameters, their temperature dependences, and the changes occurring during deformation. These values are given partly in Appendix C and partly in the present chapter where they bear on the comparisons of the experimental results with the theoretical calculations. The forbidden band width (energy gap) has been obtained from investigations of the fundamental absorption edge, magneto-optical measurements, photoconductivity, photomagnetic effect, photovoltaic effect in a p − n junction, recombination radiation, temperature dependence of the Hall effect, electrical properties under pressure, and the thermal conductivity due to ambipolar diffusion. The most accurate values of the forbidden band width at low temperatures have been obtained by measurements of the optical absorption in strong magnetic fields. A study of the ultraviolet reflection and of the photoelectric emission has yielded energy gaps between bands located far from the Fermi level. The values of the effective electron and hole masses have been obtained from investigations of the thermoelectric power, Nernst − Ettingshausen effect, thermoelectric power in a magnetic field, cyclotron resonance, infrared reflection, fundamental absorption in a magnetic field, and Shubnikov − de Haas, de Haas − van Alphen, and Faraday effects. The spectroscopic splitting factors have been deduced from the magneto-optical absorption and the Shubnikov − de Haas effect.

10. The forbidden band width depends nonmonotonically on the atomic number of group VI elements: PbSe has the smallest forbidden band and PbS the largest. The forbidden band width of PbTe − PbSe and PbSe − PbS alloys is intermediate between the values of the forbidden band widths of the pure substances used to make these alloys.

11. The Shubnikov−de Haas effect in all the investigated materials, the thermoelectric power of n- and p-type PbTe in strong magnetic fields, the Nernst − Ettingshausen effect of n-type PbTe, the magnetic susceptibility of n- and p-type PbSe, the magnetoacoustical oscillations in p-type PbSe, and the cyclotron resonance in n-type PbTe all indicate a considerable nonparabolicity of the conduction and valence bands.

12. Measurements of the Shubnikov–de Haas effect show that the longitudinal effective masses of PbSe and PbS depend on the energy in the same way as the transverse masses (the mass anisotropy coefficient is independent of the energy), while in PbTe the longitudinal effective mass depends on the energy less strongly than the transverse mass (the anisotropy coefficient decreases with increasing energy). Other experiments yield somewhat different results. For example, it follows from measurements of cyclotron resonance that the energy dependence of the longitudinal effective electron mass in PbTe is negligibly weak, and the magnetoacoustical oscillations indicate that the nature of the anisotropy of the valence band of PbSe varies with energy.

13. The dependence of the transverse effective masses on the energy becomes weaker along the series PbTe, PbSe, PbS.

14. The longitudinal effective mass decreases and the transverse effective mass increases from PbTe to PbS and the relative increase in the transverse mass is greater than that in the forbidden band width.

15. The relative changes in the forbidden band width and effective masses under pressure are approximately equal.

16. The longitudinal g factors of lead chalcogenides are considerably greater than 2, as indicated by the Shubnikov–de Haas effect and by the optical absorption in a magnetic field. The spin splitting of the Landau levels is approximately equal to half the separation between these levels.

17. The Knight shift of p-type PbTe indicates that the electron wave function at the top of the valence band has a nonzero density near Pb atoms.

18. Investigations of the fundamental absorption edge and of stimulated recombination radiation (laser emission) of $Pb_xSn_{1-x}Te$ solid solutions have indicated that the addition of SnTe to PbTe reduces the forbidden band width, and extrapolation of the dependence of the width on the composition shows that, at some composition, the conduction and valence band edges should merge.

19. Measurements of the mobility in alloys of these $A^{IV}B^{VI}$ compounds indicate that the electrons in the valence band are concentrated near chalcogens to a greater degree than the electrons in

Sec. 6.1] GENERAL ENERGY BAND STRUCTURE 267

the conduction band. Moreover, it is found that impurities occupying Pb sites in the lattice reduce the electron mobility more strongly than the impurities occupying chalcogen sites.

20. Measurements of the Hall effect, thermoelectric power, and infrared reflection indicate the existence of a second valence band in PbTe; this band has a larger effective mass and at low temperatures it is located about 0.1 eV below the edge of the upper (light-hole) band, whereas at high temperatures it lies higher than the edge of the light-hole band. A nonlinearity of the temperature dependence of the forbidden band width of PbTe is attributed to the presence of two valence bands.

B. Methods for Calculating Band Structure. Relativistic Corrections and Spin—Orbit Interaction

In this subsection, we shall summarize the information on the methods of calculating the energy band structure of a solid (including group-theoretic methods), which is necessary in order to understand the results of the calculations of the band structure of lead chalcogenides. We shall make this summary as simple as possible without attempting to be rigorous or completely general. Readers acquainted with the group theory and with methods for calculating the energy-band structure should ignore this subsection.

The energy levels and wave functions of any system are calculated using the Schrödinger equation:

$$\mathcal{H}\psi = \mathcal{E}\psi. \qquad (6.1)$$

We shall consider the one-electron problem, neglecting at first the electron spin, i.e., assuming ψ to be a function of three space coordinates r_i (i = 1, 2, 3).

The systems to be investigated, including crystal lattices, always have some symmetry, i.e., there is a set of displacements (rotations, reflections, translations, etc.) which can make a system coincide with itself. Each such displacement of the system corresponds to an orthogonal transformation of the coordinate system, relative to which the Hamiltonian of the system is invariant, i.e., the form of the Hamiltonian operator $\mathcal{H}$ in the new coordinates is identical with its form in the old coordinates. All possible symmetry transformations of a system form a group G.* We note that the group G always includes E, which is the identity transformation $r'_i = r_i$, corresponding to the absence of any displacement in the system.

A group-theoretic analysis of the symmetry and its consequences is one of the most important stages in the calculations concerned with such a complex system as a crystal. The group-theoretic methods cannot be used to find the energy levels of the system but they give absolutely reliable information on the multiplicity of degeneracy of all the possible states, splitting of the energy levels due to perturbation, symmetry of the wave functions, selection rules for the matrix elements, some quantitative relationships between the matrix elements, etc.

Let us assume that there are n linearly independent functions ψ_l ($l = 1, 2, \ldots, n$) corresponding to an n-fold degenerate level $\mathscr{E}$. After each transformation of the coordinates, we obtain, from the group G, new functions $\psi_{l'}$. Since the form of the Schrödinger equation is not affected by these coordinate transformations, the functions $\psi_{l'}$ are solutions of this equation for the same value of the energy $\mathscr{E}$, as the functions ψ_l. Consequently, each function $\psi_{l'}$ is a linear combination of functions ψ_l:

$$\psi_{l'} = \sum_{l=1}^{n} D_{l'l} \psi_l. \tag{6.2}$$

Each transformation in the group G corresponds to a particular matrix $D_{l'l}$. The complete set of n-dimensional $D_{l'l}$ matrices represents the irreducible representations of the group G and the set of n functions ψ_l is the basis of the irreducible representation. Any wave functions belonging to one level are transformed in accordance with any of the irreducible representations of the total symmetry group of the system.† Conversely, the eigenfunctions of the Hamiltonian which are transformed in accordance with one of the irreducible representations of the symmetry group belong to the same energy level.

---

*The formal mathematical definition of a group is given in several reviews and monographs [472-474], etc.

†Accidental degeneracy, not associated with any symmetry properties, will not be considered.

Sec. 6.1]   GENERAL ENERGY BAND STRUCTURE

By way of example, we shall consider the one-electron state of an atom. In this case, the Hamiltonian has completely spherical symmetry, i.e., it is invariant under rotation through any angle about an arbitrary axis passing through the center. The wave function, which depends only on the radius vector r (the s function) is also spherically symmetrical, i.e., in all the transformations of the symmetry group of the system this function retains its form. Therefore, all the matrices of an irreducible representation whose basis is the s function are identical, one-dimensional, and equal to unity. Such a representation is known as unitary. If a symmetry group were to have only the unitary representation, the wave functions would have the same symmetry as the Hamiltonian and the symmetry would not result in a degeneracy of energy levels.

Let us now consider p functions. One of the p-functions is $r_1 f(r)$, where f depends only on the radius vector. Such a function has a much lower symmetry than the Hamiltonian. It does not change its form only in the case of rotation about the r_1 axis. Rotations transforming the r_1 axis into r_2 or r_3 axes transform the function $r_1 f(r)$ into two other p-functions, $r_2 f(r)$ and $r_3 f(r)$, which are also solutions of the problem for the same value of the energy, i.e., the energy level considered is degenerate. Other linearly independent functions are not generated by rotation and, therefore, these three functions can be used as the basis of a three-dimensional representation, corresponding to a triply degenerate level.

As the basis of a representation, we can select linear combinations of wave functions belonging to different levels. In this case, we obtain a reducible representation — so called because the basis functions can be combined in such a way that the representation assumes the quasidiagonal form:

$$D = \begin{pmatrix} D_1 & 0 & \cdots & 0 \\ 0 & D_2 & \cdots & 0 \\ \cdots & \cdots & \cdots & \cdots \\ 0 & 0 & & D_m \end{pmatrix}, \qquad (6.3)$$

where D_i are the matrices of the irreducible representations. A reducible representation can thus be separated into irreducible representations.

Reducible representations appear in the analysis of the splitting of levels by a perturbation which reduces the symmetry of the system. Let us consider functions which form the basis of an irreducible representation of the symmetry group of the original high-symmetry Hamiltonian and use these functions as the basis of a representation of the symmetry group of a new Hamiltonian. The second group is a subgroup of the first and contains fewer elements than the first group. Therefore, a representation of the second group contains fewer matrices and these matrices may be reduced to the quasidiagonal form for a certain set of the wave functions of the basis which we can combine in an arbitrary manner. A representation of this type thus splits into irreducible representations of lower dimensions, and this process corresponds to the splitting of a level into a series of sublevels with lower degeneracy multiplicities.

Without knowing the form of the wave functions but knowing the representation,* in accordance with which they are transformed, we can obtain important information on the wave functions and the energy spectrum. The dimensions of the matrices of irreducible representations indicate the multiplicity of the level degeneracy. The expansion of reducible representations into irreducible forms is directly related to the splitting of the energy levels under the action of a perturbation. The law of transformation of the functions (6.2) is an algebraic expression of the symmetry of wave functions, i.e., if we know the representations, we know the symmetry properties of the wave functions. The symmetry of the wave functions governs the selection rules of the matrix elements and some relationships between those matrix elements which do not vanish. The fundamental theorem on the selection rules is formulated, for example, in [472]; we shall give here a simple example.

Let us assume that a system has a center of inversion, i.e., it is symmetrical with respect to reflection at a point. We shall also assume that the inversion operation commutes with all the other elements of the group G, i.e., the result of a consistent application of inversion and any other symmetry transformation is independent of the order in which these two operations are carried

---

*It must be mentioned that, in fact, in all the operations with representations it is sufficient to know their characters, i.e., the sums of the diagonal elements (traces) of the matrices. Tables of characters of the representations of all the point groups of symmetry (which will be discussed later) are given, for example, in [474].

Sec. 6.1] GENERAL ENERGY BAND STRUCTURE 271

out. It is then found that the wave functions are either even or odd, i.e., they are either invariant under inversion or they reverse their sign. When the representation is known, all the transformation properties are known, including the nature of the change in the function under inversion, i.e., its parity. Let us assume that we are interested in a matrix element of some odd operator, for example, one of the momentum components

$$P_i = \int \psi_1^* \hat{p}_i \psi_2 d\mathbf{r}. \qquad (6.4)$$

We shall consider a specific case by assuming that the function ψ_2 is even and then $\hat{p}_i \psi_2$ is an odd function. In this case, the matrix element of Eq. (6.4) does not vanish only when ψ_1^* is an odd function. The matrix element of the momentum is thus equal to zero when the parity of the functions ψ_1^* and ψ_2 is the same, i.e., when the integrand $\psi_1^* \hat{p}_i \psi_2$ is not invariant under inversion.

The selection rules may be formulated in a more general form in the following way: the integral over all space is not equal to zero only when the representation used to transform the integrand splits into irreducible representations, one of which is the unitary representation. Naturally, in the example given here, the selection rule can be obtained quite easily without using representations but, in more complex cases, the group theory formalism facilitates the solution of the problem.

Let us consider an electron in the periodic field of a crystal. In this case, the potential energy U in the Hamiltonian

$$\mathcal{H} = \frac{\hat{p}^2}{2m_0} + U \qquad (6.5)$$

is invariant under translations by multiples of the lattice vector and under a particular set of rotations and reflections (the point group of a crystal) and combinations of point transformations with translations. The translational symmetry is taken into account by the fact that the wave functions have the form of the Bloch functions:

$$\psi = u_\mathbf{k}(\mathbf{r}) e^{i(\mathbf{k}\mathbf{r})}. \qquad (6.6)$$

The periodic function $u_\mathbf{k}(\mathbf{r})$ need be considered only within one unit cell. Its symmetry and the associated degeneracy of the energy bands at a point $\mathbf{k}$ are governed by the point group.* Substituting Eq. (6.6) into Eqs. (6.1) and (6.5), we obtain the following equation for the function $u_\mathbf{k}(\mathbf{r})$, corresponding to a given value of the wave vector $\mathbf{k}$:

$$(\mathcal{H} + \mathcal{H}_\mathbf{k}) u_\mathbf{k} = \mathcal{E}_\mathbf{k} u_\mathbf{k}, \qquad (6.7)$$

where $\mathcal{H}$ is the Hamiltonian of Eq. (6.5), which is independent of $\mathbf{k}$, and $\mathcal{H}_\mathbf{k}$ is an operator of the form

$$\mathcal{H}_\mathbf{k} = \frac{\hbar}{m_0}(\mathbf{k}\hat{\mathbf{p}}) + \frac{\hbar^2 k^2}{2 m_0}. \qquad (6.8)$$

Because of the presence of $\mathcal{H}_\mathbf{k}$ in Eq. (6.7), the effective Hamiltonian $\mathcal{H} + \mathcal{H}_\mathbf{k}$ is invariant only under those symmetry transformations of a crystal which do not alter the vector $\mathbf{k}$ or which do not alter it basically — for example, when this vector is altered into a reciprocal lattice vector. These transformations form the group of the vector $\mathbf{k}$. The functions $u_\mathbf{k}$ are thus transformed not in accordance with the irreducible representations of the point symmetry group of the crystal but in accordance with the irreducible representations of the group of the vector $\mathbf{k}$, which is also a point group. A consequence of the symmetry transformations contained in the group of the vector $\mathbf{k}$ is the degeneracy of bands at a given point $\mathbf{k}$. The presence of symmetry transformations of a crystal, which are not included in the group of the vector $\mathbf{k}$ and which transform the point $\mathbf{k}$ into an equivalent point of the Brillouin zone produces a degeneracy of a different type. All the vectors $\mathbf{k}$ obtained from one vector by these transformations are called the star of the vector $\mathbf{k}$. The energy levels at the points belonging to the same star coincide and the function $\mathcal{E}(\mathbf{k})$ has the full point symmetry of the crystal.

There are altogether 32 point groups, each of which has a finite number of different irreducible representations, and the parameters which represent these representations (they are called the characters) are known and tabulated. By way of example, let us

---

*For the sake of simplicity, we shall assume that the crystal has no screw axis or glide planes, as is the case in lead chalcogenide crystals.

Sec. 6.1]   GENERAL ENERGY BAND STRUCTURE   273

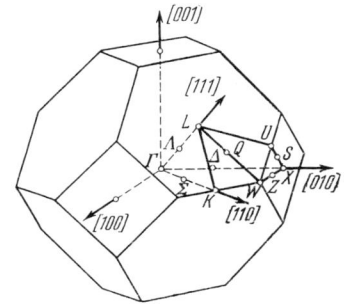

Fig. 6.1. Brillouin zone and symmetry points for an fcc lattice.

consider a crystal of the lead chalcogenide type. Such a crystal consists of two fcc sublattices. The Pb atoms are located at the sites of one of these sublattices, while the sites of other sublattices (shifted relative to the lead sublattice by half a principal diagonal of the cubic cell) contain the chalcogen atoms. The symmetry of this crystal is characterized by a group of all the symmetry transformations of a cube (m3m) or Q_h (in the Schönflies notation). The Brillouin zone and the symmetry points of the zone for an fcc lattice are shown in Fig. 6.1.

$\mathcal{H}_k = 0$ at $k = 0$ (point Γ in the Brillouin zone) and the symmetry group of the vector $k = 0$ (the group of the point Γ) is identical with the total point group of the symmetry of the crystal. There are 10 irreducible representations of this group which are denoted by the symbols Γ_i [474]. They include a one-dimensional unitary representation Γ_1 whose basis is a function corresponding to a nondegenerate level and which transforms into itself by the operations of the symmetry group of a cube. The surfaces $\psi = $ const of such a function have the cubic symmetry. This function is similar to the atomic s function.

The functions which are the basis of a three-dimensional representation Γ_{15} are similar to the atomic p functions. They can be selected so that operations of the symmetry group of a cube transform them as functions $r_i f(r)$ $(i = 1, 2, 3)$. For example, in the case of inversion $(r'_i = -r_i)$, all three functions reverse their sign, i.e., they are odd. A rotation by 120° about a principal diagonal of a cube, corresponding to a cyclic permutation of the coordinates, produces a corresponding permutation of the functions and the transformation matrix is of the form

$$\begin{pmatrix} 0 & 1 & 0 \\ 0 & 0 & 1 \\ 1 & 0 & 0 \end{pmatrix}.$$

The surfaces $\psi = \text{const}$ have a much lower symmetry than the cubic cell of a crystal. They have fourfold symmetry axes, coinciding with the r_i axis. The representation Γ_{15} applies to the transformations not only of the wave functions but also to the components of the momentum operator.

The symmetry group of a vector **k**, corresponding to the edge (boundary) of the Brillouin zone along the [111] direction (the L-point group), contains a smaller number of symmetry transformations and is denoted by the symbols ($\bar{3}$m) or D_{3d}. It is important to note that the L-point group includes the inversion operation. Inversion transfers each L-point to the opposite face of the Brillouin zone boundary in the **k** space, and the vectors which represent the position of the L-point before and after inversion differ by one reciprocal lattice vector.

Six representations of the L-point group are denoted by L_1, L_2, L_3, L_1', L_2', and L_3' [474]. The dimensions of these representations give the possible multiplicity of the level degeneracy at the point L of the Brillouin zone without allowance for the spin. The L_1, L_2, L_1', and L_2' representations are one-dimensional and the unitary representation L_1 correspond to a function which is invariant under the symmetry transformations of the group of the point L. The representations L_3 and L_3' are two-dimensional and correspond to doubly degenerate levels. The representations with unprimed subscripts correspond to even functions and those with primed subscripts correspond to odd functions. The definite parity of the functions is a consequence of the fact that the L-point group includes the inversion transformation, which commutes with all other point transformations.

So far, we have ignored the electron spin. In the simplest case, ignoring the interaction of the magnetic field generated by the orbital motion of an electron with the spin magnetic moment (the spin−orbit interaction), the wave function can be regarded as depending only on the space coordinates. Thus, having determined the level scheme, we may assume that at each level there are two electrons with opposite spins. This automatically takes into account the additional double degeneracy due to spin. The situation changes appreciably when the spin−orbit interaction is taken into account.

Sec. 6.1]   GENERAL ENERGY BAND STRUCTURE   275

The spin–orbit interaction, like the concept of the electron spin itself, is introduced consistently in the relativistic quantum theory. The transformation of the four-component Dirac equation gives a two-component Hamiltonian which differs from the Hamiltonian of Eq. (6.5) by the presence of terms of the order of v^2/c^2:

$$\mathcal{H} = \frac{\hat{p}^2}{2m_0} + U - \frac{\hat{p}^4}{8m_0^3 c^2} + \frac{\hbar^2}{8m_0^2 c^2}\nabla^2 U + \frac{\hbar}{4m_0^2 c^2}(\hat{\sigma}\,[(\nabla U)\cdot\hat{p}]). \quad (6.9)$$

The first term represents the kinetic energy and the second represents the potential energy of an electron. The third term appears due to the dependence of the mass on the velocity; the fourth represents a relativistic correction to the potential energy, which has no classical analog. The third and fourth relativistic terms affect only the space components of the wave functions but not the spin components. The first four terms in Eq. (6.9) are invariant under the symmetry transformations of the operator U and in an analysis of the symmetry properties we can combine them into one term. The two relativistic corrections do not give rise to the splitting of the energy levels but are important in calculations of the mutual positions of the levels, which will be considered later.

The fifth term in Eq. (6.9), which describes the spin–orbit interaction, contains the spin operator $\hat{\sigma}$, which acts on the spin variable on which the wave function depends. In this case, the Hamiltonian is not invariant under the transformations of the spatial coordinates alone but is invariant under the simultaneous symmetry transformations of the space and spin coordinates.

The spin part of the wave function, which must be included when the spin–orbital interaction is considered, has the following important property: when the spin coordinates are rotated through an angle of 2π, it changes its sign and only double application of the transformation, denoted by the symbol $\bar{E}$, restores the original function. Each transformation of the coordinates in accordance with the symmetry group G is equivalent to the successive application of an ordinary transformation and a transformation $\bar{E}$. However, the wave functions are transformed in a different manner by these two operations. Each element in the group G thus corresponds to two matrices, i.e., the representation is double-valued.*

---
*The double-valued representations of the group G formally correspond to the single-valued representations of some double group which is obtained from the group G by the addition of the element $\bar{E}$.

Thus, when spin is taken into account, the wave functions are transformed by the double-valued representations of the symmetry group. Since the dimensions of the single- and double-valued representations are, in general, different, the spin – orbital interaction may split the energy levels. The space components of the wave functions, transformed in accordance with the double-valued representations, may consist of functions transformed in accordance with different single-valued representations, i.e., the spin – orbit interaction mixes states with different space symmetries.

One important point must also be made. In the presence of an external magnetic field, the Schrödinger equation is symmetrical with respect to time reversal. To explain the consequences of such symmetry, we shall introduce a time reversal operator $\hat{T}$. The function $\hat{T}\psi$ is obtained from ψ by complex conjugation and spin flip (reversal):

$$\hat{T}\left[\psi_1(\mathbf{r})\uparrow + \psi_2(\mathbf{r})\downarrow\right] = -\psi_1^*(\mathbf{r})\downarrow + \psi_2^*(\mathbf{r})\uparrow.$$

The symmetry with respect to time reversal makes the functions ψ and $\hat{T}\psi$ correspond to the same energy eigenvalue. Hence, it follows that if the two sets of functions ψ_l and $\hat{T}\psi_l$ are the bases of different irreducible representations, these two complex-conjugate representations can be considered from the physical point of view as a single representation with doubled dimensions; then the wave functions of the two sets refer to the same energy level. Such an additional degeneracy, resulting from the symmetry with respect to time reversal, is frequently called the Kramers degeneracy.

The symmetry with respect to time reversal gives rise to an n-fold degeneracy of all the levels in systems with an inversion center; here, n is an even number in spite of the fact that the symmetry group may have one-dimensional representations. This means that for each one-dimensional representation there is a corresponding complex conjugate representation which is different and the level is doubly degenerate.

All that we have said about the spin – orbit interaction, double-valued representations, and the Kramers degeneracy applies also to crystals. In contrast to finite systems, we must now consider the energy $\mathscr{E}$ and the degeneracy at a given point of the Brillouin zone $\mathbf{k}$. Each operation of the symmetry group of a crystal transforms the Bloch function $\psi_\mathbf{k}$ into a function, corresponding to the same value of the energy and wave vector, which is obtained from $\mathbf{k}$ by such an operation. Therefore, the function $\mathscr{E}(\mathbf{k})$ has, as already pointed out, the complete point symmetry of the crystal. Moreover, the time reversal operator $\hat{T}$ transforms the Bloch function $\psi_\mathbf{k}$ into another Bloch function corresponding to $-\mathbf{k}$ and, therefore, $\mathscr{E}(\mathbf{k}) = \mathscr{E}(-\mathbf{k})$, even if the crystal has no inversion center.

A band at an arbitrary point $\mathbf{k}$ may be nondegenerate, even when the spin is allowed for, if the crystal is not symmetrical with respect to inversion. If the symmetry group of the crystal contains the inversion operation $\hat{I}$, then each band is at least doubly degenerate at any point $\mathbf{k}$ because the application of the two operations $\hat{I}$ and $\hat{T}$ transforms a Bloch function $\psi_\mathbf{k}$ into a new Bloch function $\hat{T}\hat{I}\psi_\mathbf{k}$, which is linearly independent of $\psi_\mathbf{k}$, but corresponds to the same values of the wave vector $\mathbf{k}$ and energy $\mathscr{E}$.

The double-valued representations of the L-point group, which are used to transform the wave functions when the spin is allowed for, are denoted by the symbols L^4, L^5, L^6, $L^{4'}$, $L^{5'}$, and $L^{6'}$.* The parity of the corresponding wave functions is denoted in the same way as the parity of the functions transformed in accordance with the single-valued representations: the prime is used to denote the odd functions. The L^6 and $L^{6'}$ representations are two-dimensional and correspond to levels which are doubly degenerate with respect to the spin. The L^4, L^5, $L^{4'}$, and $L^{5'}$ are one-dimensional but L^5 and $L^{5'}$ are obtained, respectively, from L^4 and $L^{4'}$ by complex conjugation and, therefore, the corresponding energy levels are doubly degenerate (the Kramers degeneracy) and these representations can be considered as two-dimensional representations L^{45} and $L^{45'}$. Consequently, all the levels are found to be doubly degenerate which is due to the presence of an inversion center in the total symmetry group of the crystal.

There is a definite correspondence between the wave functions transformed in accordance with the single-valued and double-valued representations. Let us consider, for example, the space wave functions transformed in accordance with the representations L_1 and L_3 (for the sake of brevity, we shall use the terms "the L_1

---

*This notation differs from that employed by Koster [474] in that the symbol L^i is used in place of L_i^+, and $L^{i'}$ is used in place of L_i^-.

function" and the "L_1 level"). The space wave function L_1, multiplied by the spin functions, yields two linearly independent wave functions, transforming in accordance with the L^6 representation. We shall denote these wave functions by L_1^6. From the two L_3 wave functions we can thus form four functions, two of which are transformed in accordance with the L^6 representations and the other two in accordance with the L^4 and L^5 representations (L_3^6 and L_3^{45}). Physically, this means that the L_3 level, which is doubly degenerate without allowance for the spin (and quadruply degenerate when the spin is allowed for) may be split into two doubly degenerate levels L_3^6 and L_3^{45} by the spin−orbit interaction (the spin−orbit splitting). However, the L^6 wave function may, in general, be formed from the L_1 and L_3 wave functions simultaneously, i.e., it may be represented in the form of a linear combination of L_1^6 and L_3^6 (the spin−orbit mixing). The spin−orbit mixing is appreciable in those cases when the spin−orbit splitting of the L_3 level is of the same order as (or larger than) the gap between the L_1 and L_3 levels.

To determine the mutual positions of the energy levels at a given point in the **k** space, we must solve the Schrödinger equation using the Hamiltonian of Eq. (6.9), in which U represents the potential energy of an electron in the crystal lattice field (the crystal potential). The selection of the crystal potential with an accuracy sufficient for the narrow forbidden band of lead chalcogenides is the main difficulty of the theory. Various methods of energy band structure calculations are discussed in Ziman's book [478]. We shall now list the methods which are employed in band structure calculations of lead chalcogenides.

The simplest methods of calculating the energy band structure are the nearly-free (weakly bound) electron method and the tight-binding method (linear combination of atomic functions). These methods are well known and we shall not consider them in detail.
In the cell (cellular) method, a crystal is split into elementary cells. The solutions in each of the cells are constructed from atomic functions and linear combinations of these functions have to match at a finite number of points on the cell boundaries.

In the augmented-plane-wave (APW) method, each atom is surrounded by a sphere in which the potential is assumed to be equal to the potential of a free atom. Between these spheres the potential is assumed to be constant. The combinations of atomic functions within each sphere are made to match the plane waves outside the sphere and the functions obtained in this way, known as the augmented plane waves, are used to find the solution to the problem.

In the orthogonalized-plane-wave (OPW) method the solution is obtained by combining a plane wave with atomic functions describing the inner state of the atom, which corresponds to deep narrow bands in a crystal. These combinations, known as the orthogonalized plane waves, have coefficients which are selected so that the functions obtained are orthogonal to the atomic wave functions.

A development of the OPW method is the pseudopotential approach. In this, the solution of the Schrödinger equation by means of orthogonalized plane waves is reduced to the solution of a new Schrödinger equation which differs from the initial one because the effective Hamiltonian has, instead of U, some other operator which is known as the pseudopotential and depends on the inner atomic functions. The pseudopotential differs from the potential U in that it is close to zero in the inner (core) part of the atom.

Finally, we shall consider briefly one other method of calculating the band structure, known as the semiempirical approach. In this approach, the wave functions and the energy levels are not calculated at the points where there are extrema but preliminary considerations (which follow from the symmetry properties, experimental data, and calculations carried out by the simple nearly-free-electron and tight-binding methods) are used to select a small number of possible variants of the mutual positions of levels at a given point in the **k** space. The next step is to calculate the effective masses (cf. §6.2A and §6.2B), g factors, etc.; the distances between the levels and the matrix elements used in the calculations are used as adjustable parameters. Comparison of the results obtained with the experimental data makes it possible to select (from several variants) the level distribution which best fits the data and to determine the adjustable parameters.

Some adjustable parameters are also used in the APW and the pseudopotential methods. For example, none of the methods yields, with reasonable accuracy, such a narrow forbidden band as that of lead chalcogenides and, therefore, the forbidden band width is usually employed as the adjustable parameter which has to be determined from the experimental data.

C. Application of Band Structure Calculation Methods to Lead Chalcogenides

In the band structure calculations we start from the results obtained by the nearly-free-electron and tight-binding methods [439, 470, 475] or from the symmetry considerations. Applying the nearly-free-electron model to an fcc lattice and taking into account the spin−orbit interaction, it is found that there is a group of six doubly degenerate bands near the forbidden band at the point L: three of these bands lie above and three below the forbidden band. Other bands are separated from this band group by relatively large gaps. The double Kramers (spin or time reversal) degeneracy of all these bands is a consequence of the fact that the crystal lattice has an inversion center. This degeneracy applies throughout the Brillouin zone.

In the absence of the spin−orbit interaction, the group of six bands, mentioned above, can be described by space wave functions which are the bases of single-valued irreducible representations of the symmetry group of the point L, denoted by the symbols L_1, $L_{2'}$, L_3, and $L_{3'}$ (see preceding subsection). The symmetry properties of these wave functions near the lattice sites are given in Table 6.1.

When the spin−orbit interaction is taken into account, the wave functions should be regarded as functions of the space and spin coordinates. They are bases of the double-valued irreducible representations L^{45}, $L^{45'}$, L^6, and $L^{6'}$. The nondegenerate (without allowance for spin) L_1 and $L_{2'}$ states transform into the L^6 and $L^{6'}$ states, respectively, and we shall denote these states by L_1^6 and $L_2^{6'}$. The doubly degenerate states split into two components: L_3 becomes L_3^6 and L_3^{45}, while $L_{3'}$ becomes $L_{3'}^{6'}$ and $L_{3'}^{45'}$.

TABLE 6.1. Symmetry Properties of Wave Functions near Lattice Sites

State	Cation (Pb)	Anions (Te, Se, S)	State	Cation (Pb)	Anions (Te, Se, S)
L_1	s-type	p-type	L_3	d-type	p-type
$L_{2'}$	p-type	s-type	$L_{3'}$	p-type	d-type

TABLE 6.2. Estimates
of Spin—Orbit Splittings of L_3 and $L_{3'}$
States in PbS, PbSe, and PbTe
(in eV) [475]

State	PbS	PbSe	PbTe
L_3	0.06	0.24	0.56
$L_{3'}$	0.96	0.97	0.98

The spin—orbit splitting of the L_3 and $L_{3'}$ states may be estimated from the values of the spin—orbit splitting of the atomic p and d levels of Pb, Te, Se, and S by assuming that, in any compounds of elements of groups IV and VI, electrons in both bands spend 40% of their time near the group IV atoms and 60% of their time near the group VI atoms. Rough estimates obtained in this way [475] are given in Table 6.2. They are in satisfactory agreement with the results of more rigorous calculations [469, 476]. It is evident from Table 6.2 that the spin—orbit splitting in lead chalcogenides is several times larger than the forbidden band width. The large spin—orbit splitting in lead chalcogenides is due to the presence of heavy atoms, particularly those of Pb.

In addition to the splitting of the levels which are doubly degenerate without allowance for the spin, the spin—orbit interaction mixes states with similar energies and the wave functions of these states transform in accordance with the same double-valued representation, but they have different space symmetries. In other words, the states are described by linear combinations of L_1^6 with L_3^6 and of $L_{2'}^6$ with $L_{3'}^6$. The values of the coefficients in these linear combinations depend strongly on the energy gaps between the bands. The spin—orbit mixing alters considerably the selection rules for the matrix elements and is important in the effective mass calculations. We shall consider this problem again later.

Since transitions between the top of the uppermost valence band and the bottom of the lowest conduction band are allowed, it follows that the corresponding states have opposite parities. The

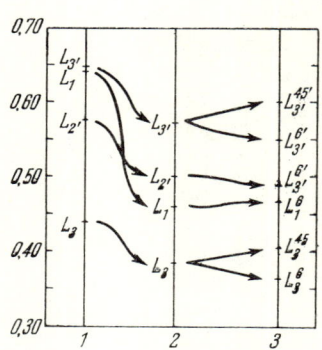

Fig. 6.2. Results of calculations of the energy level positions at the point L of the Brillouin zone of PbTe [518]. The calculations were made using the APW method: 1) distribution of the levels without the relativistic corrections; 2) with the relativistic corrections but without the spin – orbit interaction; 3) with the relativistic corrections and with the spin – orbit interaction.

tight-binding method [470, 475] shows that the $L_{3'}$ states have higher energies than the L_3 states.

Conklin et al. [467, 476] calculated theoretically the band structure of PbTe by the augmented-plane-wave (APW) method, generalized by the inclusion of the relativistic terms. The same method was employed to calculate the band structure of PbSe and PbS [468]. Since there are no reliable data on the ionicity of PbTe, Conklin et al., used solutions obtained for neutral atoms. No self-consistent calculation of the constant potential between spheres was carried out and this potential was regarded as a free parameter to be determined from the experimental data.

The considerable importance of the two spin-independent relativistic terms in the Hamiltonian of Eq. (6.9) in the calculations of the band structure of semiconductors, particularly those containing such heavy elements as lead, was pointed out first in [476, 477]. Johnson et al. [476] used the APW method to calculate the structure of PbTe, while Herman et al. [477] estimated the relativistic corrections for solids from the values of these corrections for atoms. It is possible to make these estimates and those of the spin – orbit splitting (given earlier) because the relativistic terms are large only near the lattice sites, where the wave functions are close to the atomic functions.

Figure 6.2 shows the band positions at the point L near the forbidden band in PbTe, obtained by the APW method. The initial positions were found by solving the nonrelativistic Schrödinger equation and then relativistic corrections were added, ignoring the spin – orbit interaction and, finally, the spin – orbit interaction was included. It is evident from this figure that two spin-independent relativistic corrections mix considerably (by several electron volts)

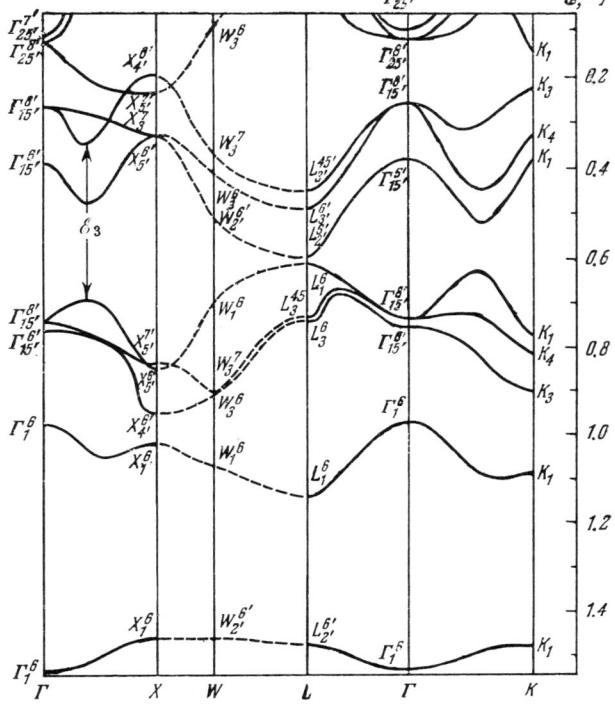

Fig. 6.3. Band structure of PbS, calculated by the pseudopotential method [470].

the energy levels and this alters the mutual positions of the bands. The largest relativistic displacement is found for the state L_1 whose wave function has a nonzero value near the lead atoms. Without allowance for the relativistic corrections, the results do not agree with the experimental data. The relativistic corrections for PbSe and PbS are smaller than those for PbTe.

Initially, without reference to the experimental data, Conklin et al. [467] found that the L_1^6 level lies 0.31 eV above $L_{2'}^6$ and therefore it belongs to the conduction band. However, the existence of the Knight shift in p-type PbTe indicates that the wave function describing the top of the valence band is of the s-type near the lead atoms (cf. §5.1). The only function in the group considered which satisfies this requirement is the L_1^6 function. Therefore, the interatomic potential has to be modified in such a way as to interchange the positions of the L_1^6 and $L_2^{6'}$ levels and to yield a forbidden band width of 0.19 eV, found experimentally from the magneto-optical measurements (cf. §5.4).

The general band structure of PbTe, obtained by Conklin et al. by the APW method [467], is close to the structure found by the pseudopotential method and shown in Fig. 6.5. Calculations of the effective masses, g factors, energy gaps between the bands, and deformation potentials, obtained from the band structures of PbTe, PbSe, and PbS calculated by the APW method [467, 468, 234], give values in agreement with the experimental data. This agreement provides confirmation of the correctness of the results obtained by the APW method. A detailed comparison of the theoretical and experimental values will be given later.

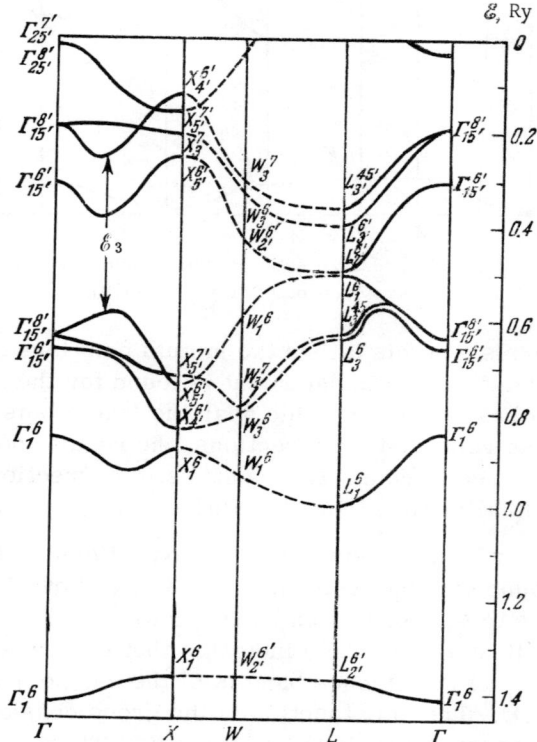

Fig. 6.4. Band structure of PbSe calculated by the pseudopotential method [470].

Sec. 6.1] GENERAL ENERGY BAND STRUCTURE

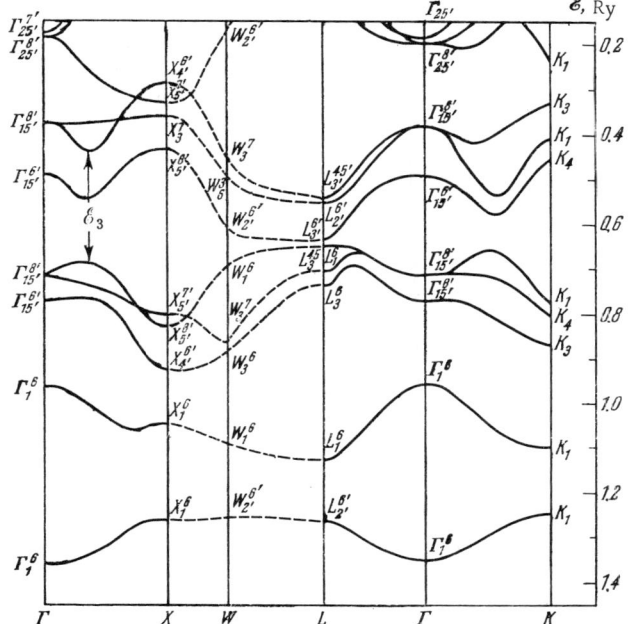

Fig. 6.5. Band structure of PbTe calculated by the pseudopotential method [470].

Another method which has been used successfully in calculations of the band structure of lead chalcogenides is the pseudopotential approach. The pseudopotential, derived by Kleinman and Lin [469, 470] contains five adjustable parameters whose values are found from the experimental data. As in the APW calculations, the theoretical values of the energy gaps, effective masses, g factors, and the pressure dependence of the forbidden band width were compared with the experimental values in order to check the theory.

The energy levels were found at the Γ, L, X, K, and W symmetry points of the Brillouin zone (these points are located, respectively, at the center of the zone, on the boundary along the <111>, <100>, and <110> directions, and at a vertex of the polyhedron bounding the Brillouin zone), as well as at several other points along the directions already mentioned (except for the <110> direction in PbSe). The results of calculations of the band structures of all three chalcogenides are given in Figs. 6.3-6.5. It is evident from

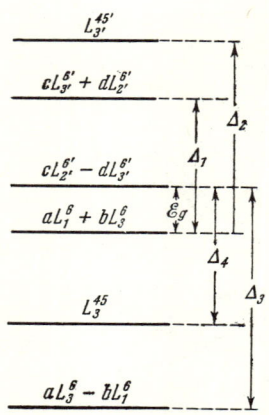

Fig. 6.6. Schematic distribution of the energy levels at the L point of the Brillouin zone near the forbidden band of lead chalcogenides, calculated by Lin and Kleinman [470]. Each level is doubly degenerate. The energy gaps $\Delta_1 - \Delta_4$ and the spin – orbit mixing parameters a-d are listed in Table 6.3.

these figures that the band structures of these three compounds are similar. The distribution of the levels at the point L near the forbidden band is shown in Fig. 6.6. The parameters used in this figure (describing the spin – orbit mixing of the wave functions and the energy gaps between bands) are listed in Table 6.3. It is clear from the reported results that the top of the valence band is described, in agreement with the Knight shift measurements, by the wave function L_1^6 with a small admixture of the L_3^6 function. The bottom of the conduction band is described by a linear combination $cL_{2'}^{6'} - dL_{3'}^{6'}$; for PbTe we have $c < d$, so that the contribution of $L_{3'}^{6'}$ predominates in the combined wave function, but for PbSe and PbS we find that $c > d$ and therefore $L_{2'}^{6'}$ predominates in these two compounds. This explains the large difference between the anisotropy coefficients of the effective masses of these three semiconductors (cf. §6.2A).

The forbidden band width $\mathscr{E}_g$ is an adjustable parameter in theoretical calculations; it is found experimentally from magneto-optical measurements (cf. §5.4C). The other energy gaps were calculated by Lin and Kleinman [470]. All these other gaps are sev-

TABLE 6.3. Parameters Governing Wave Functions and Mutual Positions of Energy Bands at Point L (Δ and $\mathscr{E}_g$ are given in eV) [470]

Compound	a	b	c	d	$\mathscr{E}_g$	Δ_1	Δ_2	Δ_3	Δ_4
PbTe	0.974	0.227	0.668	0.744	0.190	1.30	1.47	1.40	0.96
PbSe	0.999	0.041	0.864	0.504	0.165	1.52	2.01	1.99	1.88
PbS	0.999	0.032	0.881	0.473	0.286	1.75	2.30	2.03	1.96

TABLE 6.4. Composition of Wave Functions (in %)

State	Pb	Te	Plane waves	State	Pb	Te	Plane waves
$L_{3'}$	56	13	31	L_3	6	74	20
$L_{2'}$	33	19	48	L^6	37	45	18
L_1	39	43	18	$L^{6'}$	46	16	38

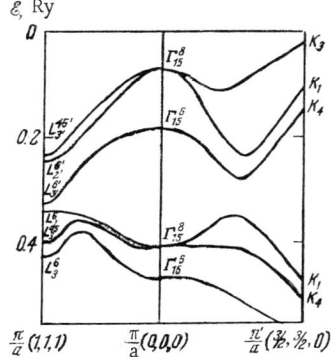

Fig. 6.7. Distribution of the energy bands of PbTe near the forbidden band along the <111> and <110> directions [470].

eral times larger than the forbidden band width. The values of the energy gaps found by the APW method [479] differ slightly from those listed in Table 6.3.

No bands with energy close to the uppermost valence band are found at the point Γ but both calculation methods indicate that there is a maximum along the <110> directions within the Brillouin zone of PbTe and the gap between this maximum and the top of the uppermost valence band, located at the point L, is very small and, therefore, cannot be calculated with sufficient accuracy [467, 470]. It is possible that this maximum represents the second valence band of PbTe, discovered in measurements of the Hall effect, thermoelectric power, and infrared reflection. The second valence band can be seen clearly in Fig. 6.7, which shows (on an enlarged scale) the energy structure near the forbidden band of PbTe.

These calculations make it possible to determine the electron density in the PbTe lattice. Conklin et al. [467] separated the integral $\int |\psi|^2 d\mathbf{r}$ for their wave functions into three parts: spheres surrounding the Pb and Te nuclei and a part assigned to the space between these spheres. Table 6.4 shows the values obtained for the space wave functions. Bearing in mind that the wave function L^6, corresponding to the top of the valence band, is formed from the function L_1 with an admixture of L_3, and the wave function of an electron at the bottom of the conduction band is formed from the functions $L_{2'}$ and $L_{3'}$, we find that the composite wave functions of elec-

TABLE 6.5. Energies of Experimentally Observed
Reflection Peaks [119] and Energy Gaps
Corresponding to Transitions at Symmetry
Points, Calculated by the Pseudopotential
Method [470] for PbTe, PbSe, and PbS (in eV)

Peak No.	Experiment			Transition type	Theory		
	PbTe	PbSe	PbS		PbTe	PbSe	PbS
$\mathscr{E}_1$	1.24	1.54	1.83	$L_1^6(2) - L_*^{6'}$*)	1.30	1.52	1.77
$\mathscr{E}_2$	2.45	3.12	3.67	$L_3^6(1) - L_*^{6'}$*)	2.52	3.35	3.50
$\mathscr{E}_3$	3.5	4.5	5.3	$\Delta_1^6(3) - \Delta_5^6(2)$	3.59	4.61	5.29
$\mathscr{E}_4$	6.3	7.1	8.1	$\Gamma_1^6(1) - \Gamma_{15'}^{6'}(2)$	6.30	7.33	8.15
$\mathscr{E}_5$	7.8	9.1	9.8	$\Gamma_1^6(1) - \Gamma_{15'}^{8'}(2)$	7.82	8.97	9.85
$\mathscr{E}_6$	11.2	12.5	13.9	$X_{5'}^{7'}(1) - X_2^7(1)$	11.36	13.20	13.91

*)$L_*^{6'}$ denotes $L_2^{6'}$ for PbTe and $L_3^{6'}$ for PbSe and PbS.

trons near the conduction and valence band edges are given by the results which are also listed in Table 6.4. The spin—orbit mixing parameters used in these calculations were taken from the work of Lin and Kleinman [470] (cf. Table 6.3).

It follows from the values obtained that the contribution of the atomic Te functions to the wave function of electrons in the valence band is approximately three times larger than to the wave function of electrons in the conduction band. This result is in qualitative agreement with the experimental observation that when one of the group VI elements in lead chalcogenides is replaced by another element of the same group, the hole mobility decreases much more than the electron mobility (§3.1A), although the conclusion that holes move mainly in the Te sublattice is too sweeping.

The wave function of an electron in the conduction band includes a 46% contribution from the atomic wave functions of Pb, while the contribution from the atomic functions of Te is approximately three times smaller. This explains the observation that the impurities which occupy the Pb sites in the lattice reduce the electron mobility much more than the impurities which occupy the Te sites.

In spite of the fact that the contribution of the atomic wave functions of chalcogens to the $L^{6'}$ function, corresponding to the bottom of the conduction band, is relatively small, it accounts for the anomalous dependence of the forbidden band width of lead chalcogenides on the atomic number [479] since one-electron energies corresponding to the outer s shells of chalcogens depend nonmonotonically on the atomic number.

The cell method was used by Nikiforov [480] to calculate the potential and some energy levels in the electron spectrum of PbSe. His calculation was carried out for the high-symmetry points of the Brillouin zone without allowance for the spin − orbit interaction.

In addition to the methods already described, the semiempirical method was also employed to calculate the band structure of lead chalcogenides [292, 297, 323, 439, 475]. The most definite conclusions about the band structure of lead chalcogenides at the L point near the forbidden band were obtained by this method by Cuff et al. [439], who used the experimental results obtained mainly from the Shubnikov − de Haas effect. Cuff et al., concluded that the only variant of the band positions which is in agreement with the experimental data is that shown in Fig. 6.6; this also agrees with the results found by the APW and pseudopotential methods. An allowance for the spin − orbit mixing explains the observed variation of the effective masses along the PbTe, PbSe, PbS series (details are given in §6.2A).

D. Comparison of Calculated Values of Energy Gaps with Experimental Data

The theoretically calculated values of the energy gaps between levels at various points of the Brillouin zone can be compared with the results of measurements of the ultraviolet reflection. As demonstrated in §2.1E, the reflection peaks correspond to the energies of allowed (by the selection rules) vertical electron transitions at those points at which the density of states has maxima.

The experimental values of the photon energies at which Cardona and Greenaway [119] found reflection peaks in lead chalcogenides are listed in Table 6.5 together with the calculated values of the energy gaps found by the pseudopotential method [470]. Table 6.5 gives also the type of transition which, according to Lin and

TABLE 6.6. Energies of Experimentally Observed Maxima of $\varepsilon_2 \mathscr{E}^2$ [119] and Energy Gaps, Corresponding to Transitions at Symmetry Points, Calculated by the APW Method for PbTe (in eV) [467]

Experiment		Theory		Experiment		Theory	
Peak No.	PbTe	Transition type	PbTe	Peak No.	PbTe	Transition type	PbTe
$\mathscr{E}_1$	1.2	$L_1^6(2) - L_{2'}^{6'}$	1.23	$\mathscr{E}_4$	6.2	$X_{5'}^{6'} - X_3^7$	6.18
$\mathscr{E}_2$	2.0	$L_3^{45} - L_{2'}^{6'}$	2.10	$\mathscr{E}_5$	7.25	$\Gamma_{15'}^{8'}(1) - \Gamma_{25'}^{8'}$	6.98
$\mathscr{E}_3$	3.2	$L_3^6 - L_{3'}^{45'}$	3.8	$\mathscr{E}_6$	10.3	$\Gamma_1^6(1) - \Gamma_{15'}^{8'}$	10.0

Kleinman [470], corresponds to the experimental values of $\mathscr{E}_1 - \mathscr{E}_6$. The agreement between the experimental and theoretical values is good for all the peaks with the exception of the broad and weak peak $\mathscr{E}_6$ of PbSe.

We must mention that the theory predicts a considerably larger number of energy gaps, at which reflection peaks should be observed, than the number of such peaks found in the experimental spectra. Obviously, some peaks overlap and cannot be resolved. For example, the $\mathscr{E}_1 = 1.2$ eV peak of PbTe, which was attributed by Lin and Kleinman [470] and Conklin et al. [467] to a transition of the $L_1^6 - L_{2'}^{6'}$ type, can be, as pointed out in [198], a pair of unresolved peaks at 1.0 and 1.5 eV due to the transitions from the spin–orbit split states L_3^{45} and L_3^6 to the bottom of the conduction band $L_{3'}^{6'}$.

Although the band structure schemes obtained by the APW and pseudopotential methods are similar, the APW and pseudopotential interpretations of the experimentally observed ultraviolet reflection peaks are quite different for all the peaks, with the exception of $\mathscr{E}_1$. Table 6.6 gives the results of calculations of the energy gaps by the APW method for PbTe [467]. For the sake of comparison, Table 6.6 lists the energies at which maxima of the function $\varepsilon_2 \mathscr{E}^2$ are observed (ε_2 is the imaginary component of the permittivity and $\mathscr{E}$ is the energy of a transition). These values, obtained from an analysis of the ultraviolet reflection data [119], are close to the values at which the reflection peaks are observed but they should correspond more accurately to the transition energies at the symmetry points of the Brillouin zone.

Nikiforov [480] attributes the peak of PbSe at $\mathscr{E}_4 = 7.1$ eV to transitions from the Γ_{15} level of the valence band to the Γ_1 level of the conduction band; the corresponding transition energy is, according to the cell method calculations, about 8 eV. The peak at $\mathscr{E}_5 = 9.1$ eV is, according to Nikiforov, due to transitions from the Γ_1 level of the valence band to the Γ_{15} level of the conduction band and the energy gap between these levels is approximately 9 eV. This interpretation of the $\mathscr{E}_5$ peak agrees with that given by Lin and Kleinman.

In view of the ambiguity of the interpretation of the ultraviolet reflection data, we cannot regard the agreement between the theoretical conclusions and the experimental data as a sufficient proof of the correctness of the theory, but this agreement does suggest that the theory is worth further examination.

Experimental information on the energy bands far from the Fermi level is also provided by measurements of the electron photoemission spectra. The results of such measurements for PbTe and PbS, presented in §2.1E, were used by Lin and Kleinman to check their own calculations [470].

A doublet in the photoelectron spectrum of PbTe was attributed to transitions from two valence bands at the point L, located 0.7 and 1.2 eV below the uppermost valence band. The calculations of Lin and Kleinman [470] indicated the existence of the L_3^{45} and L_3^6 bands, split by the spin – orbit interaction and located 0.77 and 1.21 eV below the L_1^6 valence band. The corresponding energy gaps, calculated by Conklin et al. [439] are 0.8 and 1.4 eV (these values are given in [198]).

The transitions causing this doublet take place from the two valence bands found by Lin and Kleinman to a band which has a maximum at the point L and which lies, according to the experimental data, about 8.3 eV above the uppermost valence band. The calculations of Lin and Kleinman [470] indicate the existence of the $L_2^{6'}$ level, lying 8.8 eV above the uppermost valence band and transitions to this state from the L_3 state are allowed by the selection rules.

The state which lies, according to the measurements, 2.4 eV above the top of the valence band and which is responsible for a maximum in the photoelectron spectrum may be identified with the calculated state $X_5^{6'}$ [470], lying 2.44 eV below the maximum of the

TABLE 6.7. Energy Gaps Between Anion and Cation s States in Crystals and Free Atoms (in Ry) [470]

States	PbS	PbSe	PbTe
$\Gamma_1(2) - \Gamma_1(1)$	0.564	0.563	0.393
$\mathscr{E}_s^{Pb} - \mathscr{E}_s^{VI}$	0.641	0.606	0.371

uppermost valence band. But it is not clear why there is no peak corresponding to transitions from the $X_5^{7'}$ state.

PbS exhibits maxima corresponding to levels separated by 2.0 and 3.0 eV from the top of the valence band. The first of these maxima may be attributed to transitions from the L_3 levels; the splitting of these levels by the spin−orbit interaction is 0.067 eV [470], which agrees with an experimentally obtained estimate according to which the splitting should not exceed 0.15 eV. The calculated depth of the lower of these levels, L_3^6, is 1.74 eV]470]. The second maximum of lead sulfide may be attributed to indirect transitions from the $X_5^{7'}$ and $X_5^{6'}$ states, lying 2.98 and 3.13 eV above the valence band top.

In the tight-binding approximation, one of the Γ_1 states is described by an anion s function and the other such states is described by an s function of the lead atom. Therefore, the gaps between these two levels, calculated for lead chalcogenides [470], can be compared with the corresponding gaps between the energy levels of free atoms. It is evident from Table 6.7 that the agreement between these two sets of energy gaps is good.

Recent electro-optical measurements, carried out by Aspnes and Cardona [542] on PbTe, PbSe, and PbS (absorption of light quanta of 0.25- to 0.65-eV energy and reflection of 0.6- to 5.5-eV quanta in an electric field), gave results which were in agreement with the theoretical calculations of Lin and Kleinman [470].

6.2. Structure of the Energy Band Edges

A. Effective Masses

The dependence $\mathscr{E}(k)$ near an extremum is represented by an effective mass. The effective mass is calculated using Eqs.

Sec. 6.2] STRUCTURE OF THE ENERGY BAND EDGES

(6.7) and (6.8). Let us assume that an extremum α of a nondegenerate band is located at a point $\mathbf{k}_0$ (the spin degeneracy in the calculation of the effective masses is unimportant and therefore we shall neglect it). A periodic function $u_{\mathbf{k}_0}^\alpha(\mathbf{r})$ and the energy at this extremum $\mathscr{E}_{\mathbf{k}_0}^\alpha$ are the solutions of an equation which is analogous to Eq. (6.7):

$$(\mathscr{H} + \mathscr{H}_{\mathbf{k}_0}) u_{\mathbf{k}_0} = \mathscr{E}_{\mathbf{k}_0} u_{\mathbf{k}_0}. \tag{6.10}$$

Apart from $u_{\mathbf{k}_0}^\alpha$, Eq. (6.10) has other eigenfunctions and all the solutions $u_{\mathbf{k}_0}^{\alpha'}$ ($\alpha' = 1, 2, \ldots, \alpha, \ldots$) form a complete system of orthonormalized functions. The corresponding eigenvalues $\mathscr{E}_{\mathbf{k}_0}^{\alpha'}$ are equal to the energies in all bands at $\mathbf{k} = \mathbf{k}_0$. We shall now consider the point $\mathbf{k} = \mathbf{k}_0 + \mathbf{k}'$, located near the point $\mathbf{k}_0$. The equation for the wave function and the energy at the point $\mathbf{k}$ is of the form

$$(\mathscr{H} + \mathscr{H}_{\mathbf{k}} + \mathscr{H}_{\mathbf{k}_0, \mathbf{k}'}) u_{\mathbf{k}_0 + \mathbf{k}'} = \mathscr{E}_{\mathbf{k} + \mathbf{k}'} u_{\mathbf{k} + \mathbf{k}'}, \tag{6.11}$$

$$\mathscr{H}_{\mathbf{k}, \mathbf{k}'} = \frac{\hbar^2 k'^2}{2 m_0} + \frac{\hbar}{m_0}(\mathbf{k}' \hat{\mathbf{p}}) + \frac{\hbar}{m_0}(\mathbf{k}' \mathbf{k}_0). \tag{6.12}$$

If $\mathbf{k}'$ is a sufficiently small quantity, the operator $\mathscr{H}_{\mathbf{k}_0, \mathbf{k}'}$ can be regarded as a small perturbation compared with the operator $\mathscr{H} + \mathscr{H}_{\mathbf{k}_0}$, and the dependence of the energy correction $\mathscr{E}_{\mathbf{k}_0 + \mathbf{k}'} - \mathscr{E}_{\mathbf{k}_0}$ on $\mathbf{k}'$ determines the effective mass. The first term in Eq. (6.12) represents the energy correction, which is a quadratic function of $\mathbf{k}'$, at least in the first approximation of the perturbation theory. In the presence of only that term, the effective mass would be equal to the free-electron mass m_0.

The sum of the second and third terms in Eq. (6.12) give the correction to the energy only in the second approximation of the perturbation theory, because the correction in the first approximation should be linear in $\mathbf{k}'$, which is impossible in the presence of an extremum at the point $\mathbf{k}_0$. The third term does not give the correction even in the second approximation because of the orthogonality of the functions $u_{\mathbf{k}_0}^{\alpha'}$. Thus, the deviation of the effective mass from the free-electron mass is completely determined by the term proportional to $\mathbf{k}' \cdot \hat{\mathbf{p}}$ and therefore this method of calculating the effective mass is known as the $\mathbf{k} \cdot \hat{\mathbf{p}}$ perturbation method.

Reducing the effective mass tensor to the diagonal form, we obtain an expression for the three components of the effective mass

$$\frac{1}{m_i^{*\alpha}} = \frac{1}{m_0} + \frac{2}{m_0^2} \sum_{\alpha' \neq \alpha} \frac{|\langle u_{k_0}^{\alpha} | \hat{p}_i | u_{k_0}^{\alpha'} \rangle|^2}{\mathscr{E}_{k_0}^{\alpha} - \mathscr{E}_{k_0}^{\alpha'}}. \quad (6.13)$$

In the derivation of the expression for the effective mass, we have ignored the relativistic terms, including the spin−orbit interaction. However, the use of the Hamiltonian with the relativistic terms, given by Eq. (6.9), in the $\mathbf{k} \cdot \hat{\mathbf{p}}$ perturbation method yields only small corrections to the effective mass. For example, in a calculation of the effective mass of PbTe it was found [479] that the contribution of the spin−orbit term to the matrix elements governing the value of the effective mass is 10^6 times smaller and the contribution of other relativistic terms is 10^2 times smaller than the main contribution due to the momentum operator. Thus, the relativistic terms, which are important in the determination of the mutual positions of the bands at a given point in the Brillouin zone, have practically no direct effect on the effective masses.

Each term in the sum on the right-hand side of Eq. (6.13) can be arbitrarily regarded as the consequence of the interaction of the bands α and α'. The term "interaction" is used here in its purely mathematical sense.

The interaction between these two bands is represented by the matrix elements of the momentum operator and by the differences between band energies. The presence of the energy terms in the denominators of the sum means that the bands located far from the one being considered can be ignored in the summation. We can also ignore the terms with the matrix elements equal to zero due to the symmetry properties. In particular, the selection rules permit us to ignore the interaction between those states whose wave functions retain their parity under the inversion transformation (cf. §6.1B). Since optical dipole transitions are also described by the matrix elements of the momentum operator, the selection rules for optical transitions are related directly to the selection rules for the terms in the sum given by Eq. (6.13).

The simple form of the perturbation theory is applicable only when the correction to the energy is considerably smaller than the distance to the nearest unperturbed level contributing to the sum in Eq. (6.13). When we are considering the effective mass of electrons whose energy (measured from the band edge) is comparable with the distance to the nearest band at the point $\mathbf{k}_0$, we must allow for the interaction with the neighboring bands even in the zeroth approximation of the perturbation theory, as is done for a degenerate unperturbed level. In such cases, the function $\mathscr{E}(\mathbf{k}') = \mathscr{E}_{\mathbf{k}_0+\mathbf{k}'} - \mathscr{E}_{\mathbf{k}_0}$ deviates from the quadratic form. The band nonparabolicity will be considered in the next subsection.

To calculate the effective masses, it is essential to know the energy gaps between relatively closely spaced bands and the corresponding matrix elements. Such calculations for lead chalcogenides have been carried out by the APW [467, 468, 479] and pseudopotential [469, 470] methods. Comparison of the results of the calculated effective masses with the experimental values provides an important check of the band structure calculations carried out by these two methods.

In addition to such calculations, several investigators [230, 286, 292, 297, 439] considered the influence of the symmetry properties on the effective masses in lead chalcogenides and derived expressions in which the matrix elements and energy gaps are regarded as parameters to be determined from the experimental data. The determination of the effective masses is an essential step in the calculation of the band structure by the semiempirical method [439].

It follows from symmetry considerations that the constant-energy surfaces near the point L, where the principal extrema of the conduction and valence bands of lead chalcogenides are located, should be in the form of ellipsoids of revolution with their axes directed along <111>. These ellipsoids have two components of the effective mass: longitudinal and transverse.

As pointed out in §6.1C, lead chalcogenides have a relatively isolated group of six energy bands near the forbidden band at the L point and we need consider only these six bands in calculations of the effective masses. Since optical transitions between the principal extrema of the conduction and valence bands are allowed, it follows that the interaction of the bands separated by the forbidden gap makes a contribution at least to one of the components of the effective mass.

TABLE 6.8. Values of Matrix Elements Calculated by the Pseudopotential Method (in $Ry^{1/2}$) [470]

Compound	$M_1 \langle L_1^6 - L_2^{6'} \rangle$	$M_2 \langle L_3^6 - L_3^{6'} \rangle$	$M_3 \langle L_1^6 - L_3^{6'} \rangle$	$M_4 \langle L_1^6 - L_3^{45'} \rangle$	$M_5 \langle L_{2'}^{6'} - L_3^{6} \rangle$	$M_6 \langle L_{2'}^{6'} - L_3^{45} \rangle$	$M_7 \langle L_3^{45} - L_3^{6'} \rangle$	$M_8 \langle L_3^{6} - L_3^{45'} \rangle$
PbTe	0.410	0.306	0.802	0.810	0.779	0.781	0.639	0.674
PbSe	0.464	0.319	0.857	0.866	0.788	0.786	0.678	0.731
PbS	0.467	0.329	0.902	0.901	0.848	0.857	0.717	0.767

Pikus and Bir [292] pointed out that, in the absence of the spin–orbit mixing of wave functions, the contribution of the interaction between the uppermost valence band and the lowest conduction band should be equal to zero for at least one of the components of the effective masses. Since the top of the valence band is described mainly by the wave function L_1^6, the interaction of the two nearest bands should contribute either only to the transverse effective mass, if the bottom of the conduction band corresponds to the function $L_{3'}^{6'}$, or only to the longitudinal component, if the bottom is described by the function $L_{2'}^{6'}$.

In fact, the wave function of electrons at the bottom of the conduction band is a linear combination of the $L_{2'}^{6'}$ and $L_{3'}^{6'}$ functions. Therefore, the interaction between the nearest bands contributes to the longitudinal and transverse components of the effective mass. Since the forbidden band width is smaller than other gaps between the bands at the point L, we may assume that this interaction makes the principal contribution to the effective masses. An analysis of the dependence of the effective mass and forbidden band width on the temperature and pressure led Moizhes [230, 286] to the conclusion that both components of the effective masses of lead chalcogenides are governed by the interaction of the two bands separated by the forbidden gap.

This conclusion was essentially confirmed by later calculations of the effective masses which indicated that the interaction of the nearest bands makes the greatest contribution to both components of the effective masses [439, 470, 479]. This explains, in particular, the approximate equality of the effective electron and hole masses in all three chalcogenides. The greatest difference (by a factor of 1.5) is found between the longitudinal effective masses of electrons and holes in PbTe. This difference is mainly associated with the contribution of the free-electron mass to the effective masses which have opposite signs for electrons and holes. The differences between the other effective masses are of the order of 10-20%. Although the interaction between the uppermost valence band and the lowest conduction band makes the greatest contribution to the effective electron and hole masses, the contributions of other bands are not small and should be allowed for in the calculations. Expressions for the effective masses, obtained by the $\mathbf{k} \cdot \hat{\mathbf{p}}$ perturbation method taking into account the interaction of all six bands have the following form [470]:

$$\frac{m_0}{m^*_{\parallel p}} = -1 + \frac{4}{3}\left(\frac{|acM_1 - bdM_2|^2}{\mathscr{E}_g} + \frac{|adM_1 + bcM_2|^2}{\Delta_1}\right), \quad (6.14)$$

$$\frac{m_0}{m^*_{\perp p}} = -1 + \frac{2}{3}\left(\frac{|adM_3 + bcM_5|^2}{\mathscr{E}_g} + \frac{|acM_3 - bdM_5|^2}{\Delta_1} + \frac{|aM_4|^2 + |bM_8|^2}{\Delta_2}\right), \quad (6.15)$$

$$\frac{m_0}{m^*_{\parallel n}} = 1 + \frac{4}{3}\left(\frac{|acM_1 - bdM_2|^2}{\mathscr{E}_g} + \frac{|bcM_1 + adM_2|^2}{\Delta_3}\right), \quad (6.16)$$

$$\frac{m_0}{m^*_{\perp n}} = 1 + \frac{2}{3}\left(\frac{|bcM_5 + adM_3|^2}{\mathscr{E}_g} + \frac{|acM_5 - bdM_3|^2}{\Delta_3} + \frac{|cM_6|^2 + |dM_7|^2}{\Delta_4}\right). \quad (6.17)$$

Here, M_1-M_8 are quantities which are proportional to the matrix elements of the components of the momentum operator, obtained ignoring the spin–orbit mixing. These quantities, calculated by the pseudopotential method [470], are listed in Table 6.8. It follows from this table that the values of the matrix elements M_1-M_8 vary little from one chalcogenide to another, like the matrix elements which occur in Kane's theory [205] for compounds of the $A^{III}B^V$ group. Table 6.8 gives also the states to which these matrix elements refer.

The energy gaps Δ_1-Δ_4 and the spin–orbit mixing parameters a-d are shown in Fig. 6.6 and their values are listed in Table 6.3 (§6.1C). These values vary considerably along the PbTe, PbSe, PbS series and this variation governs the difference between the effective masses of lead chalcogenides.

The effective masses calculated using Eqs. (6.3)-(6.6) are listed in Table 6.9 together with the experimental values obtained from measurements at low temperatures. The agreement between the theoretical and experimental values can be regarded as satis-

TABLE 6.9. Theoretical and Experimental Values of Effective Masses and g Factors of PbTe, PbSe, and PbS at Low Temperatures

Compound	Determination method	$m^*_{\parallel p}/m_0$	$m^*_{\parallel n}/m_0$	$m^*_{\perp p}/m_0$	$m^*_{\perp n}/m_0$	$|g_{\parallel p}|$	$|g_{\parallel p}|$
PbTe	[470] (theor., pseudopotential method)	0.20	0.16	0.037	0.033	39.51	47.67
	[467] (theor., APW method)	0.43	0.24	0.034	0.031	31.36	29.16
	[439] (exp., Shubnikov – de Haas eff.)	0.31	0.24	0.022	0.024	51	45
	[450] (exp., magneto-optics)	0.25	0.25	0.028	0.028		
	[455] (exp., cyclotron resonance)	0.38	0.24	0.031	0.024		
PbSe	[470] (theor.)	0.059	0.054	0.058	0.057	20.12	21.80
	[439] (exp.)	0.068	0.070	0.034	0.040	32	27
	[450] (exp.)	0.078	0.078	0.049	0.049	22*	22*
PbS	[470] (theor.)	0.101	0.086	0.085	0.077	10.81	10.09
	[439] (exp.)	0.105	0.105	0.075	0.080	13	12
	[450] (exp.)	0.110	0.110	0.100	0.100	8.5*	10*

*An asterisk is used to denote those values of the g factors along the [100] direction which are related to $g_\parallel$ and $g_\perp$ by Eq. (5.4).

factory. The pseudopotential method [470] overestimates somewhat the values of the transverse effective masses and underestimates the longitudinal masses, i.e., it gives a lower mass anisotropy coefficient than that obtained from the measurements. The agreement would be better if the matrix elements M_2, M_3, and M_5 were somewhat larger and M_1 was somewhat smaller. The matrix elements calculated for PbTe by the APW method [479] differ from those obtained by the pseudopotential method in exactly this way and therefore Pratt and Ferreira [479] found, by the APW method, more correct values for the electron and hole anisotropy coefficients of PbTe.

The spin−orbit mixing of the states makes it possible to account for the variation of the effective masses along the PbTe, PbSe, PbS series [439]. The selection rules and the values of the matrix elements depend considerably on the spin−orbit mixing. It is evident from Table 6.3 that the energy gaps Δ_1-Δ_4 between the energy bands at the point L increase monotonically from PbTe to PbS. This reduces the spin−orbit mixing. Consequently, there is an increase in the value of c, which governs the contribution of the function $L_{2'}^{6'}$ to the states at the bottom of the conduction band, which results − according to Eqs. (6.3) and (6.5) − in a reduction of the longitudinal effective masses. A reduction of the contribution of the function $L_{3'}^{6'}$ (the quantity d) increases the transverse effective masses by an amount greater than the increase in the forbidden band width.

This reduces the anisotropy coefficients along the PbTe, PbSe, PbS series. It is found [439] that the relatively small change in the separation between levels corresponding to the mixed states, may give rise to a large change in the components of the effective mass and in the anisotropy coefficient; for example, a change in the energy gaps by a factor less than two alters the anisotropy by more than one order of magnitude. It is interesting to note that in the tight-binding approximation the constant-energy surfaces near the extrema of the conduction and valence bands of PbS are spherical [471].

B. Energy Band Nonparabolicity. Kane and Cohen Models

When the energy of an electron, measured from the band edge, is comparable with the energy gap between bands (in lead chalcogenides the smallest gap is the forbidden band), the dependence of the energy on the crystal momentum is nonquadratic and the effective masses are functions of the energy. As pointed out in the preceding subsection, the interaction of the nearest bands should be allowed for in calculations of the nonparabolicity in the zeroth approximation of the $\mathbf{k} \cdot \hat{\mathbf{p}}$ perturbation theory. The interaction of the nearest bands is then allowed for exactly and it contributes to the dependence of the energy on the crystal momentum not in the form of a series in powers of the momentum, but in the form of a finite expression. Calculations of this type were carried out first by Kane [205] for InSb.

Let us consider the band nonparabolicity in the simplest case, which applies directly to lead chalcogenides. Let us assume that the valence and conduction band extrema, separated by an energy gap $\mathscr{E}_g$, are located at the same point in the $\mathbf{k}$ space, for example, at the point L. Let us assume also that the extrema are not degenerate (no allowance is made for the spin). The axes of the ellipsoidal constant-energy surfaces are then directed along the vector $\mathbf{k}_0$. We shall assume that the dependence of the energy on the momentum in the transverse direction ($\mathbf{k}' \perp \mathbf{k}_0$) is governed only by the interaction between these two bands.

The periodic components of the wave functions at the extrema of the conduction and valence bands will be denoted by u_1 and u_2, respectively. The wave functions at the point $\mathbf{k}_0 + \mathbf{k}'$ will be taken in the form of a linear combination $c_1 u_1 + c_2 u_2$, where the coefficients c_1 and c_2 depend on $\mathbf{k}'_\perp$. In this case, the perturbation Hamiltonian has the same form as Eq. (6.12) except that the last term is dropped. The next steps in the calculation are the same as those in the case of an unperturbed doubly degenerate energy level except that the two eigenvalues of the unperturbed Hamiltonian are different:

$$\langle u_1 | \mathscr{H} + \mathscr{H}_{\mathbf{k}_0} | u_1 \rangle = 0, \quad \langle u_2 | \mathscr{H} + \mathscr{H}_{\mathbf{k}_0} | u_2 \rangle = -\mathscr{E}_g$$

(the energy is measured from the lower edge of the conduction band).

We thus obtain, as usual, a secular equation for the energy as a function of $\mathbf{k}'_\perp$:

$$\begin{vmatrix} -\mathscr{E} + \dfrac{\hbar^2 k'^2_\perp}{2m_0} & \dfrac{\hbar k'_\perp}{m_0} \langle u_1 | \hat{p}_\perp | u_2 \rangle \\ \dfrac{\hbar k'_\perp}{m_0} \langle u_2 | \hat{p}_\perp | u_1 \rangle & -(\mathscr{E} - \mathscr{E}_g) + \dfrac{\hbar^2 k'^2_\perp}{2m_0} \end{vmatrix} = 0. \quad (6.18)$$

Sec. 6.2] STRUCTURE OF THE ENERGY BAND EDGES 301

This equation is quadratic in $\mathscr{E}$ and it has two solutions which give the electron energies in the conduction and valence bands:

$$\mathscr{E} = \frac{\hbar^2 k_\perp'^2}{2m_0} - \frac{\mathscr{E}_g}{2} \pm \left(\frac{\mathscr{E}_g^2}{4} + P_\perp^2 k_\perp'^2\right)^{1/2}, \qquad (6.19)$$

$$P_\perp \equiv \left|\frac{\hbar}{m_0} \langle u_1 | \hat{p}_\perp | u_2 \rangle\right|. \qquad (6.20)$$

The components of the effective mass can be conveniently determined applying Eq. (3.13) to most of the available experimental data. Then, if the effective mass is considerably smaller than the free-electron mass, we can drop the first term from Eq. (6.19) and obtain the following expression for the transverse effective electron and hole masses:

$$m_\perp^* = \frac{\hbar^2 \mathscr{E}_g}{2P_\perp^2}\left(1 + \frac{2\mathscr{E}}{\mathscr{E}_g}\right). \qquad (6.21)$$

This simple method is very similar to the theory developed by Kane [205] for InSb. However, there are differences between our calculation and Kane's theory. One difference is that in InSb the extrema of the conduction and valence bands are located at the center of the Brillouin zone and the constant-energy surfaces are spherical in the zeroth approximation of the $\mathbf{k} \cdot \hat{\mathbf{p}}$ perturbation theory. Secondly, the top of the valence band of InSb is doubly degenerate without allowance of spin and, moreover, there is a third valence band which is split-off by the spin−orbit interaction. Thirdly, Kane developed his theory further, taking into account the contribution of the bands lying further away and using higher approximations of the perturbation theory. For these three reasons, the formulas derived by Kane [205] are more cumbersome but, in the limiting cases of strong and weak spin−orbital interactions, Kane's formulas for the conduction band have a form which differs only slightly from the relationships (6.19) and (6.21).

We have mentioned earlier that lead chalcogenides have six closely spaced bands, the interaction between which must be allowed for in the zeroth approximation of the $\mathbf{k} \cdot \hat{\mathbf{p}}$ perturbation theory. Such a calculation was carried out by Dimmock and Wright [475]. Allowance for the interaction between the six closely spaced bands yielded a secular equation of the sixth degree, whose solution does not give, in general, simple expressions relating the energy to the momentum.

Therefore, it is usual to employ formulas obtained for two simple models which are special cases of the six-band model. In the derivation of these models, we shall exploit the fact that the energy gaps between the edges of the conduction and valence bands and the other bands are several times greater than the forbidden band width.

The first of these models is a generalization of the Kane model for the conduction band of InSb to the case of ellipsoidal constant-energy surfaces. In this model, the longitudinal and transverse effective masses of electrons and holes are governed by the interaction of the lowest conduction band with the highest valence band and the contributions of other bands are assumed to be negligibly small. In order to obtain simple expressions, it is usual to neglect also the contribution of the free-electron mass to the effective mass, which is permissble when $m^*_\perp \ll m_0$ and $m^*_\parallel \ll m_0$. The second inequality is satisfied poorly by PbTe, in which the contribution of the free-electron mass is responsible for the difference between the longitudinal electron and hole masses near an extremum. In the nonparabolicity region, where the effective masses are larger than they are near the band edges, the contribution of the free-electron mass is even more important. Therefore, the simple model is definitely inaccurate as far as PbTe is concerned. However, the two inequalities are satisfied quite well by PbSe and PbS.

In this modified Kane model, the effective masses and the mass anisotropy coefficients of electrons and holes are equal and the relationship between the energy and crystal momentum is given by the formula

$$\mathscr{E}(\mathscr{E} + \mathscr{E}_g) = \mathscr{E}_g \left(\frac{\hbar^2 k_\perp^2}{2m^*_{\perp 0}} + \frac{\hbar^2 k_\parallel^2}{2m^*_{\parallel 0}} \right). \tag{6.22}$$

The constant-energy surfaces for any value of the energy are ellipsoids and the anisotropy coefficient is independent of the energy, i.e., the longitudinal and transverse components of the effective mass vary with the energy in the same manner.

The density of states, considered as a function of the energy, is described by the following formula of the Kane model:

$$\rho(\mathscr{E}) = \frac{2^{1/2} m_{d0}^{*3/2}}{\pi^2 \hbar^3} \mathscr{E}^{1/2} \left(1 + \frac{2\mathscr{E}}{\mathscr{E}_g} \right) \left(1 + \frac{\mathscr{E}}{\mathscr{E}_g} \right)^{1/2}. \tag{6.23}$$

Sec. 6.2]   STRUCTURE OF THE ENERGY BAND EDGES   303

The cyclotron effective mass can be obtained from Eq. (5.3) bearing in mind that $m_\perp^*$ depends on the energy in accordance with the relationship

$$m_\perp^* = m_{\perp 0}^* \left(1 + \frac{2\mathscr{E}}{\mathscr{E}_g}\right). \tag{6.24}$$

Using Kane's model, we can determine the density-of-states effective mass m_d^* using Eqs. (4.20) and (4.21); this mass depends on the energy in the same way as the transverse effective mass, i.e., in accordance with Eq. (6.24).

Expressions for the electrical conductivity, Hall effect, and thermoelectric power in the Kane model are given in [250, 332, 333, 481], for the Nernst–Etingshausen effect in [250, 481], for the magnetic susceptibility in [333], for the thermoelectric power in strong magnetic fields in [335], for the Faraday effect in [443], and for the tunnel effect in a magnetic field in [323] (cf. also Appendix A). These expressions were used to analyze the experimental data [323, 332-336, 430]. In the derivation of the expressions which require knowledge of the energy dependence of the relaxation time, it was assumed that

$$\frac{1}{\tau} \propto \left[\mathscr{E}\left(1 + \frac{\mathscr{E}}{\mathscr{E}_g}\right)\right]^{-r'} \rho(\mathscr{E}), \tag{6.25}$$

where the parameter r' is equal to zero for the scattering by the acoustical phonons.* All these effects are expressed in terms of integrals of the type

$$^n\mathscr{L}_k^m(\zeta^*, \beta) \equiv \int_0^\infty \left(-\frac{\partial f}{\partial z}\right) z^n (z + \beta z^2)^m (1 + 2\beta z)^k \, dz,$$

where ζ^* is the reduced Fermi level; $\beta \equiv k_0 T/\mathscr{E}_g$; $z = \mathscr{E}/k_0 T$; f is the Fermi function; n, m, and k are integers and half-integers. These integrals are tablulated by Zawadzki et al. [482]. The relationship between the chemical potential and the carrier density is given by the formula

$$n = \frac{(2m_{d0}^* k_0 T)^{3/2}}{3\pi^2 \hbar^3} \, {}^0\mathscr{L}_0^{3/2}. \tag{6.27}$$

---
*As pointed out in §3.1A, this approximation is quite rough and Eq. (6.25) should include a factor reflecting the dependence of the square of the matrix element on the energy in the nonparabolic region. Formulas for the relaxation time in the nonparabolic case, obtained allowing for this factor, are given in Appendix A.

The second model, which was used to analyze the experimental data for PbTe [112, 113, 335, 336, 438], is a special case of the Cohen model [483]. It is derived from the six-band model [475] on the assumption that the interaction between the lowest conduction band and the highest valence band determines the transverse effective mass. The values of the longitudinal electron and hole masses are assumed to be affected by the interaction with further bands and by the conduction from the free-electron mass, but the dependences of these quantities on the energy can be neglected. The transverse effective mass is assumed to be considerably smaller than the free-electron mass. Instead of Eq. (6.22), we now obtain

$$\left(\mathscr{E} - \frac{\hbar^2 k_\parallel^2}{2m^*_{\parallel n}}\right)\left(\mathscr{E} + \mathscr{E}_g + \frac{\hbar^2 k_\parallel^2}{2m^*_{\parallel p}}\right) = \mathscr{E}_g \frac{\hbar^2 k_\perp^2}{2m^*_{\perp 0}}. \qquad (6.28)$$

In this model, the transverse electron and hole masses are equal. For the sake of simplicity, it is usual to assume that the longitudinal masses are also equal, although it does not follow from this model. Such an approximation simplifies the formulas quite considerably without introducing serious errors [112].

The constant-energy surfaces in the Cohen model, sufficiently far from the extrema, are not ellipsoids but they can be roughly approximated by ellipsoids with anisotropy coefficients smaller than the values near the band edge [339, 439]. The anisotropy coefficient varies with the energy so that it is inversely proportional to the transverse effective mass, which depends on the energy in accordance with Eq. (6.24), as in the Kane model.

In the Cohen model, the energy dependence of the density of states and the dependence of the carrier density on the Fermi energy are given by the formulas

$$\rho(\mathscr{E}) = \frac{2^{1/2} m_{d0}^{*3/2}}{\pi^2 \hbar^3} \mathscr{E}^{1/2} \left(1 + \frac{2\mathscr{E}}{\mathscr{E}_g}\right), \qquad (6.29)$$

$$n = \frac{(2m_{d0}^* k_0 T)^{3/2}}{2\pi^2 \hbar^3} [F_{1/2}(\zeta^*) + 2\beta F_{3/2}(\zeta^*)], \qquad (6.30)$$

where $F_n(\zeta^*)$ are the Fermi integrals and $\beta = k_0 T/\mathscr{E}_g$. The expressions for the electrical conductivity and the Hall effect are given in [484], for the thermoelectric power in a strong magnetic field in [485], and for the infrared reflection in [112, 113]. All these quantities, except the thermoelectric power in a strong magnetic field, are expressed in terms of integrals of the type given by Eq. (6.26). The formulas for the cyclotron effective masses and the extremal cross sections of the nonellipsoidal constant-energy surfaces are derived in [475].

The results of the experimental investigations of the thermomagnetic effects, Shubnikov – de Haas, infrared reflection, and magnetic susceptibility, discussed in earlier chapters, indicate that the nonparabolicity of the conduction and valence bands can be described with satisfactory accuracy by simple two-band models. It follows from the Shubnikov – de Haas effect (cf. §5.2C) that the anisotropy coefficients of PbSe and PbS are independent of the energy but in the case of PbTe this coefficient decreases with increasing energy; for this reason, the Kane model represents well the band structures of PbSe and PbS but is only a rough approximation for PbTe. The observation that the anisotropy coefficient of PbTe decreases with increasing energy more slowly than the reciprocal of the transverse effective mass indicates that the Cohen model, like the Kane model, describes only approximately the nonparabolicity of PbTe and that in fact the band structure of lead telluride is intermediate between the structures given by the Kane and Cohen models.

However, investigations of the cyclotron resonance in n-type PbTe show that the anisotropy coefficient and the reciprocal of the transverse effective mass have the same energy dependences, which corresponds to the Cohen model. Investigations of the magnetoacoustical oscillations show that the constant-energy surfaces of p-type PbSe are not ellipsoidal, which cannot be explained using the Kane model. Thus, further studies are needed before the final decision can be reached about the applicability of these simple nonparabolic models to the band structures of lead chalcogenides.

We should not expect exact agreement between the theoretical values of the effective masses, calculated using one of these two models, with the experimental data, because the influence of the bands lying further away from the forbidden gap is not negligibly small. The agreement between these models and the experimental data is improved if the forbidden bandwidth $\mathscr{E}_g$, which is one of the most important theoretical parameters, is replaced by a quantity $\mathscr{E}_{gi}$, which we have called already the effective width of the interaction forbidden band.

In spite of some influence of the bands distant from forbidden band, the quantity $\mathscr{E}_{gi}$, which describes the nonparabolicity, differs little from the true forbidden band width, which can be determined most accurately by the optical and magneto-optical methods. Our theoretical estimates show that the influence of the lower valence bands on the effective mass and nonparabolicity is largely compensated by the influence of the upper conduction bands. A considerable scatter is found in the experimental values of the effective width of the interaction forbidden band obtained by various methods [112, 113, 335, 336, 438, 439, 530-534].

We must bear in mind that the Kane and Cohen models give the same energy dependence for the transverse effective mass, which — according to Eqs. (2.22) and (3.12) — has a stronger influence on the average values of the effective masses than the longitudinal component. Therefore, in measurements of the quantities depending on the effective masses averaged over various directions (these quantities include the carrier mobility, Hall effect, thermoelectric power, Nernst — Ettingshausen effect, etc.), we can analyze the experimental data, at least in the case of PbTe, using both models and selecting a suitable value of $\mathscr{E}_{gi}$. The data available at present indicate that the Kane model is to be preferred for PbSe and PbS. The formulas for the Kane model are given in Appendix A. The effects which depend on the shape of the constant-energy surfaces (for example, cyclotron resonance and various quantum oscillations in magnetic fields) can be used to determine the validity of either of these models under given conditions.

The use of simple nonparabolic models in calculations of the g factors gives results which are not in agreement with the experimental data. This is because the interaction between further bands affects much more the spin splitting of the Landau levels in a magnetic field, and the two-band models are, strictly speaking, inapplicable.

C. Values of the g Factors

Because of the spin — orbit interaction, the g factor of electrons in a solid may differ from the g factor of a free electron, which is equal to 2. A calculation of the g factor given in [486] yields expressions similar to Eq. (6.13) for the effective masses, i.e., the g factor is expressed in terms of the energy gaps between bands and the matrix elements of the momentum operator. Like the effective masses, the g factor in the case of a narrow forbidden band and a strong spin–orbit interaction may be governed mainly by the interaction between the highest valence band and the lowest conduction band. In this case, the g factor is considerably larger than 2 and if the principal extrema of the conduction and valence bands are degenerate only in respect of the spin, the g factor is related to the cyclotron effective mass [given by Eq. (5.3)] by the simple expression

$$g \frac{m_c^*}{m_0} = 2. \qquad (6.31)$$

The case considered corresponds to the Kane model. However, it is important to stress that, in contrast to the Kane model of InSb, we are assuming that there is no degeneracy of the conduction and valence band edges, except for the spin degeneracy. Equation (6.31) applies also when the energy of carriers is comparable with half the width of the forbidden band, i.e., it is valid also in the nonparabolic region and in this case the g factor depends on the energy in accordance with the law

$$g = g_0 \left(1 + \frac{2\mathscr{E}}{\mathscr{E}_g}\right)^{-1}, \qquad (6.32)$$

which follows from Eqs. (6.31), (5.3), and (6.24). We should consider also the case corresponding to the Cohen model when only the longitudinal g factor is governed by the interaction of the two nearest bands. Then, Eq. (6.31) is satisfied only by the longitudinal g factor and is of the form

$$g_\parallel \frac{m_\perp^*}{m_0} = 2. \qquad (6.33)$$

It follows from Eq. (6.31) that the spin splitting of the Landau levels is equal to the separation between these levels. Investigations of the Shubnikov–de Haas effect [439] demonstrated that the spin splitting in lead chalcogenides is equal to only one-half the separation between the Landau levels, i.e., $gm_c^*/m_0 \approx 1$. This is explained by the fact that near the forbidden band at the L point there are actually six closely spaced bands and the interaction between these bands yields lower values of the g factor than the interaction between two bands [475]. Measurements of the g factor provide the most rigorous check of the two-band model, since the relationship used in such a check, given by Eq. (6.31), has no adjustable parameters.

The longitudinal g factors were calculated in [439, 468, 470, 479] using the values of the energy gaps and matrix elements calculated by the APW [468, 479], pseudopotential [470], and semi-empirical [439] methods. Dimmock and Wright [475] considered the influence of various mutual positions of the six bands at the L point on the value of the g factor. Expressions for the longitudinal and transverse g factors for all six bands were obtained by Mitchell and Wallis [471].

We shall compare first the theoretical and experimental values of the longitudinal g factor, which determines the spin splitting of the Landau levels when the magnetic field and the ellipsoid axis lie along the same [111] direction. Calculations of the longitudinal g factors of electrons and holes near the principal extrema were carried out by Lin and Kleinman, using the following formulas [470]:

$$g_{\parallel p} = 2 \left[a^2 - b^2 - \frac{2}{3} \left(\frac{|bcM_5 + adM_3|^2}{\mathscr{E}_g} + \frac{|acM_3 - bdM_5|^2}{\Delta_1} - \frac{|aM_4|^2 - |bM_8|^2}{\Delta_2} \right) \right], \quad (6.34)$$

$$g_{\parallel n} = 2 \left[c^2 - d^2 + \frac{2}{3} \left(\frac{|bcM_5 + adM_3|^2}{\mathscr{E}_g} + \frac{|acM_5 - bdM_3|^2}{\Delta_3} + \frac{|cM_6|^2 - |dM_7|^2}{\Delta_4} \right) \right]. \quad (6.35)$$

The notation used for the energy gaps, matrix elements, and spin−orbit mixing parameters is the same as in Eqs. (6.14)-(6.17) for the effective masses and the numerical values of these quantities, found by the pseudopotential method [470], are listed in Tables 6.3 and 6.8.

The absolute values of the longitudinal g factors calculated theoretically by Lin and Kleinman [470] are in satisfactory agree-

Sec. 6.2]   STRUCTURE OF THE ENERGY BAND EDGES   309

ment with the values determined by Cuff et al. [439] from an experimental investigation of the Shubnikov – de Haas effect (cf. Table 6.9). Lin and Kleinman's theoretical values obey even better the experimentally determined relationship $g_\| m^*_\perp/m_0 \approx 1$, since their somewhat high calculated values of the transverse effective masses are compensated by the underestimated values of the longitudinal g factors. The magneto-optical measurements yield the values of the g factors somewhat greater than the calculated values, as shown in Table 6.9.

The determination of the signs of the g factors is a fairly complex problem, whose theoretical aspects were tackled by Mitchell and Wallis [471]. The pseudopotential method [470] gives positive signs for the longitudinal g factors of electrons and negative signs for holes, in agreement with the magneto-optical investigations [127, 450] (cf. §5.4C).

We shall now consider the transverse g factor, which governs the spin splitting in a magnetic field perpendicular to the ellipsoid axis. A theoretical analysis [471, 475] showed that the transverse g factor differs from 2 only when a given state is related to some other state by the matrix elements of the operators $\hat{p}_\|$ and $\hat{p}_\perp$ simultaneously. If we neglect the spin – orbit mixing, we find that none of the pairs of states in the considered group of six levels has nonzero matrix elements $\hat{p}_\|$ and $\hat{p}_\perp$ simultaneously and the transverse g factors of all bands are equal to 2. However, when the spin – orbit mixing is allowed for, the matrix elements of the longitudinal and transverse components of the momentum operator are finite for states separated by the forbidden band. In the two-band model case, we obtain a formula which follows from Eq. (6.31):

$$g_\perp \frac{(m^*_\| m^*_\perp)^{1/2}}{m_0} = 2.$$

An expression for the transverse g factors in the six-band model is given by Mitchel and Wallis [471]. No measurements of the transverse g factors of lead chalcogenides have yet been reported.

6.3. Dependence of the Band Structure Parameters on Temperature, Deformation, and Composition

A. Influence of Pressure on the Forbidden Band Width and Effective Masses. Deformation Potentials

Having calculated the dependence of the forbidden band width on the lattice constant by the pseudopotential method, Lin and Kleinman [470] derived the change in the forbidden band width due to the isotropic deformation caused by the application of hydrostatic pressure. The calculated values of the derivatives $\partial \mathscr{E}_g/\partial P$, which are $-6.64 \cdot 10^{-6}$ eV $\cdot$ cm$^2 \cdot$ kg^{-1} for PbTe and $-5.44 \cdot 10^{-6}$ eV $\cdot$ cm$^2 \cdot$ kg^{-1} for PbS, are in satisfactory agreement with the results of electrical and optical measurements under hydrostatic pressure: $-(8 \pm 1) \cdot 10^{-6}$ eV $\cdot$ cm$^2 \cdot$ kg^{-1} for PbTe and $-(7.5 \pm 2) \cdot 10^{-6}$ eV $\cdot$ cm$^2 \cdot$ kg^{-1} for PbS (cf. §3.3A). The errors in the experimental data are estimated from the differences between the results of measurements reported by various investigators. The values obtained from the piezoresistance investigations are also in agreement with the above values, although the scatter of the results reported by different workers is quite considerable.

The result obtained for PbTe by Ferreira [234] is also close to the experimental data. Ferreira's calculations were carried out by the generalization of the APW method to a deformed crystal. Ignoring the spin − orbit mixing, the calculated pressure dependence of the forbidden band width is $\partial \mathscr{E}_g/\partial P = -9 \cdot 10^{-6}$ eV $\cdot$ cm$^2 \cdot$ kg^{-1} on the assumption that the compressibility ξ is $2.3 \cdot 10^{-6}$ cm^2/kg [283]. The maximum spin − orbit mixing, after which the L^6 state still remains the upper valence state, gives the value $-7 \cdot 10^{-6}$ eV $\cdot$ cm$^2 \cdot$ kg^{-1}.

Experiments show that the energy gap between the highest valence band and the second valence band of PbTe varies as rapidly with the applied pressure as the forbidden band width but the variation is of the opposite sign, i.e., the second valence band moves away from the edge of the highest valence band so that the gap between the conduction band and the second valence band is not greatly affected. The pressure coefficient of the gap between the highest and second valence bands $\partial \Delta \mathscr{E}/\partial P$ is $(6-7) \cdot 10^{-6}$ eV $\cdot$ cm$^2 \cdot$ kg^{-1}.

Ferreira's calculations [234] give not only the pressure dependence of the forbidden band width of PbTe but also the values of the deformation potentials Ξ_u and Ξ_d separately [cf. Eq. (3.54)]. The values of the shear deformation potential constants Ξ_u for electrons and holes obtained from measurements of the piezoresistance are lower than Ferreira's values and agree only in the order of magnitude (cf. §3.3B). This discrepancy is attributed by Ferreira to the errors in analysis of the experimental data, resulting from the unjustified assumptions that the statistics is nondegenerate and that the pressure-induced changes in the effective mass and relaxation time do not vary strongly with temperature.

According to Ferreira's calculations, the absolute values of Ξ_d are lower than those of Ξ_u, which is not in agreement with the experimental data: measurements of the carrier mobility yield a values of Ξ (cf. §3.1B), which is several times larger than Ξ_u (cf. §3.3B) and consequently represents basically the constant Ξ_d. Thus, it follows from the experimental data that $|\Xi_d|$ is several times larger than $|\Xi_u|$.

The change in the forbidden band width under pressure should, according to Eqs. (6.14)-(6.17), result in corresponding changes in the effective masses. If both components of the effective mass were completely determined by the interaction between the lowest conduction band and the highest valence band, the effective masses near the extrema would vary with pressure proportionally to the forbidden band width, i.e., we would observe the relationship $\partial \ln \mathscr{E}_g / \partial P = \partial \ln m^* / \partial P$.

In fact, we may expect the relative change in the effective masses with pressure differ somewhat from the relative change in the forbidden band width for the following reasons. First of all, the values of the effective masses, particularly the longitudinal component in the case of PbTe, are governed not only by the interaction of the bands separated by the forbidden gap, but also by the contribution of the free-electron mass and by the interaction with other bands. The energy gaps separating these bands are several times larger than the forbidden band width and their relative change with increasing pressure is smaller so that the corresponding contributions to the effective masses depend weakly on the pressure.

Secondly, when the samples have sufficiently high carrier densities, the measurements yield the effective mass not near an extremum but at the Fermi level where the effective mass depends less strongly on the forbidden band width. It follows from Eqs. (6.14)-(6.17) and (6.24) that when the effective mass is governed by the interaction of the bands separated by $\mathscr{E}_g$, the dependence of the mass on the forbidden band width at energies $\mathscr{E} > 0$ is given by the formula

$$m^* \propto \mathscr{E}_g + 2\mathscr{E} \qquad (6.36)$$

so that when $\mathscr{E} > 0$ a change in $\mathscr{E}_g$ produces a smaller relative change in m* than at $\mathscr{E} = 0$.

Thirdly, the matrix elements which govern the interaction between the bands in lead chalcogenides may depend appreciably on the pressure due to the spin−orbit mixing, and this dependence is similar to the strong dependence of the matrix elements along the PbTe, PbSe, PbS series, which results in corresponding changes in the effective masses (§6.2A).

The experimental data on the pressure dependences of the effective masses and the forbidden band width reported by different workers are not in very good agreement and are insufficiently accurate for a quantitative check of the conclusions reached in the preceding paragraphs. The ratio $(\partial \ln \mathscr{E}_g/\partial P)/(\partial \ln m^*/\partial P)$ ranges from 1 to 1.5, but the second of these values is not sufficiently reliable to conclude that the transverse effective mass is governed entirely by the interaction of the nearest bands while the longitudinal mass is affected also by the interaction of the other bands. In general, the experimental data show that the relative change in the effective masses due to the application of pressure is somewhat smaller than the corresponding change in the forbidden band width and that this change in the effective masses becomes smaller at higher carrier densities, which is explained qualitatively by the interaction between bands, considered in the preceding paragraphs.

B. Temperature Dependences of the Forbidden Band Width and Effective Masses

The temperature dependence of the forbidden band width is due to the thermal expansion of a crystal and the interaction of carriers with the lattice vibrations. The derivative $\partial \mathscr{E}_g/\partial T$ can be represented in the form of the sum:

$$\frac{\partial \mathscr{E}_g}{\partial T} = \left(\frac{\partial \mathscr{E}_g}{\partial T}\right)_0 + \left(\frac{\partial \mathscr{E}_g}{\partial T}\right)_V, \qquad (6.37)$$

where $(\partial \mathscr{E}_g/\partial T)_0$ is the temperature coefficient of the forbidden band width in a lattice whose atoms are at rest (this coefficient represents the influence of the thermal expansion) and $(\partial \mathscr{E}_g/\partial T)_V$ is the temperature coefficient of $\mathscr{E}_g$ for a crystal of fixed volume (it represents changes in the interaction of carriers with the thermal vibrations).

The first term in Eq. (6.37) can be expressed in terms of the pressure dependence of $\mathscr{E}_g$ using the following thermodynamic relationship

$$\left(\frac{\partial \mathscr{E}_g}{\partial T}\right)_0 = -\frac{3\alpha_{t.e}}{\xi} \frac{\partial \mathscr{E}_g}{\partial P}, \qquad (6.38)$$

where ξ is the compressibility of a crystal and $\alpha_{t.e}$ is the linear thermal expansion coefficient. Substituting into Eq. (6.38) the values given in Appendix C, we obtain $(\partial \mathscr{E}_g/\partial T)_0 = 2 \cdot 10^{-4}$ eV/deg. Comparing this value with that deduced from the optical and photoelectric measurements, $\partial \mathscr{E}_g/\partial T = 4 \cdot 10^{-4}$ eV/deg, we find that only about half of the temperature-induced change in the forbidden band can be attributed to the thermal expansion of the lattice.*

Thus, the interaction of carriers with phonons increases the forbidden band when the temperature is increased and this effect is represented by the coefficient $(\partial \mathscr{E}_g/\partial T)_V \approx 2 \cdot 10^{-4}$ eV/deg. It is usual to estimate $(\partial \mathscr{E}_g/\partial T)_V$ by means of Fan's expression [131] obtained in the second approximation of the perturbation theory ignoring virtual transitions to other bands. This expression can give only negative values of $(\partial \mathscr{E}_g/\partial T)_V$. An allowance for all bands regarded as intermediate states in virtual transitions yields a more

---

*It must be pointed out, however, that the scatter of the experimentally obtained temperature and pressure coefficients of the forbidden band width and of the values of the elastic constants is fairly large and some authors [125] concluded that almost the whole of the change in the forbidden band with temperature is due to the thermal expansion.

complex expression [229] and it is found that the interaction between the valence and conduction bands makes a positive contribution to the temperature coefficient of $\mathscr{E}_g$. Calculation of the temperature dependence of the forbidden band width carried out by Keffer et al. [543] gave a value of $(\partial\mathscr{E}_g/\partial T)_V$ which was in agreement with the experimental value both in respect of its sign and order of magnitude.

The temperature dependence of the depth of the second valence band in PbTe can be expanded into two components in a similar manner. The corresponding values are $\partial\Delta\mathscr{E}/\partial T \approx -4 \cdot 10^{-4}$ eV/deg, $(\partial\Delta\mathscr{E}/\partial T)_0 \approx -2 \cdot 10^{-4}$ eV/deg, $(\partial\Delta\mathscr{E}/\partial T)_V \approx -2 \cdot 10^{-4}$ eV/deg.

The temperature dependence of the effective mass is due to the same two effects, the thermal expansion of the lattice and its thermal vibrations. Using the plane-wave approximation for the wave functions of carriers, Ehrenreich [238] demonstrated that the interaction with phonons makes a negligibly small contribution to the temperature dependence of the effective mass. Hence, it follows that the whole of the temperature dependence of the effective mass is due to the thermal expansion of the lattice; if the effective mass is governed by the interaction of the bands separated by the forbidden gap, then the relative change in the effective mass near a band edge is equal to the relative change in the forbidden band width due to the thermal expansion alone:

$$\left(\frac{\partial \ln m^*}{\partial T}\right)_0 = \left(\frac{\partial \ln \mathscr{E}_g}{\partial T}\right)_0. \tag{6.39}$$

However, it follows from the expressions obtained using the periodic part of the Bloch functions [229] that the temperature dependence of the effective mass is affected by the interaction with phonons and therefore has a term related to the temperature dependence of the energy gaps, in particular, the temperature dependence of the forbidden band width. Ravich [487] demonstrated that this term is the main one in semiconductors with a narrow forbidden band (such as InSb and PbTe), in which the effective masses are governed by the interaction of the valence and conduction band.*

---

*One of the results reported in [487] was the conclusion that the phonon contribution to the temperature dependence of the effective mass is solely due to the nonpolar interaction and that the polar interaction in semiconductors with a small polaron coupling constant α_p [cf. Eq. (3.22)] gives rise to small changes in the band structure parameters, even in those cases when the polar interaction dominates carrier scattering.

The relative changes in the effective mass near a band edge and the changes in the forbidden band width due to the interaction with phonons are related by an equality similar to that given by Eq. (6.39):

$$\left(\frac{\partial \ln m^*}{\partial T}\right)_V = \left(\frac{\partial \ln \mathscr{E}_g}{\partial T}\right)_V. \tag{6.40}$$

Adding Eqs. (6.39) and (6.40), we obtain

$$\frac{\partial \ln m^*}{\partial T} = \left(\frac{\partial \ln m^*}{\partial T}\right)_0 + \left(\frac{\partial \ln m^*}{\partial T}\right)_V = \frac{\partial \ln \mathscr{E}_g}{\partial T}. \tag{6.41}$$

The quantity $(\partial \ln m^*/\partial T)_0$ can be found in the same way as $(\partial \mathscr{E}_g/\partial T)_0$, using the formula

$$\left(\frac{\partial \ln m^*}{\partial T}\right)_0 = -\frac{3\alpha_{t.e}}{\xi} \frac{\partial \ln m^*}{\partial P}. \tag{6.42}$$

Comparing the derivatives calculated from Eq. (6.42) with the values of $\Delta = \partial \ln m^*/\partial \ln T$ found from measurements of the thermoelectric power, Nernst – Ettingshausen effect, and Faraday effect, we find that about half the temperature-induced change in the effective mass is due to the thermal expansion of a crystal and the other half should be attributed to the influence of the lattice vibrations.

It is interesting to compare the values of Δ, given in Appendix C, with the derivatives $\partial \ln \mathscr{E}_g/\partial \ln T$, whose 300°K values are 0.3 for PbS, and 0.4 for PbSe and PbTe. The values of Δ are approximately equal to $\partial \ln \mathscr{E}_g/\partial \ln T$ for PbTe and PbSe, as expected on the basis of Eq. (6.41), while in the case of PbS this approximate equality applies only to the order of magnitude.

In semiconductors with a narrow forbidden band, such as lead chalcogenides, the temperature dependence of the forbidden band width is relatively strong and consequently the temperature dependence of the effective mass is also strong. The relative change in the deformation potential due to increasing temperature is not related to the change in the forbidden band and is considerably smaller than the relative changes in the forbidden band width and effective mass. To account for the difference between the temperature dependence of the deformation potential and the corresponding dependences of the forbidden band width and effective mass, we can use the same considerations as in the explanation of the difference between the relative changes of the same quantities under the influence of pressure (cf. §3.3A).

We must bear in mind that the experimental data used to determine the temperature dependence of the effective mass are normally analyzed ignoring the nonparabolicity, which should have some effect on the results because of the redistribution of carriers in a nonparabolic band. Moreover, in the nonparabolic region, the effective mass depends less on the forbidden band width and Eq. (6.41) should be replaced by a more general relationship:

$$\frac{\partial \ln m^*}{\partial T} = \left(\frac{\partial \ln m^*}{\partial \mathscr{E}_g}\right)_T \frac{\partial \ln \mathscr{E}_g}{\partial T}, \qquad (6.43)$$

where the effective mass, considered as a function of $\mathscr{E}_g$ is given by a formula of the (6.36) type. An analysis of the experimental data on the thermoelectric power and other effects, taking into account the nonparabolicity of n-type PbTe [334, 337] (cf. §4.1B and D), shows that the effective mass at the bottom of the conduction band varies so that the relative change in the forbidden band width is equal to the relative change in the effective mass, in agreement with the theoretical predictions.

Summarizing this discussion, we may conclude that the pressure and temperature dependences of the effective mass and the forbidden band width are in agreement with the assumption that the effective mass is governed by the interaction of the conduction and valence bands and that about half of the temperature-induced change in the forbidden band width and effective mass is due to the thermal expansion of the lattice while the other half is due to the interaction of carriers with phonons.

All this applies in the range of temperatures in which the forbidden band width varies linearly with temperature. Investigations of the fundamental absorption edge, Hall effect, and thermal conductivity of PbTe at temperatures above 400°K show that the heavy-hole band edge rises above the light-hole band edge (Fig. 3.10) and the forbidden band width is practically independent of temperature (Fig. 4.20a). The effective mass of holes in PbTe near 400°K increases strongly with increasing temperature, and above 400°K it is not directly related to the forbidden band width, because at high

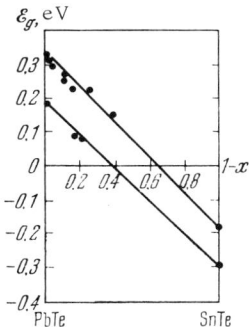

Fig. 6.8. Forbidden band width of $Pb_xSn_{1-x}Te$ solid solutions as a function of the composition [137]. The straight lines are drawn through the experimental points obtained from investigations of the optical absorption at 300°K (the upper line) and recombination radiation at 12°K (the lower line). The points for SnTe were obtained from measurements of the tunnel effect at 300 and 4.2°K.

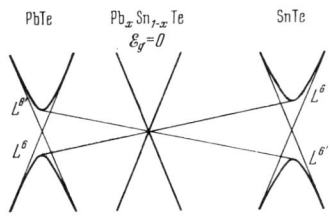

Fig. 6.9. Schematic representation of the conduction and valence bands of PbTe, SnTe and a PbTe – SnTe alloy in which the forbidden band vanishes [137].

temperatures the principal extrema of the conduction and valence bands are at different points in the k space. PbSe and PbS also exhibit deviations from the linear temperature dependence of the forbidden band width above ~400°K. However, an anomalous temperature dependence of the Hall coefficient has not been observed for p-type PbSe and p-type PbS and therefore the problem of the existence of the second valence band and of the temperature dependence of the forbidden band width of lead selenide and sulfide at high temperatures requires further investigations.

C. Forbidden Band Width of Solid Solutions

Optical investigations (cf. §2.1) indicate that the forbidden band width of PbTe – PbSe and PbSe – PbS alloys is a monotonic function of the composition. Such a dependence on the composition is attributed to the similarity of the band structures of all three lead chalcogenides.

The dependence of the forbidden band width on the composition is more complex for $Pb_xSn_{1-x}Te$ alloys. Investigations of the optical absorption at room temperature (cf. §2.1) and recombination (laser) radiation at low temperatures (cf. §2.2D) show that when SnTe is added to PbTe, the forbidden band width decreases. If we extrapolate the obtained linear dependence of the forbidden band width on the composition to solid solutions with higher concentrations of SnTe, we find that at some value $x = x_{cr}$, which depends on temperature, the forbidden band width vanishes (Fig. 6.8). This means that the top of the valence band coincides with the bottom of the conduction band, as shown in Fig. 6.9.

If we assume that the observed relationship applies at lower values of x, we find that the bands diverge again but their positions are interchanged, i.e., at compositions corresponding to $x < x_{cr}$, the top of the valence band is described by the wave function $L^{6'}$ while the bottom of the conduction band is described by L^6, which is the reverse of the situation for $x > x_{cr}$ (similar to the case of PbTe).

Dimmock et al. [137] observed this relationship and provided a qualitative theoretical explanation of this effect. Their explanation is based on the assumption that the relativistic corrections for the materials considered affect considerably the mutual positions of the energy levels. Since the relativistic terms are large only near the lattice sites, where the wave functions are close to the atomic functions, we can estimate the relativistic shifts of the band edges in a crystal using known differences between the relativistic corrections for the corresponding atomic levels of Pb and Sn as well as employing the data on the composition of the wave functions obtained by the APW method (cf. §6.1C). Estimates show that the upward relativistic shift of the L^6 level, relative to the $L^{6'}$ level, is 0.76 eV in the transition from PbTe to SnTe. Such a shift is sufficient for the coincidence of the conduction and valence bands at some intermediate composition [to explain the observed relationships, the relative shift of the L^6 and $L^{6'}$ levels should be approximately equal to $\mathscr{E}_g(\text{PbTe}) + \mathscr{E}_g(\text{SnTe})$ 0.5 eV].

Band inversion should give rise to different signs of the quantities representing the dependence of the forbidden band width on the temperature and pressure, on both sides of x_{cr}.

Investigations of the fundamental absorption edge, laser emission, and photoelectric phenomena in $Pb_xSn_{1-x}Te$ and $Pb_xSn_{1-x}Se$ solid solutions [544-549] have confirmed the band inversion in these compounds and its consequences. When $x < x_{cr}$, the forbidden band width has a negative temperature coefficient [547, 548]. Thermomagnetic investigations carried out by the authors of this book and their colleagues have shown that the effective mass and nonparabolicity of $Pb_xSn_{1-x}Te$ solid solutions are in agreement with Fig. 6.9: the effective mass and the effective width of the interaction forbidden band decreases when SnTe is added to PbTe.

6.4. Carrier Scattering Mechanisms

Analysis of the experimental data on the transport phenomena and the theoretical calculations show that the principal scattering mechanisms in lead chalcogenides are the scattering by the acoustical and optical phonons, collisions between carriers, and scattering by impurities. The contributions of the polar scattering by the optical phonons and of the collisions between carriers have been deduced reliably using formulas free of any adjustable parameters. The contribution of the acoustical scattering is represented by the deformation potential constant, which is selected so as to make the experimental value of the mobility equal to the theoretical value (at the same temperature and for the same carrier density). All subsequent calculations of the transport coefficients can be carried out, in a wide range of temperatures and carrier densities, without introducing new adjustable parameters.

We shall now consider the role of each of the scattering mechanisms in greater detail (cf. also §3.1, §3.2D, §4.1E, §4.2A).

The assumption that carriers are scattered by the long-wavelength acoustical vibrations explains, when the temperature dependence of the effective mass is taken into account, the experimentally observed temperature dependence of the mobility above 100°K (the $T^{-5/2}$ law). Scattering by the acoustical phonons can be represented by the deformation potential constants, which are 20-25 eV for n- and p-type samples of all three lead chalcogenides. The relaxation time for the acoustical scattering is almost isotropic. In calculations of the transport phenomena we must take into account the band nonparabolicity as well as the dependence of the matrix element of the interaction of carriers with the acoustical phonons on the energy in the nonparabolic region (the formula for the relaxation time of lead chalcogenides is given in Appendix A). The nonparabolicity results in a considerably deviation of the scattering parameter $r = \partial \ln \tau / \partial \ln \xi$ from $-1/2$.

The polar scattering by the long-wavelength longitudinal optical phonons gives rise, at some temperatures, to approximately the same temperature dependence as the acoustical scattering. For example, the ratio of the mobility governed by the acoustical and optical scattering in PbTe and PbSe is practically independent of temperature between 77 and 300°K. The polar scattering at these temperatures is stronger than the acoustical scattering for carrier densities below $4 \cdot 10^{18}$ cm^{-3} for PbTe, $9 \cdot 10^{18}$ cm^{-3} for PbSe, and $1.1 \cdot 10^{19}$ cm^{-3} for PbS; at higher carrier densities the acoustical scattering predominates.

When the temperature is increased considerably above the Debye value (above 300°K for PbTe and PbSe) the relative importance of the acoustical scattering gradually increases. Thus, at high temperatures and carrier densities, which are important in thermoelectric applications of lead chalcogenides (§7.1), the scattering by the acoustical phonons predominates.

When the temperature is reduced below 77°K, the relative contribution of the polar scattering again decreases. This explains why the $T^{-5/2}$ law for the mobility is exhibited by relatively pure samples even well below 77°K (down to 20°K for PbTe with an electron density of $2 \cdot 10^{18}$ cm^{-3}), in spite of the fact that at low temperatures the effective mass is independent of temperature.

In contrast to the acoustical scattering, whose inelasticity appears only at the lowest temperatures (below 7°K for PbTe with an electron density of $2 \cdot 10^{18}$ cm^{-3}), the polar scattering is strongly inelastic already at liquid-nitrogen temperature. However, above the Debye temperature (above 300°K for PbTe and PbSe and above 600°K for PbS) the polar scattering can be regarded as almost elastic and we can use the relaxation time concept. At lower temperatures we can use the variational method for the solution of the transport equation. The relaxation time at high temperatures is given in Appendix A.

As in the case of calculations of the scattering by the acoustical phonons, the nonparabolicity must be allowed for in determination of the relaxation time for the polar scattering in lead chalcogenides as well as in calculation of the matrix elements of the carrier – phonon interaction. Moreover, since the density of free carriers in lead chalcogenides is always high, we must take into account the scattering of the polar vibrations by free carriers. The screening not only reduces the strength of the polar scattering but also gives rise to a considerable dispersion of the longitudinal optical phonons in the range of low values of the wave number (this applies to semiconductors with a large differences between the static and high-frequency values of the permittivity). These effects have a strong influence on the magnitude and the energy dependence of the relaxation time in the polar scattering case and they are responsible for a deviation of r from 1/2.

In semiconductors with an anisotropic band structure the scattering by the optical phonons is anisotropic ($\tau_{\parallel}/\tau_{\perp}$ is of the order of 2-3 for PbTe), which reduces the effective anisotropy coefficient found from measurements of such anisotropic transport properties as the electrical resistivity and thermal conductivity in a magnetic field.

Collisions between carriers play an important role in the thermal, thermoelectric, and thermomagnetic phenomena near the liquid-nitrogen temperature. In particular, calculations show that such collisions reduce considerably the Lorenz number (approximately by a factor of 2 at carrier densities of the order of 10^{18}-10^{19} cm^{-3} for all three materials). The polar scattering, which is inelastic at low temperatures, also reduces somewhat the Lorenz number but this reduction is insufficient to explain the experimental data without assuming that collisions between carriers take place.

The selection rules for the matrix elements show that the scattering by the short-wavelength phonons accompanied by carrier transfer between equivalent extrema is not a very likely process. This conclusion has been confirmed by direct measurements of the intervalley-scattering relaxation time, carried out using an ultrasonic method at low temperatures and carrier densities. However, we cannot assume that this inelastic scattering mechanism is completely negligible; it may play some role at high carrier densities and temperatures, when high-energy carriers are available. Transitions between inequivalent extrema are important in p-type samples of PbTe, SnTe, and alloys formed from these two materials.

The scattering by ionized impurities and vacancies is the dominant mechanism at low (liquid-helium) temperatures and the main contribution is made not by the screened Coulomb potential but by the inner component of the impurity ion potential, acting over distances of the order of the lattice constant. This scattering mechanism is responsible for the temperature-independent mobility in degenerate samples. Estimates of the scattering by lattice inhomogeneities due to shear deformation and stretching by impurity ions show that such scattering is not very important.

The scattering by dislocations and other lattice defects may be important at low temperatures and it may give rise to a scatter of the mobility values. Measurements have shown that an increase in the dislocation density reduces strongly the carrier mobility.

The scattering of carriers by the surface and by grain boundaries is observed sometimes in thin films as well as in polycrystalline samples.

Calculations have been made of various transport coefficients (mobility, Lorenz number, Hall coefficient, thermoelectric power, and the Nernst – Ettingshausen coefficient) and of their dependences on temperature, carrier density, and magnetic field, using a single adjustable parameter (the deformation potential constant). The scattering by the acoustical phonons, polar scattering, collisions between carriers, and scattering by impurities (at the lowest temperatures) have been allowed for in these calculations. Such calculations make it possible to explain, on the basis of a unified theory, a large number of experimental observations of various transport phenomena in a wide range of temperatures, carrier densities, and magnetic fields. In this way, we obtain detailed and reliable information on the scattering mechanisms in lead chalcogenides, which is still lacking even for such classical semiconductors as germanium and indium antimonide.

CHAPTER VII

APPLICATIONS OF LEAD CHALCOGENIDES

7.1. Thermoelectric Properties

The first investigations of the thermoelectric properties of lead chalcogenides were carried out by Ioffe, Maslakovets et al. [207, 488-490]. The outcome of these investigations was the use of lead sulfide in the first thermoelectric generators and of lead telluride in various cooling units [490, 491]. Later it was found that alloys based on antimony and bismuth tellurides were more suitable materials for cooling purposes. Lead chalcogenides, particularly lead telluride and its alloys, were found to be the most efficient materials for thermoelectric generators working between room temperature and 600-650°C.

A. Thermoelectric Figure of Merit of Lead Chalcogenides and their Alloys

According to the theory of the thermoelectric energy conversion, the figure of merit of materials used as arms of a thermoelement (thermocouple) is

$$Z = \frac{\alpha_0^2 \sigma}{\varkappa}. \tag{7.1}$$

All quantities which appear in Z are functions of the temperature and carrier density (cf. Chaps. III and IV) and consequently Z depends strongly on the temperature and carrier density. At a given temperature, the value of Z reaches its maximum, Z_{max}, at some optimum carrier density n_0, which usually increases with increasing temperature. In view of this, semiconductors used in thermoelectric devices are always doped and the doping is usually fairly heavy.

TABLE 7.1. Comparison of Thermoelectric Figures
of Merit of Lead Chalcogenides
at Room Temperature*

Compound	Conduction type	N	$\dfrac{m^*}{m_0}$	μ, $\dfrac{cm^2}{V \cdot sec}$	$\varkappa_1 \cdot 10^3$, $\dfrac{cal}{cm \cdot sec \cdot deg}$	$\dfrac{\mu}{\varkappa_1} \cdot 10^{-5}$	Z_{max}, 10^{-3} deg^{-1}
Pb Te	n	4	0.13	1730	4.8	3.6	1.9
	p	4	0.15	840	4.8	1.8	1.2
Pb Se	n	4	0.14	1000	3.9	2.6	1.5
	p	4	0.13	1000	3.9	2.6	1.4
Pb S	n	4	0.15	620	6.0	1.0	0.7
	p	4	0.18	610	6.0	1.0	0.9

*This table lists the calculated values of Z, obtained using the approximate formula (7.2). The values of m^*/m_0 are taken from the thermoelectric power data for $n \approx 10^{19}$ cm^{-3} [218, 230, 122].

A general mathematical analysis of the dependence of Z on the temperature and carrier density is fairly complex. However, several simplifying assumptions (the presence of carriers of one sign only, the absence of degeneracy, a low value of the electronic component of the thermal conductivity, and some other assumptions), make it possible to establish a relationship between Z_{max} and the properties of a material:

$$Z_{max} \propto N \frac{m^{*3/2} \mu}{\varkappa_1} T^{3/2} \exp(r + 1/2) \qquad (7.2)$$

The majority of the assumptions made in the derivation of the above formula for Z_{max} are not satisfied exactly and therefore the expression given in Eq. (7.2) can be used only in qualitative estimates. However, in spite of the approximate nature of this relationship, it is very useful in comparative estimates and in analysis of the potentialities of thermoelectric materials. The values of the properties which govern Z_{max} are given in Table 7.1 for all three compounds of the PbS group at $T = 300°K$. It is evident from the values in Table 7.1 that the high thermoelectric figures of merit of lead chalcogenides are due to the many-ellipsoidal nature of their energy spectrum (N = 4, cf. §4.1) and low values of the lattice thermal conductivity ($\sim 5 \cdot 10^{-3}$ cal · cm^{-1} · sec^{-1} · sec^{-1}) and relatively high carrier mobilities (~ 1000 cm^2 · V^{-1} · sec^{-1}).

The low values of $\varkappa_l$ are, to a considerable degree, due to the large masses of the atoms forming lead chalcogenides while the relatively high values of the mobility are due to the low values of the effective carrier masses and the relatively weak ionic component of binding (cf. §1.1).

The high value of the permittivity ε_0 (cf. Appendix C) is also a favorable factor, which contributes to the high value of the thermoelectric figure of merit of lead chalcogenides. The high value of the permittivity is responsible for a considerable reduction in the scattering cross section of electrically charged impurity centers so that the scattering by ionized impurities has little effect. This aspect is particularly important in the case of materials for thermoelectric generator applications in which the optimum carrier density usually considerably exceeds 10^{19} cm^{-3}.

Comparison of the properties of the three compounds of the PbS group shows that the differences between their thermoelectric properties at $T = 300°K$ are governed mainly by the ratio of the carrier mobility and the lattice thermal conductivity. The most favorable value of this ratio is obtained for n-type lead telluride, which has a higher carrier mobility than other lead chalcogenides.

At high temperatures, another fundamental parameter becomes important. This is the forbidden band width which governs the onset of the intrinsic conduction. The appearance of carriers of opposite sign much reduces the value of Z because of the reduction of the thermoelectric power and the rise of the thermal conductivity due to ambipolar carrier diffusion (cf. §4.2B). From this point of view, a favorable factor is the positive temperature coefficient of the forbidden band width of all compounds of the PbS group (cf. Appendix C). This increase of the forbidden band width with increasing temperature delays the onset of intrinsic conduction and makes it possible to reach higher values of Z_{max}.

The largest value of $\mathscr{E}_g$ among lead chalcogenides is that of lead sulfide, but this compound is considerably inferior to lead selenide and telluride in respect of the other thermoelectric parameters (cf. Table 7.1). Lead telluride has the second widest forbidden band and the highest value of the ratio $\mu/\varkappa_l$. The combination of these two factors makes lead telluride the most promising thermoelectric material among the three chalcogenides.

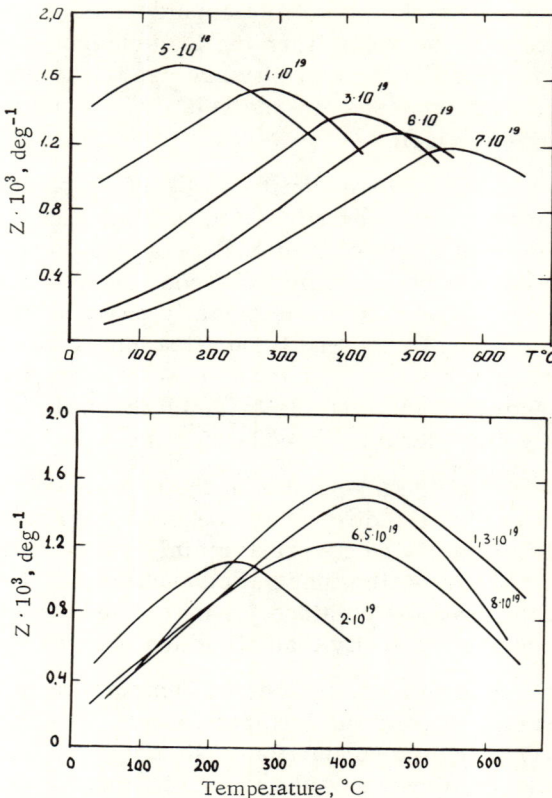

Fig. 7.1. Thermoelectric figure of merit of n-type (a) and p-type (b) lead telluride.

The essential experimental data on the thermoelectric figure of merit of lead chalcogenides are reported in the papers of Ioffe [488], Wright [493], Iordanishvili [401, 503], and others [494, 495, 550]. The figure of merit of lead telluride has been investigated most thoroughly. Figure 7.1 shows the temperature dependence of Z, at 20-650°C obtained for n- and p-type PbTe by Stavitskaya [550]. Iodine was used as a donor impurity and sodium was used as an acceptor. The maximum value of the thermoelectric figure of merit of n-type PbTe, $Z_{max} = 1.7 \cdot 10^{-3}$ deg^{-1}, was reached at T $\approx$ 150°C and it corresponded to $n_{opt} = 5 \cdot 10^{18}$ cm^{-3}. When the carrier density was increased, the maximum of Z for n-type PbTe shifted in the direction of higher temperatures and its absolute value decreased.

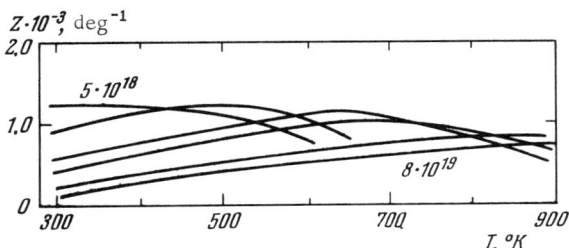

Fig. 7.2. Thermoelectric figure of merit of n-type lead selenide.

At $T \approx 500°C$ the value of Z_{max} for n-type PbTe did not exceed $1.3 \cdot 10^{-3}$ deg^{-1}.

The temperature dependence of Z_{max} of p-type PbTe was more complex (Fig. 7.1b). At $T \approx 200°C$, Z_{max} of p-type PbTe corresponded to $n_{opt} \approx 2 \cdot 10^{19}$ cm^{-3} and its value did not exceed $1.2 \cdot 10^{-3}$ deg^{-1}. However, when the optimum value of the carrier density was increased further, Z_{max} increased to $1.6 \cdot 10^{-3}$ deg^{-1} and the position of the maximum of Z in the range of carrier densities $6 \cdot 10^{19}$–$1.5 \cdot 10^{20}$ cm^{-3} remained practically constant and was observed at $T \approx 400°C$. Such a temperature dependence of Z_{max} was in good agreement with the complex structure of the valence band of PbTe (§3.2B) and corresponded to an increase of the importance of the heavy holes when the temperature and the degree of doping were increased.

The temperature dependences of the thermoelectric figure of merit of n-type PbSe (Fig. 7.2) are similar to the corresponding dependences of n-type PbTe, but the selenide has lower values of Z_{max} and a broader maximum [401]. The figure of merit of p-type lead selenide is considerably lower than the corresponding figure of merit of p-type PbTe, especially if we take into account that up to $p = 1 \cdot 10^{20}$ cm^{-3} the influence of the heavy-hole band is unimportant and the figure of merit of p-type PbSe does not increase with increasing temperature.

Lead sulfide has the lowest value of the figure of merit ($\sim 0.8 \cdot 10^{-3}$ deg^{-1}). However, the cheapness of this material and its natural occurrence in nature (in the form of galenite) may recommend it for some applications.

Higher values of the thermoelectric figure of merit have been reported for solid solutions based on lead chalcogenides. It has long been known that the thermoelectric figure of merit of n-type PbTe – PbSe solid solutions is high. In this case, the figure of merit increases because of an increase in $\mu/\varkappa_l$ (which is approximately 40% higher compared with lead telluride) and the appearance of additional scattering by neutral impurity centers. The stronger scattering of phonons, compared with electrons, is due to the difference between their wavelengths and it was predicted theoretically by A. V. and A. F. Ioffe [489]; it forms the basis of one of the main methods for increasing the efficiency of thermoelectric materials. However, the gain in $\mu/\varkappa_l$ usually decreases with increasing temperature because of an increase in the importance of the phonon scattering mechanism and a corresponding reduction in the importance of the scattering by impurities. At temperatures above 400°C, the value of $\mu/\varkappa_l$ for n-type PbTe – PbSe solid solutions approaches the value of $\mu/\varkappa_l$ of n-type PbTe. For this reason, PbTe – PbSe solid solucially as the formation of a solid solution reduces the forbidden band width.

Similar results are obtained for n-type PbTe – SnTe solid solutions.

The greatest changes in the thermoelectric properties are observed in p-type alloys. Obviously, in this case, the introduction of impurities alters the positions of the heavy-hole band (cf. §3.2B) and this results in a considerable change in the thermoelectric properties. An excellent example of such a change is provided by p-type PbTe – SnTe solid solutions [255] in which the presence of 1-3% of SnTe affects the carrier-density dependence of the thermoelectric power. However, changes in the band structure in these solid solutions are unfavorable from the point of view of the thermoelectric figure of merit since the introduction of SnTe strongly reduces the hole mobility.

The figure of merit of p-type PbTe can be increased considerably by adding $AgSbTe_2$. Changes in the nature of the temperature dependences of the thermoelectric power and electrical conductivity, observed in this case, result in an increase of the maximum values of $\alpha_0^2 \sigma$. Moreover, the introduction of $AgSbTe_2$ molecules into PbTe gives rise to a strong phonon scattering and a sharp fall in $\varkappa_l$. All this increases the maximum values of Z of p-type PbTe – $AgSbTe_2$ solid solutions (up to $2 \cdot 10^{-3}$ deg^{-1} at T = 600°C [496]).

An increase in the figure of merit of lead chalcogenides can be expected also, as indicated by Eq. (7.2), when the nature of the scattering mechanism is altered and the parameter r is increased (for example, by introducing impurity ions). However, experiments carried out in order to test this method of improving the figure of merit were not successful [222]. This is because the potentials of the ions in lead chalcogenides differ very considerably from the Coulomb potential because of screening (cf. §3.1C).

We can expect an increase in the figure of merit of lead chalcogenides due to the strong band nonparabolicity and the increase in the effective mass with increasing temperature and carrier density. According to Eq. (7.2), the effective mass appears explicitly in Z_{max}, which is proportional to $(m^*)^{3/2}$, and implicitly through the carrier mobility. For the scattering by the acoustical vibrations, $\mu \propto (m^*)^{-5/2}$ and $Z_{max} \propto 1/m^*$. However, if we establish a scattering mechanism in which the carrier mobility depends weakly on the effective mass or increases with increasing m^* (for example, the scattering by grain boundaries), we can use the characteristic features of the band structure of lead chalcogenides to increase their thermoelectric figure of merit.

A comparative analysis of the figures of merit of materials based on lead chalcogenides and on antimony and bismuth chalcogenides shows that, at temperatures of 200-600°C, the most efficient materials are lead telluride and its alloys doped to give a carrier density $n > 10^{19}$ cm^{-3}. Since the temperature dependence of the figure of merit of lead telluride alloys has a fairly narrow maximum, it is convenient to use a variable carrier density, which gives the optimum values at different parts of a thermoelement arm in accordance with the temperature distribution along the arm. In such a case, the average value of the figure of merit of a thermoelement consisting of n-type and p-type PbTe arms is about $1.5 \cdot 10^{-1}$ deg^{-1} in the interval between room temperature and 900°K. The efficiency of such a thermoelement, calculated from the formula

$$\eta = \frac{\Delta T}{T_{cold}} \frac{\sqrt{1+ZT}-1}{\sqrt{1+ZT}+\frac{T_{cold}}{T_{hot}}},$$

amounts to about 17%

Table 7.2. Calculated Efficiencies of Thermoelectric Generators Based on PbTe [492]

No.	Nature of elements	T_{hot}, °C	T_{cold}, °C	Efficiency calc. from props. of material, %
1	n- and p-type PbTe with gradual variation of optimum carrier density	593	38	11.7
2	Dopant compositions: n — 0.055% PbI_2 p — 1.0% Na			10.28
3	Two dopants in each arm: n — 0.055% PbI_2 below 371°C n — 0.10% PbI_2 above 371°C p — 0.3% Na below 315°C p — 1.0% Na above 315°C			11.2

However, a continuous variation of the carrier density along a thermoelement arm is fairly difficult to achieve technically and therefore, in practical applications, it is usual to employ a stepwise change in the carrier density by using two or three sections with different carrier densities [492]. Comparisons of the efficiencies of thermoelements at different optimum carrier densities is given in Table 7.2.

B. Thermoelectric Generators Based on Lead Telluride

In practical applications of lead telluride in thermoelectric generators, we encouter a number of technical problems associated with the physicochemical properties of this compound. One such problem is the protection of the thermoelement materials from oxidation and evaporation. The oxidation of lead telluride, which gives rise to acceptor centers (cf. §1.4), may alter strongly all the thermoelectric properties and even change the type of conduction and give rise to unwanted p–n junctions in the n-type arm. The formation of an oxide film has a particularly strong effect in the presence of microcracks or a considerable porosity of a sample and it is manifested by a strong increase of the resistance of a thermoelement. The placing of the thermoelements in evacuated containers prevents oxidation but it is undesirable for other reasons: lead telluride

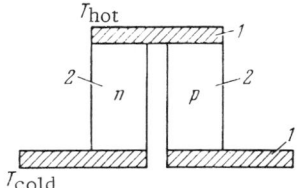

Fig. 7.3. Construction of a thermoelement: 1) bridging plates; 2) working thermoelectric material.

has a relatively high vapor pressure at T > 500°C (cf. §1.2). Strong evaporation of PbTe may result in the appearance of shunting bridges and, during long service life, may alter the geometrical dimensions of the thermoelement and, consequently, the electrical characteristics of the generator as a whole. Lead telluride is usually protected from oxidation and evaporation by protective greasing and coating. There are several USA and Japanese patents dealing with the coating of thermoelements with ceramic or silicone materials [497, 498, 555]. The suggested materials ensure that the coating is vacuum-tight and their linear expansion coefficients match those of the working material of the thermoelement.

An important problem is the selection of the bridging materials and the methods (i.e., materials and methods for joining the p- and n-type arms of a thermoelement to ensure minimum thermal and electrical resistance between them: cf. Fig. 7.3). The bridging materials, like the protective coatings, should have a linear expansion coefficient close to that of the working material of the thermoelement ($\sim 18 \cdot 10^{-6}$ deg^{-1}) and they should not interact with the thermoelement arms because this might alter considerably the electrical properties of the arm materials. In the case of lead telluride, we can use pure iron or lightly alloyed steels as the bridging materials [494, 499, 500]. The bridging method is governed, to a considerable degree, by the technique used in the fabrication of thermoelement. In thermoelectric generators, it is usual to employ polycrystalline thermoelement arms prepared by casting, drawn under pressure, or pressed (cf. §1.2). In the latter case, the bridging connection is made by pressing a metal layer to the ends of the arms. When cast samples are used, the bridging is achieved either by diffusion or by pressure contacts [500, 501]. The pressure-contact method is the simplest but the contact resistance is not less than 20% of the total resistance [502, 503]. The joining of iron to n-type PbTe by the diffusion method is carried out by heating the contact at $T = 858 \pm 2°C$ for 20-30 min under a load of 500 g/cm^2. The direct joining of Fe with p-type PbTe is difficult and therefore it is usual to employ an intermediate p-type SnTe layer, which penetrates

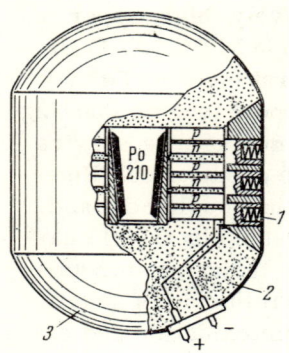

Fig. 7.4. SNAP-III thermoelectric generator based on PbTe: 1) spring; 2) thermal insulation; 3) radiator.

iron satisfactorily (at $T \approx 850°C$). The connection between p-type SnTe and p-type PbTe is made by diffusion at $T = 808°C$ [500].

The temperature range in which lead telluride is employed in thermoelectric generators is usually limited to 600-650°C on the high-temperature side and to $T = 30-300°C$ on the low-temperature side. The upper limit is governed primarily by the physicochemical properties of lead telluride (its strong volatility, oxidation, and deterioration of mechanical properties at high temperatures), while the lower limit is set by the availability of more efficient low-temperature materials. We have mentioned in the preceding subsection that at $T < 200°C$ lead telluride is less efficient than alloys based on antimony and bismuth tellurides, but if cascading is impossible, one can use lead telluride right down to room temperature. In view of its limited working temperature range, lead telluride is employed in thermoelectric generators in which heat sources maintain the hot-junction temperature, which is not higher than 700°C.* Such heat sources include:

1. Radioactive heat sources (Po^{210}, Pu^{238}, Sr^{90}), which are employed, for example, in American thermoelectric generators of the SNAP-III type in which the junction temperatures are $T_{hot} = 590°C$, $T_{cold} = 200°C$ (Fig. 7.4). Generators of this type have a power output of the order of 10-100 W and an efficiency up to 5-6% [502, 551].

2. Heat from a nuclear reactor, transferred to the hot junction by means of a suitable heat carrier (for example, a potassium − sodium mixture) [503, 506]. In this case, the hot-junction temperature reaches 600-700°C. The cold-junction temperature should not

---

*In some cases, the p-type arms of the thermoelements used in the working temperature range of lead telluride are made of alloys of p-type GeTe, which have practically the same figure of merit as p-type PbTe but have a higher vapor pressure and a lower melting point [493]; the use of germanium telluride improves the mechanical properties of thermoelements.

TABLE 7.3. Thermoelectric Generators Based on PbTe [503, 504, 506]

Type, company	Heat source	Output, W	T_{cold}, °C	T_{hot}, °C	Eff., %	Applications
SNAP III	Po^{210}	3-5	200	590	5.8	Space vehicles
SNAP VII	Pu^{238}	—	—	—	4	Naval applications
Neptune	Sr^{90} Reactor heat	15,000	—	570	—	Submarine power supply (design)
3-M	Propane – butane	10-50	149	593	1.6	Radio power supply; illumination of railroad cars
WEC	Propane	340	125	435	—	Radio power supply
B-1	Propane	100	50	600	2.4	—

be lower than 300°C because at lower temperatures the properties of lead telluride may be affected by the reactor radiation. The design power output of such devices is 1-10 kW.

3. Heat of chemical reactions (burning of propane, butane, etc.). Characteristics of some PbTe generators using the heat of combustion are listed in Table 7.3. The efficiency of such devices is usually low (2-3%) since the gas burners employed to heat the hot junction have an efficiency not greater than 40% [505, 552].

Comparison of the efficiency of actual PbTe generators and of the efficiency calculated from the properties of PbTe (Tables 7.3 and 7.2, respectively) shows that the losses in currently produced thermoelements are very considerable. These losses are mainly due to the still unsatisfactory methods of bridging and heat transfer, and also to the partial matching of the parameters of the working material with a given temperature distribution. Therefore, further improvements in thermoelement technology should result in a considerable increase in the efficiency of these generators. Under laboratory conditions, it is already possible to reach generator efficiencies approaching 10-15%, which are close to the efficiencies calculated from the properties of the thermoelement materials [495, 503].

In spite of their relatively low efficiencies, lead telluride generators are widely used in a number of special applications, particularly in space exploration [553, 554]. The absence of moving parts, reliability of operation, long service life, and small size make thermoelectric generators the best power supply units for unpopulated and remote areas as well as for small unattended units in terrestrial or space applications (Table 7.3).

7.2. Infrared Technology

A. Photoresistors

Photoresistors made of lead chalcogenides are used widely in the detection and measurement of infrared radiation (cf. reviews [146, 508-513]). The development of photoconducting infrared detectors in the late thirties and forties was a great step forward in the technique of infrared detection. Lead chalcogenide photoresistors have not lost their importance even now, although devices made of $A^{III}B^{V}$ semiconductors have been successfully developed in recent years and they have a number of advantages over lead chalcogenide photoresistors.* Lead sulfide photoresistors are used in high-resolution spectroscopes.

The sensitivity of lead chalcogenide photoresistors is fairly high. For example, a lead sulfide photoresistor is approximately 100 times more sensitive to monochromatic radiation then thermal detectors such as bolometers and thermopiles. Lead chalcogenide devices have a fast response and their time constant is considerably shorter than that of modern thermal detectors. Lead chalcogenide photoresistors, produced on a mass scale, are cheap because of the easy availability of these compounds and the simplicity of the fabrication techniques. These photoresistors have a low ohmic resistance, small size, are simple to use, and have a very long service life.

The properties of photoresistors made of the three chalcogenides are essentially similar. The most widely used are lead sulfide photoresistors which are better detectors in the special region near $2\,\mu$. PbSe detectors, used at longer wavelengths, were

---

*InAs is used at the same wavelengths as PbS, while InSb is used in the same spectral range as PbSe and PbTe.

Fig. 7.5. Lead sulfide photoresistor produced by British Thomson-Houston.

developed later; current models have good properties and are used widely. PbTe photoresistors have been developed least of all since they have approximately the same spectral sensitivity characteristic as PbSe devices but a lower sensitivity threshold (detectivity).

The intrinsic photoconductivity is used in lead chalcogenide photoresistors, because the forbidden bands of these materials are relatively narrow. The sensitive elements of photoresistors are usually polycrystalline films about 1 μ thick with a grain size of about 0.1 μ; recently, single-crystal epitaxial films [41] and thin pure single-crystal photoresistors [36, 514] were described. The photoconductivity mechanism in films is not yet clear; it is usual to attribute the photoconductivity to the carrier-density and barrier mechanisms. These two mechanisms, their role in the photoconductivity, the influence of oxygen on the photoconductivity,

Fig. 7.6. Lead sulfide photoresistor of the 61SV type produced by the Mullard Ltd.

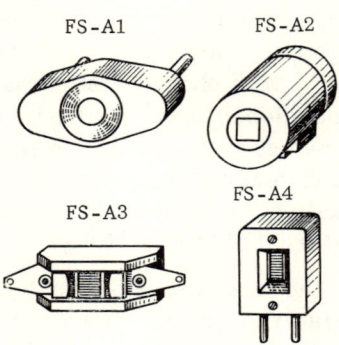

Fig. 7.7. Lead sulfide photoresistors of the FS-A series, manufactured in the Soviet Union.

as well as the properties of lead chalcogenides manifested in the photoconductivity, are all considered in § 2.2.

Thin photoconducting lead chalcogenide films are deposited on glass or a similar substrate by evaporation (sublimination) or by a chemical method (cf., for example, [508]). In the chemical method, photoconducting films are precipitated from a solution in the presence of an oxidizing agent. Such films are coated with a protective lacquer before exposure to air. This method was used by British Thomson-Houston (England), Eastman Kodak (USA), and other firms, in the manufacture of uncooled photoresistors (Fig. 7.5).

Photosensitive films can be prepared also by sublimation in vacuum followed by heating in a low-pressure oxygen atmosphere, or by direct evaporation at a low oxygen pressure, which is then pumped down. Figure 7.6 shows an uncooled lead sulfide photoresistor enclosed in an evacuated glass envelope which was prepared by the evaporation method and marketed by Mullard Ltd. (England). Cooled photocells require Dewar flasks.

The outside of some Soviet lead sulfide photoresistors of the FS-A series is shown in Fig. 7.7. The most popular are the FS-A1 and FS-A4 devices [509]. The fabrication technology was developed by Kolomiets [515], who also investigated the properties of these photoresistors.

The photocells prepared by the evaporation method are more sensitive at room temperature than those prepared by the chemical method. However, the chemical method is simpler and therefore it is used more frequently in industry, while the evaporation method is usually employed in scientific research establishments.

Typical parameters of lead chalcogenide photoresistors, collected from the catalogs of various firms by Kruse et al. [510], are listed in Table 7.4. It must be stressed that the properties of manufactured infrared detectors are being continuously improved and

TABLE 7.4. Typical Parameters of Lead Chalcogenide Photoresistors [510]

Compound	T, °K	λ_m, μ	$\lambda_{1/2}$, μ	$D^*_{500°K}$, cm·cps$^{1/2}$·W^{-1}	$D^*_{\lambda m}$, cm·cps$^{1/2}$·W^{-1}	Frequencies (in cps) for values of D^* and D^*_λ	Time constant, μsec	Calculated optimum modulation frequency, cps	Resistance per square, MΩ
PbS	295	2.1	2.5	$4.5 \cdot 10^8$	$1.0 \cdot 10^{11}$	90	250	640	1.5
	195	2.5	3.0	$4.0 \cdot 10^9$	$1.7 \cdot 10^{11}$	1,000	455	350	4
	77	2.5	3.3	$4.0 \cdot 10^{10}$	$8.0 \cdot 10^{10}$	90	455	350	5
PbSe	295	3.4	4.2	$3.0 \cdot 10^7$	$2.7 \cdot 10^{8}$ *	90	4	40,000	0.005
	195	4.6	5.4	$7.5 \cdot 10^8$	$6.0 \cdot 10^9$	900	125	1,270	40
	77	4.5	5.8	$2.2 \cdot 10^9$	$1.1 \cdot 10^{10}$	90	48	3,300	5
PbTe	77	4.0	5.1	$3.8 \cdot 10^8$	$2.7 \cdot 10^9$	90	25	6,500	32

* Johnson et al. [516] have described a more sensitive uncooled PbSe photoresistor with a detectivity $D^*_{\lambda m} = 10^{10}$ cm·cps$^{1/2}$·W^{-1} at $\lambda_m = 4$ μ at a modulation frequency of 10 kc. The theoretical limit of a device with $\lambda_{1/2} = 4.5$ μ is $2 \cdot 10^{11}$ cm·cps$^{1/2}$·W^{-1}.

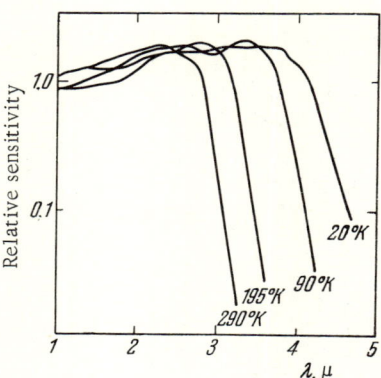

Fig. 7.8. Sensitivity spectrum of a lead sulfide photoresistor (according to Moss).

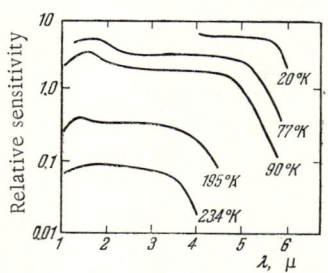

Fig. 7.9. Sensitivity spectrum of a lead telluride photoresistor (according to Moss).

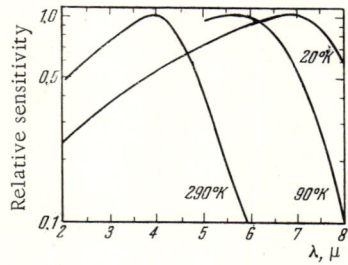

Fig. 7.10. Sensitivity spectrum of a lead selenide photoresistor (according to Moss).

some of the data given here may be out of date by the time this book appears.

PbS photoresistors are widely used at wavelengths of 2-3 μ. The spectral characteristics of lead sulfide resistors at various temperatures are shown in Fig. 7.8. They all are in the form of a curve with a maximum. The wavelength $\lambda_{1/2}$, at which the detectivity is equal to half its maximum value, is close to the fundamental absorption edge and is directly related to the forbidden band width by the formula

$$\lambda_{1/2} \approx 1.24/\mathscr{E}_g, \qquad (7.3)$$

where $\lambda_{1/2}$ is in microns and $\mathscr{E}_g$ in electron volts. Table 7.4 lists the values of the wavelength $\lambda_{1/2}$ as well as the wavelength λ_m at which the detectivity reaches its maximum.

Lead sulfide photoresistors may exhibit some sensitivity in the wavelength range of the order of 10 μ [190] due to their impurity photoconductivity. This makes it possible to detect infrared radiation of sources kept at a temperature of 40°C, in spite of the fact that the energy of radiation emitted by such sources is negligibly small in the wavelength range 0-3 μ.

PbSe and PbTe photoresistors are used mainly in the wavelength range 4-6 μ. Their spectral characteristics are shown in Figs. 7.9 and 7.10.

Cooling of lead chalcogenide photoresistors increases the long-wavelength limit of their sensitivity, in accordance with the temperature coefficient of the forbidden band width $\partial \mathscr{E}_t/\partial T \approx 4 \cdot 10^{-4}$ eV/deg. When cooled with liquid hydrogen, PbSe detectors are sensitive up to 9 μ.

An important characteristic of photoconducting radiation detectors is their detectivity D^*, governed by their voltage sensitivity and the noise level per unit frequency interval. The main form of noise in lead chalcogenide photoresistors is the current noise. This noise appears during the flow of a constant current through a photoresistor and is related to the complex structure of polycrystalline photoconducting films. The power of this noise is approximately proportional to the square of the current. For a given resistor and a fixed value of the current, the low-frequency noise power, taken per unit frequency interval, is approximately inversely proportional to the frequency. The temperature dependence of the current noise is relatively weak. Apart from the current noise, the detectivity is limited by the fluctuations of the radiation emitted by the ambient medium, which depend strongly on the temperature of the medium. The thermal and recombination-generation noise mechanisms are relatively unimportant in polycrystalline lead chalcogenide films.

Table 7.4 gives the detectivity for white light, whose composition is governed by the temperature of the light source, as well as for monochromatic radiation of wavelength corresponding to the sensitivity maximum.

Single-crystal photoresistors can have a detectivity not lower than that of polycrystalline samples provided the carrier density in single crystals is comparable with the intrinsic carrier density at room temperature. Prior et al. [36, 514] developed a method for preparing and etching such pure PbSe single crystals by means of which thin plates (several microns thick) with a low surface recombination velocity can be produced.

In most single-crystal samples, the main form of noise is of thermal origin. The current noise is absent even at currents which heat the sample. Therefore, the signal-to-noise ratio increases as the current rises until heating causes the signal to decrease. This makes it possible to increase the detectivity by means of a substrate which increases the rate of heat loss. The highest detectivity reached in lead selenide single-crystal photoresistors, at room temperature, at a wavelength of 4.4 μ, is $3.7 \cdot 10^8$ cm $\cdot$ cps$^{1/2} \cdot$ W^{-1}. The optimum thickness of single-crystal plates lies within the limits 10-20 μ.

The detectivity of many PbS photoresistors increases when they are cooled. Lead sulfide photoresistors work quite well at room temperature as well as at "dry ice" and liquid-nitrogen temperatures. The detectivity of PbSe and PbTe increases considerably when the temperature is lowered. They are usually employed at low temperatures although the characteristics of lead selenide photoresistors are satisfactory also at room temperature [516]. To increase the detectivity, it is necessary to cool not only the photoconducting film but also the surrounding medium in order to reduce the fluctuations of the background radiation.

The rate of response of photoconducting radiation detectors is represented by their time constants. Fast resonpse is required in recording rapid processes. The time constant governs also the optimum frequency of the radiation flux modulation, which is used in order to record the photosignal conveniently by means of an ac amplifier. The time constant of uniform photoconducting films is equal to the carrier lifetime τ_0 and, in the presence of carrier capture, it is equal to the photoconductivity lifetime τ_{pc} (cf. §2.2A). The situation is not so simple for a film with a complex structure or in the case of carrier capture resulting in nonlinear effects. For example, cooled lead chalcogenide photoresistors frequently have two time constants of very different values and this results in a more rapid decrease of the sensitivity with increasing frequency than that usually observed.

The time constants of various lead chalcogenide photoresistors range from several microseconds to several milliseconds. The time constant of lead sulfide photoresistors is proportional to their detectivity (the McAllister relationship [508]). This relationship is due to the photoconductivity being proportional to the carrier lifetime [Eq. (2.29)]. Therefore, lead sulfide photoresistors with the largest time constants exhibit a higher detectivity than fast-response resistors. The time constant is increased by cooling those photoresistors whose detectivity increases at lower temperatures. Table 7.4 gives typical values of the time constants. It is evident from this table that PbSe and PbTe devices have a much faster response than PbS photoresistors.

The voltage sensitivity of photoresistors decreases with increasing frequency of the modulation of the radiation, proportionally to $(1+\omega^2\tau_0^2)^{-1/2}$. Since the noise voltage is proportional to $\omega^{-1/2}$, the detectivity depends on the modulation frequency in accordance with the law

$$D^* \propto \left(\frac{\omega}{1+\omega^2\tau_0^2}\right)^{1/2}. \qquad (7.4)$$

This function has a maximum at $\omega\tau_0 = 1$. The calculated optimum modulation frequencies [510] are listed in Table 7.4.

Typical room-temperature resistances of photoconducting films are 10^5-10^6 Ω [508] when the electrodes are 1 mm long, as required for spectroscopic purposes. The resistance increases strongly at low temperatures. The resistance of polycrystalline films is governed not by the initial properties of the material but by the structure of the film. It is independent of the photoresistor area for films of given shape. Table 7.4 lists the resistances for square-shaped films.

An important characteristic of photoresistors used in space exploration is their service life under radioactive radiation conditions. Tests have shown [517] that the service life of lead sulfide photoresistors in cosmic space is about 3 years.

B. Photodiodes

Semiconductor photodetectors can be in the form of photodiodes whose action is based on the photovoltaic effect in a p−n junction. Theoretical estimates of the threshold sensitivity of PbS photodiodes, obtained by Hilsum and Simpson [518], show that a photodiode is an order of magnitude more sensitive than a single-crystal photoresistor. The sensitivity of a p−n junction in lead sulfide at room temperature is comparable with the sensitivity of evaporated PbS films, but the p−n junction has the advantage that the time constant is reduced to about 1 μsec. Moreover, in contrast to photo-

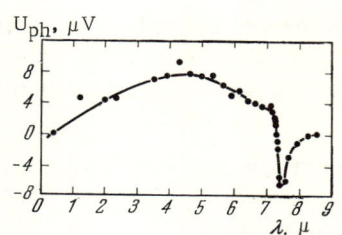

Fig. 7.11. Open-circuit photovoltage spectrum of a lead selenide photodiode at 77°K, obtained for an incident radiation density at 500 μW/cm^2 [163].

resistors, photodiodes operating at low (liquid-nitrogen) temperatures do not need to have a low impurity concentration. Photodiodes are usually operated with a reverse biased p−n junction but they can be used also under open-circuit conditions (in this case, no external power supply is needed).

The first single-crystal PbS diode, prepared by Starkiewicz et al. [162], had an active area of 8·1 mm^2, a sensitivity threshold of 6·10^{-9} W, a detectivity maximum at a wavelength of 2.5 μ, a wavelength $\lambda_{1/2}$ of about 3 μ, and a time constant shorter than 1 μsec.

Kimmitt and Prior [163] prepared a surface p−n junction in a PbSe single crystal, grown from the vapor phase, and investigated its photoelectric properties at 77°K and other temperatures. The experimental value of the detectivity was found to be $D^*_{\lambda_m} = 3 \cdot 10^8$ cm·cps$^{1/2}$·W^{-1} at a wavelength $\lambda_m = 4$ μ. An analysis of the results obtained showed that by using a low-noise amplifier and perfect contacts with this junction a detectivity of about 7·10^9 cm·cps$^{1/2}$·W^{-1} could be obtained. This value represents the limit set by the radiation noise background, corresponding to room temperature. The thermal noise is the principal type of noise in an ideal device of this type.

The effective quantum efficiency of Kimmit and Prior's PbSe photodiode was close to unity. The differential resistance of the junction (0.034 Ω·cm^2) was several orders of magnitude smaller than that expected from theoretical estimates. A better p−n junction, having a higher resistance, would make it possible to obtain a higher sensitivity and to reduce the role of the series resistance of the front layer of the photodiode. This p-type layer was several tenths of a micron thick. A silver contact was attached at one end of the front layer and a narrow illuminated strip was located near the contact.

The sensitivity spectrum of this PbSe photodiode recorded at 77°K is given in Fig. 7.11. The long-wavelength limit of the sensitivity was 7 μ and the maximum voltage sensitivity was reached at 4 μ. The maximum photovoltage was 8 μV for an incident radiation density of 500 μW/cm^2.

C. Lasers

The development of the laser technology has opened up new applications for lead chalcogenides. The potential applications of lasers are very extensive. Already they are widely used as light sources in physics experiments. The use of lasers in communications technology should make it possible to obtain channels with an enormous information capacity; there are indications of possible applications of lasers in other branches of technology. The radiation emitted by a laser has a high monochromaticity, a strong directionality, and high energy density. Lasers are of special interest as sources of coherent light.

Junga et al. [178] were the first to draw attention to the possible use of lead chalcogenides as active materials for infrared lasers. Their theroetical estimates showed that lead chalcogenides are more suitable for infrared lasers, working in the temperature range 2-110°K, than $A^{III}B^{V}$ compounds, in particular InSb, and the reasons for their superiority are as follows.

First, the conduction and valence bands of lead chalcogenides have higher densities of states due to the many-ellipsoid structure as well as to their larger effective masses. Therefore, spontaneous optical transitions are more likely and radiative recombination is stronger than in $A^{III}B^{V}$ materials.

Secondly, the interband impact recombination processes in lead chalcogenides are not of great importance because of the approximate equality of the effective electron and hole masses and because of the high value of the permittivity. The various recombination mechanisms have been described in §2.2B. Estimates obtained by Baryshev and Shtivel'man [184] show that the relationship between the radiative and the interband impact recombination in lead chalcogenides is more favorable for their use as laser materials than is the case for $A^{III}B^{V}$ compounds. The experiments of Washwell and Cuff [174] demonstrated that, even in the spontaneous recombination radiation region, the carrier recombination is mainly radiative at suitably low temperatures and carrier densities. The estimates of Junga et al. [178] showed that the interband impact recombination in lead chalcogenides exceeds the radiative recombination only at very high (metallic) carrier densities, while in InSb the interband impact recombination appears at carrier densities of the order of 10^{15} cm^{-3}. This means also that the interband impact recombination in lead chalcogenides is not important at the high injection levels employed in lasers.

Thirdly, the carrier mobility in lead chalcogenides is fairly high and this reduces the loss of photons due to their absorption by free carriers.

Two methods of establishing a high injection level (population inversion), required for the generation of laser radiation, were used in lead chalcogenides [178-182]: injection through a p−n junction [178-181] and bombardment with fast (19-60 keV) electrons [182]. In the injection lasers (laser diodes) the radiation emerged from a plate along a direction parallel to the p−n junction plane, while in the electron-bombardment lasers the radiation emerged at right angles to the electron beam. The passage of a current and the bombardment with electrons were applied in pulses sufficiently short to prevent the heating of the sample.

The laser radiation spectra, their dependencs on the temperature, pressure, and magnetic field, the dependence of the intensity and spetral composition of the radiation on the current density, as well as the role of the resonator, have all been discussed in §2.2D. The reflecting resonator surfaces are usually obtained by cleavage along the $\{100\}$ natural growth planes of the chalcogenide crystals. The use of PbTe and PbSe as the laser materials has made it possible to extend the spectral range of the laser radiation in the direction of wavelengths longer than those emitted by InSb lasers. The wavelength of the laser emission of PbSe at 12°K is 8.5 μ (12 μ under a pressure of 5000 kg/cm^2), which is in the spectral region where the atmosphere has a high transmission coefficient (the so-called "atmospheric window" at 8-14 μ) [181].

The strong dependence of the forbidden band width on the temperature, deformation, and magnetic field, as well as the dependence of the forbidden band of solid solutions on their composition, makes it possible to vary the frequency of the laser emission and thus cover a considerable range of frequencies. For example, $Pb_xSn_{1-x}Te$ lasers emit wavelengths up to 15-16 μ at 12°K [137] (the optical properties and the forbidden band width of these solid solutions are discussed in §2.1C, §2.2D, and §6.3C). Acoustic waves, which deform lead chalcogenide crystals [196], can be used in frequency modulation of laser radiation.

7.3. Strain Gauges

The relatively strong piezoresistance of lead chalcogenides (cf. §3.3B) makes it possible to use these materials in measurements of mechanical stresses (by means of strain gauges). A particularly high transverse piezoresistance of PbTe was reported by Ilisavskii [307]: the constant π_{44} is $185 \cdot 10^{-12}$ cm^2/dyn for p-type samples and $107 \cdot 10^{-12}$ cm^2/dyn for n-type samples. The disadvantage of PbTe as a semiconducting strain-gauge material is its low mechanical strength.

Arsen'eva-Geil' [308, 310] detected a high strain sensitivity of pressed PbS samples, which made it possible to use this material to prepare strain gauges for measurements of deformations. To measure bending strain, PbS samples are cemented to a steel substrate. In measurements of hydrostatic pressure, the sample is suspended, by conductors welded to the sample, in a chamber which is filled with oil under pressure. A PbS strain gauge was used by Orlov [519] to investigate volume changes in blood vessels.

A strain gauge is characterized by a strain sensitivity (known as the gauge factor), which is given by the formula

$$K_s = \frac{\Delta R/R_0}{\varepsilon}, \qquad (7.5)$$

where R_0 is the initial resistance of the gauge; ε is the relative deformation (strain); and ΔR is the change in the resistance caused by this deformation. The gauge factor of a PbS gauge is 2-3 orders of magnitude higher than that of wire strain gauges. An increase of the gauge factor was achieved by a heat treatment [309] similar to that which is used in the preparation of sensitive PbS photoresistors and which is assumed to produce p−n junctions at grain boundaries forming a photosensitive film. A change in the barrier height due to deformation is evidently the cause of the strong effect of strain on the pressed PbS powder. Such heat treatment made it possible to obtain high values of the gauge factor, amounting to several thousand [309].

PbS strain gauges exhibit a linear dependence of the change in the resistance on the deformation (i.e., they have a constant gauge factor), a fast response, and no hysteresis (in the majority of cases) [310]. The disadvantages of strain gauges of this type are the poor reproducibility of the results, instability, "ageing" phenomena, dependence on atmospheric conditions, etc. For these reasons, PbS strain gauges have not found wide acceptance. However, all these disadvantages can be eliminated and the advantages of lead sulfide strain gauges over wire gauges, particularly the high value of the gauge factor, suggest that the use of this material for strain measurement is promising.

APPENDIX A

FORMULAS FOR A NONPARABOLIC BAND

We shall now give the formulas for a nonparabolic Kane band (cf. §6.2B), for which the dispersion law is described by Eq. (6.22), the dependence of the density of states on the energy by Eq. (6.23), and the dependence of the carrier density on the Fermi energy by Eq. (6.27). We shall use the same notation as in Eqs. (6.22) and (6.26).

I. Effects Independent of the Relaxation Time

1. The thermoelectric power in a strong magnetic field:

$$\alpha_\infty = \frac{k_0}{e}\left(\frac{{}^1\mathscr{L}_0^{3/2}}{{}^0\mathscr{L}_0^{3/2}} - \zeta^*\right). \tag{A.1}$$

2. The Faraday effect:

$$\vartheta_F = \frac{2\pi n e^3 dH}{c^2 \mathscr{N} \omega^2 m_F^{*2}}, \quad \frac{1}{m_F^{*2}} = \frac{K+2}{3K m_{\perp 0}^2}\frac{{}^0\mathscr{L}_{-2}^{3/2}}{{}^0\mathscr{L}_0^{3/2}}. \tag{A.2}$$

3. The contribution of free carriers to the high-frequency electric susceptibility, which governs the value of the reflection coefficient of light in the near infrared region of the spectrum:

$$\chi_{fc} = -\frac{ne^2}{\omega^2 m_\chi^*}, \quad \frac{1}{m_\chi^*} = \frac{2K+1}{3K m_{\perp 0}^*}\frac{{}^0\mathscr{L}_{-1}^{3/2}}{{}^0\mathscr{L}_0^{3/2}}. \tag{A.3}$$

II. Effects Dependent on the Relaxation Time

For each quantity we give two formulas: for an arbitrary dependence of the relaxation time on the energy $\tau(\mathscr{E})$ and for the scattering by the acoustical phonons [in the latter case we assume that the matrix element of the interaction is independent of the energy so that $\tau^{-1} \sim T\rho(\mathscr{E})$].

1. The carrier mobility:

$$\mu = \frac{e}{m_{\chi 0}^*} \frac{\int_0^\infty \left(-\frac{\partial f}{\partial z}\right) \tau(z) \frac{(z + \beta z^2)^{3/2}}{1 + 2\beta z} dz}{\int_0^\infty \left(-\frac{\partial f}{\partial z}\right)(z + \beta z^2)^{3/2} dz}, \quad (A.4)$$

$$\mu = \frac{\text{const}}{m_{\chi 0}^* m_{\alpha 0}^{*3/2} T^{3/2}} \frac{{}^0\mathscr{L}_{-2}^1}{{}^0\mathscr{L}_0^{3/2}}. \quad (A.4a)$$

The constant in the numerator of Eq. (A.4a) is independent of the temperature and the carrier density.

2. The Hall coefficient:

$$R = \frac{A}{enc},$$

$$A = \frac{3K(K+2)}{(2K+1)^2} \frac{\int_0^\infty \left(-\frac{\partial f}{\partial z}\right) \tau^2(z) \frac{(z + \beta z^2)^{3/2}}{(1 + 2\beta z)^2} dz \int_0^\infty \left(-\frac{\partial f}{\partial z}\right)(z + \beta z^2)^{3/2} dz}{\left[\int_0^\infty \left(-\frac{\partial f}{\partial z}\right) \tau(z) \frac{(z + \beta z^2)^{3/2}}{1 + 2\beta z} dz\right]^2}, \quad (A.5)$$

$$A = \frac{3K(K+2)}{(2K+1)^2} \frac{({}^2\mathscr{L}_{-1}^{1/2})({}^0\mathscr{L}_0^{3/2})}{({}^0\mathscr{L}_{-2}^1)^2}. \quad (A.5a)$$

3. The thermoelectric power:

$$\alpha_0 = \frac{k_0}{e} \frac{\int_0^\infty \left(-\frac{\partial f}{\partial z}\right) \tau(z)(z - \zeta^*) \frac{(z + \beta z^2)^{3/2}}{1 + 2\beta z} dz}{\int_0^\infty \left(-\frac{\partial f}{\partial z}\right) \tau(z) \frac{(z + \beta z^2)^{3/2}}{1 + 2\beta z} dz}, \quad (A.6)$$

$$\alpha_0 = \frac{k_0}{e} \left(\frac{{}^1\mathscr{L}_{-2}^1}{{}^0\mathscr{L}_{-2}^1} - \zeta^*\right). \quad (A.6a)$$

APPENDIX A 349

4. The Nernst – Ettingshausen effect (divided by the Hall mobility):

$$\frac{Q}{R\delta} = \frac{k_0}{e} \left\{ \frac{\int_0^\infty \left(-\frac{\partial f}{\partial z}\right) \tau^2(z) z \frac{(z+\beta z^2)^{3/2}}{(1+2\beta z)^2} dz}{\int_0^\infty \left(-\frac{\partial f}{\partial z}\right) \tau^2(z) \frac{(z+\beta z^2)^{3/2}}{(1+2\beta z)^2} dz} \right.$$

$$\left. - \frac{\int_0^\infty \left(-\frac{\partial f}{\partial z}\right) \tau(z) z \frac{(z+\beta z^2)^{3/2}}{1+2\beta z} dz}{\int_0^\infty \left(-\frac{\partial f}{\partial z}\right) \tau(z) \frac{(z+\beta z^2)^{3/2}}{1+2\beta z} dz} \right\}, \quad \text{(A.7)}$$

$$\frac{Q}{R\delta} = \frac{k_0}{e} \left(\frac{{}^1\mathscr{L}_{-4}^{1/2}}{{}^0\mathscr{L}_{-4}^{1/2}} - \frac{{}^1\mathscr{L}_{-2}^1}{{}^0\mathscr{L}_{-2}^1} \right) \quad \text{(A.7a)}$$

5. The electronic component of the thermal conductivity:

$$\varkappa_e = L_0 T \sigma,$$

$$L_0 = \left(\frac{k_0}{e}\right)^2 \left\{ \frac{\int_0^\infty \left(-\frac{\partial f}{\partial z}\right) \tau(z) z^2 \frac{(z+\beta z^2)^{3/2}}{1+2\beta z} dz}{\int_0^\infty \left(-\frac{\partial f}{\partial z}\right) \tau(z) \frac{(z+\beta z^2)^{3/2}}{1+2\beta z} dz} - \left[\frac{\int_0^\infty \left(-\frac{\partial f}{\partial z}\right) \tau(z) z \frac{(z+\beta z^2)^{3/2}}{1+2\beta z} dz}{\int_0^\infty \left(-\frac{\partial f}{\partial z}\right) \tau(z) \frac{(z+\beta z^2)^{3/2}}{1+2\beta z} dz} \right]^2 \right\},$$

(A.8)

$$L_0 = \left(\frac{k_0}{e}\right)^2 \left[\frac{{}^2\mathscr{L}_{-2}^1}{{}^0\mathscr{L}_{-2}^1} - \left(\frac{{}^1\mathscr{L}_{-2}^1}{{}^0\mathscr{L}_{-2}^1}\right)^2 \right]. \quad \text{(A.8a)}$$

III. Expressions for the Relaxation Time

In contrast to Eqs. (A.4a)–(A.8a), an allowance is made here for the dependence of the matrix element on the energy.

1. Scattering by the acoustical phonons:

$$\frac{1}{\tau(z)} = \frac{\pi k_0 T \rho(z) \Xi^2}{\hbar c_l N} \left[1 - \frac{8\beta(z+\beta z^2)}{3(1+2\beta z)^2} \right] \quad \text{(A.9)}$$

APPENDIX A

2. Polar scattering by the optical phonons at $T \gg \theta_1/2$:

$$\frac{1}{\tau(z)} = \frac{2^{1/2} k_0 T e^2 m_{d_1}^{*1/2}(\epsilon_\infty^{-1} - \epsilon_0^{-1})}{\hbar^2 (z k_0 T)^{1/2}} \frac{1 + 2\beta z}{(1 + \beta z)^{1/2}} \left\{ \left[1 - \delta \ln\left(1 + \frac{1}{\delta}\right) \right] \right.$$

$$\left. - \frac{2\beta(z + \beta z^2)}{(1 + 2\beta z)^2} \left[1 - 2\delta + 2\delta^2 \ln\left(1 + \frac{1}{\delta}\right) \right] \right\} , \quad (A.10)$$

where

$$\delta(z) = (2k\pi_{sc})^{-2} ; \qquad k^2 = \frac{2m_{d_1}^* (z k_0 T)(1 + \beta z)}{\hbar^2} ;$$

$$\pi_{sc}^{-2} = \frac{2^{5/2} e^2 m_d^{*3/2} (k_0 T)^{1/2}}{\pi \hbar^3 \epsilon_\infty} \, {}^0\mathcal{L}_1^{1/2} .$$

APPENDIX B

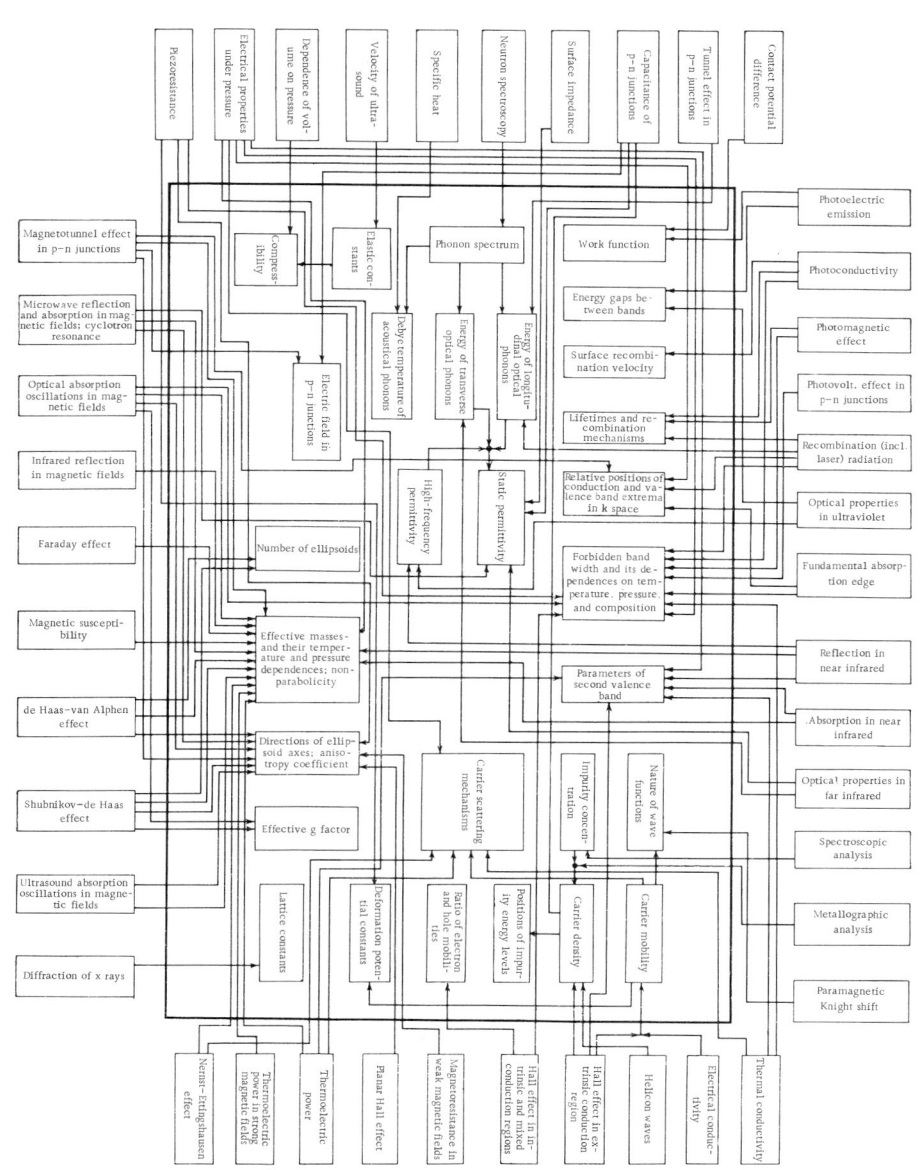

APPENDIX C

MAIN PROPERTIES OF LEAD CHALCOGENIDES

Property	PbTe	PbSe	PbS	Measurement method; remarks
a, Å	6.50	6.12	5.94	Diffraction of x rays
η, g/cm³	8.2	8.3	7.6	
T_{mp}, °C	917	1065	1114	Maximum melting point
$\mathscr{E}_g$ (300 °K), eV	0.32	0.29	0.41	[99] Fundamental absorption edge
$\partial\mathscr{E}_g/\partial T$ (77 ÷ 300 °K), eV/deg	$4 \cdot 10^{-4}$	$4 \cdot 10^{-4}$	$4 \cdot 10^{-4}$	[64]
$\mathscr{E}_g$ (77 °K), eV	0.217	0.176	0.307	[450] Absorption of light in magnetic fields
$\mathscr{E}_g$ (4.2 °K), eV	0.190	0.165	0.286	
$\partial\mathscr{E}_g/\partial P$, eV·cm²·kg⁻¹	$-8 \cdot 10^{-6}$ [289]	$-8 \cdot 10^{-6}$ [283]	$-8 \cdot 10^{-6}$ [282]	Electrical properties under pressure. Scatter of values obtained by different investigators is about 25% (cf. §3.3A)
n_i (300 °K), cm⁻³	$1.5 \cdot 10^{16}$	$3 \cdot 10^{16}$	$2 \cdot 10^{15}$	[5] Hall effect
m^* (0 °K), in m_0 units:				[439] Shubnikov – de Haas effect
$m^*_{\parallel p}$	0.31	0.068	0.105	
$m^*_{\parallel n}$	0.24	0.070	0.105	
$m^*_{\perp p}$	0.022	0.034	0.075	
$m^*_{\perp n}$	0.024	0.040	0.080	

APPENDIX C

Property	PbTe	PbSe	PbS	Measurement method; remarks
$\partial \ln m_p^*/\partial \ln T$	0.4 [341]	0.35 [230]	—	Thermoelectric power
$\partial \ln m_n^*/\partial \ln T$	0.5 [339]	0.45 [230]	0.6 [122]	
$\partial \ln m_p^*/\partial P$, cm^2/kg	$-2.1 \cdot 10^{-5}$ [285]	$-1.9 \cdot 10^{-5}$ [283]	$-1.9 \cdot 10^{-5}$ [283]	
$\partial \ln m_n^*/\partial P$, cm^2/kg	$-2.1 \cdot 10^{-5}$ [285]	$-1.8 \cdot 10^{-5}$ [283]	$-2.3 \cdot 10^{-5}$ [283]	
$\mathscr{E}_{1g}$, eV	~3	~3.5	~4	Average values obtained from piezoresistance and electrical properties under pressure (cf. §3.3)
$\|\Xi_{up}\|$, eV	~6	—	—	[297] Piezoresistance. Considerable scatter of values obtained by different investigators (cf. §3.3B)
$\|\Xi_{un}\|$, eV	~3	—	—	
$\|\Xi_p\|$, eV	~25	~23	~20	[220] Calculated from mobility values
$\|\Xi_n\|$, eV	~24	~23	~20	
c_{11}, dyn/cm^2	$10.4 \cdot 10^{11}$	—	$13.9 \cdot 10^{11}$	[301] Velocity of ultrasound. Considerable scatter of values obtained by different investigators, particularly for c_{12} (cf. §3.3B)
c_{12}, dyn/cm^2	$-0.44 \cdot 10^{11}$	—	$3.5 \cdot 10^{11}$	
c_{44}, dyn/cm^2	$1.3 \cdot 10^{11}$	—	$1.7 \cdot 10^{11}$	

Property	PbTe	PbSe	PbS	Measurement method; remarks
Velocity of sound, cm/sec	$3.6 \cdot 10^5$	—	$4.1 \cdot 10^5$	
ξ, cm^2/kg	$2.6 \cdot 10^{-6}$	$2.07 \cdot 10^{-6}$	$1.8 \cdot 10^{-6}$	[306] Dependence of volume on pressure
ε_0	400 [242, 126]	250 [128]	175 [128]	Capacitance of p – n junctions and Lyddane – Sachs – Teller formula (cf. §2.1D)
$\varepsilon_\infty = \mathcal{N}_\infty^2$	33	24	17	[128] Infrared reflection
$\hbar\omega_l$, eV	0.0136	0.0165	0.0263	[252] Tunnel effect in p–n junctions
$\hbar\omega_t$, eV	0.0039 [126]	0.0054 [121]	0.0082 [120]	Infrared reflection
μ_p (300 °K), cm$^2 \cdot$ V$^{-1} \cdot$ sec^{-1}	840	1000	620	[220] Hall mobility. Maximum values are given
μ_n (300 °K), cm$^2 \cdot$ V$^{-1} \cdot$ sec^{-1}	1730	1000	610	
μ_p (77 °K), cm$^2 \cdot$ V$^{-1} \cdot$ sec^{-1}	21600	13700	15000	
μ_n (77 °K), cm$^2 \cdot$ V$^{-1} \cdot$ sec^{-1}	31600	16500	11000	
μ_p (4,2 °K), cm$^2 \cdot$ V$^{-1} \cdot$ sec^{-1}	250000	57900	80000	
μ_n (4,2 °K), cm$^2 \cdot$ V$^{-1} \cdot$ sec^{-1}	800000	139000	68500	
$\varkappa$ (300 °K), cal$\cdot$cm$^{-1} \cdot$sec$^{-1} \cdot$deg^{-1}	$4.8 \cdot 10^{-3}$ [389]	$3.9 \cdot 10^{-3}$ [355]	$6.0 \cdot 10^{-3}$ [237]	Thermal conductivity in purest samples
θ (200 °K), deg	125	138	227	[376] Specific heat

LITERATURE CITED

1. R. P. Chasmar, Proc. Photoconductivity Conf., Atlantic City, 1954 (ed., R. G. Breckenridge), publ. J. Wiley, New York (1956), p. 463.
2. R. A. Smith, Physica, 20:910 (1954).
3. W. W. Scanlon, Solid State Physics, 9:83 (1959).
4. E. H. Putley, The Hall Effect and Related Phenomena, Butterworths, London (1960).
5. E. H. Putley, in: Materials Used in Semiconductor Devices (ed. C. A. Hogarth), Interscience, New York (1965), pp. 71-114.
6. Ya. A. Ugai, Introduction to the Chemistry of Semiconductors [in Russian], "Vysshaya shkola" (1965), p. 37.
7. L. Pauling, The Nature of the Chemical Bond and the Structure of Molecules and Crystals, 3rd ed., Cornell Univ. Press, New York (1960).
8. W. D. Lawson, J. Appl. Phys., 22:1444 (1951).
9. W. D. Lawson, J. Appl. Phys., 23:495 (1952).
10. W. D. Lawson and S. Nielsen, in: Art and Science of Growing Crystals (ed. J. J. Gilman), Interscience, New York (1963), p. 365.
11. G. P. Tilley, Brit. J. Appl. Phys., 12:524 (1961).
12. R. F. Brebrick and W. W. Scanlon, J. Chem. Phys., 27:607 (1957).
13. M. Weinstein, Trans. Met. Soc. AIME, 230:321 (1964).
14. A. S. Pashinkin and A. V. Novoselova, Zh. Neorg. Khim., 4:2659 (1959).
15. V. P. Zlomanov, B. A. Popovkin, and A. V. Novoselova, Zh. Neorg. Khim., 4:2661 (1959).
16. C. M. Hsiao and A. W. Schlechten, J. Metals, 4:65 (1952).
17. K. Sudo, J. Mining Metall. Inst. Japan, 77:844 (1958).
18. R. A. Isakova, V. N. Nesterov, and A. S. Shendyanin, Zh. Neorg. Khim., 8:18 (1963).
19. J. F. Miller and R. C. Himes, J. Electrochem. Soc., 107:915 (1960).
20. E. D. Nensberg, Izv. Akad. Nauk SSSR, Neorgan. Mat., 1:853 (1966).
21. W. M. Franklin and J. B. Wagner, Jr., J. Appl. Phys., 34:3121 (1963).
22. B. B. Houston, Jr., and M. K. Norr, J. Appl. Phys., 31:615 (1960).
23. D. Greig, Phys. Rev., 120:358 (1960).
24. J. W. Rutter and B. Chalmers, Can. J. Phys., 13:15 (1953).
25. G. W. Johnson, J. Electronics Control, 12:421 (1962).
26. P. J. Holmes, "Practical applications of chemical etching," in: The Electrochemistry of Semiconductors (ed. P. J. Holmes), Academic Press, London (1962), p. 239.
27. P. H. Schmidt, J. Electrochem. Soc., 108:104 (1961).

28. A. A. Urusovskaya, R. Tyaagaradzhan, and M. V. Klassen-Neklyudova, Kristallografiya, 8:623 (1963).
29. M. K. Norr, J. Electrochem. Soc., 109:433 (1962).
30. A. J. Crocker, Brit. J. Appl. Phys., 17:433 (1966).
31. W. W. Scanlon, Phys. Rev., 126:509 (1962).
32. V. I. Kaidanov, Author's Abstract of Dissertation for Candidate's Degree [in Russian], M. I. Kalinin Polytechnical Institute, Leningrad (1966).
33. B. M. Gol'tsman, Growth of Semiconductor Single Crystals [in Russian], LDNTP, Leningrad (1963), p. 27.
34. E. P. A. Metz, R. C. Miller, and R. Mazelsky, J. Appl. Phys., 33:2016 (1962).
35. P. M. Starik and P. I. Voronyuk, in: Growth of Crystals (ed. N. N. Sheftal'), Vol. 6B, Consultants Bureau, New York (1968), p. 91.
36. A. C. Prior, J. Electrochem. Soc., 108:82 (1961).
37. F. P. Pizzarello, J. Appl. Phys., 25:804 (1954).
38. R. H. Jones, Proc. Phys. Soc. (London), 76:783 (1960).
39. C. Fritzsche, Herstellung von Halbleitern, 2nd ed., VEB Verlag Technik, Berlin (1962).
40. S. A. Semiletov, Kristallografiya, 9:84 (1961).
41. R. B. Schoolar and J. N. Zemel, J. Appl. Phys., 35:1848 (1964).
42. J. H. Zemel, J. D. Jensen, and R. B. Schoolar, Phys. Rev., 140:A330 (1965).
43. M. Hansen and K. Anderko, Constitution of Binary Alloys, 2nd ed., McGraw-Hill, New York (1958).
44. J. Bloem, Philips Res. Rep., 11:273 (1956).
45. J. Bloem and F. A. Kröger, Z. Physik. Chem. (Frankfurt), 7:1 (1956).
46. J. Bloem and F. A. Kröger, Philips Res. Rep., 12:303 (1957).
47. J. Bloem and F. A. Kröger, Philips Res. Rep., 12:281 (1957).
48. R. F. Brebrick and W. W. Scanlon, Phys. Rev., 96:598 (1954).
49. A. E. Goldberg and G. R. Mitchell, J. Chem. Phys., 22:220 (1954).
50. R. F. Brebrick and R. S. Allgaier, J. Chem. Phys., 32:1826 (1960).
51. R. F. Brebrick, J. Phys. Chem. Solids, 18:116 (1961).
52. R. F. Brebrick and E. Gubner, J. Chem. Phys., 36:1283 (1962).
53. E. Miller, K. Komarek, and I. Cadoff, J. Appl. Phys., 32:2457 (1961).
54. H. Gobrecht and A. Richter, J. Phys. Chem. Solids, 26:1889 (1965).
55. Y. Sato, M. Fujimoto, and A. Kobayashi, Japan. J. Appl. Phys., 2:683 (1963).
56. K. Igaki and N. Ohashi, Proc. Intern. Conf. on Lattice Defects, Kyoto, 1962, in: J. Phys. Soc. Japan, 18, Suppl. 2, 143 (1963); Trans. Japan Inst. Metals, 5:94 (1964).
57. F. A. Kröger, Chemistry of Imperfect Crystals, Amsterdam (1964), p. 493.
58. M. Fujimoto and Y. Sato, Japan. J. Appl. Phys., 5:128 (1966).
59. R. F. Brebrick, J. Phys. Chem. Solids, 11:43 (1959).
60. C. Wagner, J. Chem. Phys., 18:62 (1950).
61. T. L. Koval'chik and Yu. P. Maslakovets, Zh. Tekh. Fiz., 26:2417 (1956).
62. P. M. Starik and P. I. Voronyuk, Izv. Akad. Nauk SSSR, Ser. Fiz., 28:8 (1964); Ukr. Fiz. Zh., 9:26 (1964).
63. Ya. S. Budzhak, in: Growth of Crystals (ed. N. N. Sheftal'), Vol. 6B, Consultants Bureau, New York (1968), p. 43.

LITERATURE CITED

64. A. F. Gibson, Proc. Phys. Soc. (London), B65:378 (1952).
65. F. F. Kharakhorin, D. A. Gambarova, and V. V. Aksenov, Fiz. Tverd. Tela, 7:3481 (1965).
66. V. S. Andramonov, N. S. Baryshev, and I. S. Aver'yanov, Fiz. Tverd. Tela, 4:2223 (1962).
67. B. A. Efimova, L. A. Kolomoets, and G. F. Zakharyugina, Izv. Akad. Nauk SSSR, Neorgan. Mat., 4:1157 (1968).
68. R. W. Fritts, in: Thermoelectric Materials and Devices, Reinhold, New York (1960), p. 143.
69. M. Fujimoto, J. Phys. Soc. Japan, 21:1706 (1966).
70. V. V. Agashe, B. L. Sharma, and B. K. Sachar, J. Inst. Telecommun. Eng. (India), 10:237 (1964).
71. R. F. Porter, J. Chem. Phys., 34:583 (1961).
72. J. W. Walker, J. W. Straley, and A. W. Smith, Phys. Rev., 53:140 (1938).
73. A. G. Gaydon, Dissociation Energies and Spectra of Diatomic Molecules, 3rd ed., Chapman and Hall, London (1968).
74. I. V. Korneeva, A. S. Pashinkin, A. V. Novoselova, and Yu. A. Priselkov, Zh. Neorg. Khim., 2:1720 (1957).
75. K. B. Sadykov and S. A. Semenkovich, in: Chemical Bonds in Semiconductors and Thermodynamics (ed. by N. N. Sirota), Consultants Bureau, New York (1968), p. 110.
76. B. I. Boltaks, Diffusion in Semiconductors, Infosearch, London; Academic Press, New York (1963).
77. G. Simkovich and J. B. Wagner, Jr., J. Chem. Phys., 38:1368 (1963).
78. M. S. Seltzer and J. B. Wagner, Jr., J. Phys. Chem. Solids, 28:233 (1965).
79. M. S. Seltzer and J. B. Wagner, Jr., J. Phys. Chem. Solids, 24:1525 (1963).
80. M. S. Seltzer and J. B. Wagner, Jr., J. Chem. Phys., 36:130 (1961).
81. B. I. Boltaks and Yu. Mokhov, Zh. Tekh. Fiz., 26:2448 (1956).
82. B. I. Boltaks and Yu. Mokhov, Zh. Tekh. Fiz., 28:1046 (1958).
83. F. F. Kharakhorin, D. A. Gambarova, and V. V. Aksenov, Abstracts of Papers presented at All-Union Conf. on Diffusion in Semiconductors [in Russian], Izd. "Nauka," Moscow (1964), pp. 21, 36.
84. F. F. Kharakhorin, D. A. Gambarova, and V. V. Aksenov, Izv. Akad. Nauk SSSR, Neorgan. Mat., 2:504 (1966).
85. F. F. Kharakhorin, D. A. Gambarova, and V. V. Aksenov, Izv. Akad. Nauk SSSR, Neorgan. Mat., 2:1371 (1966).
86. F. F. Kharakhorin, D. A. Gambarova, and V. V. Aksenov, Izv. Akad. Nauk SSSR, Neorgan. Mat., 1:1506 (1965).
87. N. A. Fedorovich, Fiz. Tverd. Tela, 7:1593 (1965).
88. F. F. Kharakhorin, D. A. Gambarova, and V. V. Aksenov, Izv. Akad. Nauk SSSR, Neorgan. Mat., 1:1502 (1965).
89. N. A. Fedorovich, Fiz. Tverd. Tela, 7:1594 (1965).
90. M. S. Seltzer and J. B. Wagner, Jr., J. Chem. Phys., 38:2309 (1963).
91. N. A. Fedorovich, Author's Abstract of Dissertation for Candidate's Degree [in Russian], Institute of Semiconductors, Academy of Sciences of the USSR, Leningrad (1966).

92. S. M. Klotsman, A. N. Timofeev, and N. Sh. Trantenberg, Fiz. Metallov Metalloved., 16:895 (1963).
93. Y. W. Zing, J. Phys. Chem. Solids, 18:162 (1961).
94. J. S. Anderson and J. R. Richards, J. Chem. Soc., 537 (1946).
95. D. G. Avery, Proc. Phys. Soc. (London), B64:1087 (1951); B65:425 (1952); B66:134 (1953); B67:2 (1954).
96. W. Paul, D. A. Jones, and R. V. Jones, Proc. Phys. Soc. (London), B64:528 (1951).
97. W. Paul and R. V. Jones, Proc. Phys. Soc. (London), B66:194 (1953).
98. W. W. Scanlon, Phys. Rev., 109:47 (1958).
99. W. W. Scanlon, J. Phys. Chem. Solids, 8:423 (1959).
100. M. E. Lasser and H. Levinstein, Phys. Rev., 96:47 (1954).
101. M. A. Clark and R. J. Cashman, Phys. Rev., 85:1043 (1951).
102. J. Strong, Phys. Rev., 38:1818 (1931).
103. H. Yoshinaga, Phys. Rev., 100:753 (1955).
104. S. A. Semiletov, I. P. Voronina, and E. I. Kortukova, Kristallografiya, 10:515 (1965).
105. M. L. Belle, Fiz. Tverd. Tela, 5:3282 (1963); 7:606 (1965).
106. A. K. Walton and T. S. Moss, Proc. Phys. Soc. (London), 81:509 (1963).
107. H. R. Riedl, Phys. Rev., 127:162 (1962).
108. H. R. Riedl and R. B. Schoolar, Phys. Rev., 131:2082 (1963).
109. R. B. Schoolar and J. R. Dixon, Phys. Rev., 137:A667 (1965).
110. J. R. Dixon, Bull. Am. Phys.,Soc., 6:312 (1961).
111. J. R. Dixon and H. R. Riedl, Proc. Sixth Intern. Conf. on Physics of Semiconductors, Exeter, England, 1962, Institute of Physics, London (1962), p. 179.
112. J. R. Dixon and H. R. Riedl, Physics Letters, 12:164 (1964).
113. J. R. Dixon and H. R. Riedl, Phys. Rev., 138:A873 (1965).
114. H. A. Lyden, Phys. Rev., 135:A514 (1964).
115. F. Stern, J. Appl. Phys., 32, Suppl. to No. 10, 2166 (1961).
116. J. R. Dixon and H. R. Riedl, Phys. Rev., 140:A1283 (1965); 147:670 (1966).
117. R. Geick, Physics Letters, 10:51 (1964).
118. S. Kurita, I. Nagasawa, K. Tanaka, Y. Nishina, and T. Fukuroi, Sci. Rep. Res. Inst. Tohoku Univ., A17:37 (1965).
119. M. Cardona and D. L. Greenaway, Phys. Rev., 133:A1685 (1964).
120. J. N. Zemel, Proc. Seventh Intern. Conf. on Physics of Semiconductors, Paris, 1964, Vol. 1, Physics of Semiconductors (ed., M. Hulin), Dunod, Paris; Academic Press. New York (1964), p. 1061.
121. E. Burstein, R. Wheeler, and J. N. Zemel, Proc. Seventh Intern. Conf. on Physics of Semiconductors, Paris, 1964, Vol. 1, Physics of Semiconductors (ed. M. Hulin), Dunod, Paris; Academic Press, New York (1964), p. 1065.
122. Yu. V. Mal'tsev, E. D. Nensberg, A. V. Petrov, S. A. Semiletov, and Yu. I. Ukhanov, Fiz. Tverd. Tela, 8:2155 (1966); Yu. I. Ukhanov, Author's Abstract of Dissertation for Candidate's Degree [in Russian], M. I. Kalinin Polytechnical Institute, Leningrad (1966).
123. R. N. Tauber, A. A. Machonis, and I. B. Cadoff, J. Appl. Phys., 37:4855 (1966).

124. O. V. Vakulenko, M. P. Lisitsa, and Ya. F. Kononets, Fiz. Tverd. Tela, 8:1698 (1966).
125. Ya. A. Semenov and A. Yu. Shileika, Fiz. Tverd. Tela, 6:313 (1964); 9:1810 (1967).
126. E. G. Bylander and M. Hass, Solid State Commun., 4:51 (1966).
127. E. D. Palik, D. L. Mitchell, and J. N. Zemel, Phys. Rev., 135:A763 (1964).
128. J. N. Zemel, J. D. Jensen, and R. B. Schoolar, Phys. Rev., 140:A330 (1965).
129. W. Paul, J. Appl. Phys., 32, Suppl. to No. 10, 2082 (1961).
130. T. S. Moss, Optical Properties of Semiconductors, Butterworths, London (1959).
131. H. Y. Fan, Rep. Progr. Phys., 19:107 (1956).
132. J. S. Toll, Phys. Rev., 104:1760 (1956).
133. A. F. Ioffe, Physics of Semiconductors, Infosearch, London (1960).
134. W. Paul, Private communication cited in [234].
135. J. C. Woolley and P. M. Nikolic, J. Electrochem. Soc., 112:82 (1965).
136. P. M. Nikolic, Brit. J. Appl. Phys., 16:1075 (1965).
137. J. O. Dimmock, I. Melngailis, and A. J. Strauss, Phys. Rev. Letters, 16:1193 (1966).
138. L. Esaki and P. J. Stiles, Phys. Rev. Letters, 16:1108 (1966).
139. W. G. Spitzer and H. Y. Fan, Phys. Rev., 106:882 (1957).
140. H. A. Lyden, Phys. Rev., 134:A1106 (1964).
141. W. Cochran, Physics Letters, 13:193 (1964).
142. R. H. Lyddane, R. G. Sachs, and E. Teller, Phys. Rev., 59:673 (1941).
143. N. Watanabe, Japan. J. Appl. Phys., 3:166 (1964).
144. J. C. Phillips, Phys. Rev., 133:A452 (1964).
145. M. Cardona, J. Appl. Phys., 36:2181 (1965).
146. T. S. Moss, Proc. IRE, 43:1869 (1955).
147. N. S. Baryshev and I. S. Aver'yanov, Fiz. Tverd. Tela, 4:1525 (1962).
148. J. N. Humphrey and R. L. Petritz, Phys. Rev., 105:1736 (1957).
149. V. G. Erofeichev, Fiz. Tverd. Tela, 3:3429 (1961).
150. V. G. Erofeichev and L. N. Kurbatov, in: Photoelectric and Optical Phenomena in Semiconductors [in Russian], Izd. AN UkrSSR, Kiev (1959), p. 213.
151. D. H. Roberts, in:: Solid State Physics in Electronics and Telecommunications (Proc. IUPAP Conf., Brussels, 1958, ed. by M. Desirant and J. L. Michiels), Vol. 2, Semiconductors (Part 2), Academic Press, New York (1960), p. 819.
152. G. W. Mahlman, Phys. Rev., 103:1619 (1956).
153. J. C. Slater, Phys. Rev., 103:1631 (1956).
154. R. L. Petritz, Phys. Rev., 104:1508 (1956).
155. E. S. Rittner and S. Fine, Phys. Rev., 98:545 (1955).
156. R. H. Harada and H. T. Minden, Phys. Rev., 102:1258 (1956).
157. V. F. Zolotarev and V. N. Larichev, Fiz. Tverd. Tela, 2:1741 (1960).
158. J. F. Woods, Phys. Rev., 106:235 (1957).
159. A. A. Rogachev and S. N. Chechurin, Fiz. Tverd. Tela, 4:1174 (1962).
160. D. G. Coates, W. D. Lawson, and A. C. Prior, J. Electrochem. Soc., 108:1038 (1961).
161. T. S. Moss, Proc. Phys. Soc. (London), B66:993 (1953).
162. J. Starkiewicz, G. Bate, H. Bennett, and C. Hilsum, Proc. Phys. Soc. (London), B70:258 (1957).

LITERATURE CITED

163. M. F. Kimmitt and A. C. Prior, J. Electrochem. Soc., 108:1034 (1961).
164. R. A. Laff, J. Appl. Phys., 36:3324 (1965).
165. J. Starkiewicz, L. Sosnowski, and O. Simpson, Nature, 158:28 (1946).
166. R. Ya. Berlaga, M. A. Rumsh, and L. P. Strakhov, Zh. Tekh. Fiz., 25:1878 (1955).
167. G. Schwabe, Ann. Physik, 17:249 (1956).
168. M. A. Rumsh and Yu. G. Baklagina, Zh. Tekh. Fiz., 26:1497 (1956).
169. R. Ya. Berlaga and T. T. Bykova, Fiz. Tverd. Tela, 2:3045 (1960).
170. V. K. Adamchuk, and R. Ya. Berlaga, Fiz. Tverd. Tela, 4:2382 (1962).
171. J. Tauc, Photo- and Thermoelectric Effects in Semiconductors, Pergamon Press, Oxford (1962).
172. N. F. Kovtonyuk and V. F. Fedonin, Izv. Vuzov, Fizika, No. 5, 28 (1966).
173. W. W. Scanlon, Phys. Rev., 106:718 (1957).
174. E. R. Washwell and K. F. Cuff, Proc. Seventh Intern. Conf. on Physics of Semiconductors, Paris, 1964, Vol. 4, Radiative Recombination in Semiconductors (ed. by C. Benoit a la Guillaume), publ. by Dunod, Paris; Academic Press, New York (1965), p. 11.
175. W. Van Roosbroeck and W. Shockley, Phys. Rev., 94:1558 (1954).
176. I. M. Mackintosh, Proc. Phys. Soc. (London), B69:115 (1956).
177. N. S. Baryshev, Fiz. Tverd. Tela, 3:1428 (1961).
178. F. A. Junga, K. F. Cuff, J. S. Blakemore, and E. R. Washwell, Physics Letters, 13:103 (1964).
179. J. F. Butler, A. R. Calawa, R. J. Phelan, Jr., T. C. Harman, A. J. Strauss, and R. H. Rediker, Appl. Phys. Letters, 5:75 (1964).
180. J. F. Butler, A. R. Calawa, R. J. Phelan, Jr., A. J. Strauss, and R. H. Rediker, Solid State Commun., 2:303 (1964).
181. J. F. Butler, A. R. Calawa, and R. H. Rediker, IEEE J. Quantum Electronics, QE-1:4 (1965).
182. C. E. Hurwitz, A. R. Calawa, and R. H. Rediker, IEEE J. Quantum Electronics, QE-1:102 (1965).
183. N. S. Bayshev and Z. I. Uritskii, Fiz. Tverd. Tela, 3:2861 (1961).
184. N. S. Baryshev and K. Ya. Shtivel'man, Fiz. Tekh. Poluprov., 1:172 (1967).
185. V. G. Baru and E. M. Khmel'nitskaya, Fiz. Tverd. Tela, 4:1897 (1962).
186. N. S. Baryshev and Z. I. Uritskii, Fiz. Tverd. Tela, 5:478 (1963).
187. T. S. Moss, Proc. Phys. Soc. (London), B62:741 (1949).
188. A. Smith and D. Dutton, J. Opt. Soc. Am., 48:1007 (1958).
189. A. Smith and D. Dutton, J. Phys. Chem. Solids, 22:351 (1961).
190. I. D. Konozenko, in: Photoelectric and Optical Phenomena in Semiconductors [in Russian], Izd. AN UkrSSR, Kiev (1959), p. 240.
191. L. N. Galkin and N. V. Korolev, Dokl. Akad. Nauk SSSR, 92:529 (1953).
192. G. F. J. Garlick and M. J. Dumbleton, Proc. Phys. Soc. (London), B67:442 (1954).
193. G. F. J. Garlick and R. A. Fatehally, in: Solid State Physics in Electronics and Communications (Proc. IUPAP Conf., Brussels, 1958, ed. by M. Desirant and J. L. Michiels), Vol. 4, Magnetic and Optical Properties (Part 2), Academic Press, New York (1960), p. 741.

194. G. F. J. Garlick, Proc. Seventh Intern. Conf. on Physics of Semiconductors, Paris, 1964, Vol. 4, Radiative Recombination in Semiconductors (ed. by C. Benoit a la Guillaume), publ. by Dunod, Paris; Academic Press, New York (1965), p. 3.
195. E. R. Washwell, K. F. Cuff, and L. R. Williams, Bull. Am. Phys. Soc., 10:594 (1965).
196. G. W. Pratt, Jr., and J. E. Ripper, Bull. Am. Phys. Soc., 10:84 (1965); J. Appl. Phys., 36:1525 (1965).
197. A. R. Calawa, J. F. Butler, and R. H. Rediker, Bull. Am. Phys. Soc., 10:84 (1965).
198. W. E. Spicer and G. J. Lapeyre, Phys. Rev., 139:A565 (1965).
199. R. A. Knapp, Phys. Rev., 132:1891 (1963).
200. R. M. Oman, J. Appl. Phys., 36:2091 (1965).
201. R. M. Oman, and M. J. Priolo, J. Appl. Phys., 37:524 (1966).
202. B. H. Sacks and W. E. Spicer, Bull. Am. Phys. Soc., 10:598 (1965).
203. J. M. Radcliffe, Proc. Phys. Soc. (London), A68:675 (1955).
204. R. Barrie, Proc. Phys. Soc. (London), B69:553 (1956).
205. E. O. Kane, J. Phys. Chem. Solids, 1:249 (1956).
206. E. H. Putley and J. B. Arthur, Proc. Phys. Soc. (London), B64:616 (1951).
207. E. D. Devyatkova, Yu. P. Maslakovets, L. S. Stil'bans, and T. S. Stavitskaya, Dokl. Akad. Nauk SSSR, 84:681 (1952).
208. E. H. Putley, Proc. Phys. Soc. (London), B65:388, 993 (1952).
209. E. H. Putley, Proc. Phys. Soc. (London), B68:22, 35 (1955).
210. E. D. Devyatkova, Yu. P. Maslakovets, and L. S. Stil'bans, Investigations of the Electrical Conductivity, Thermoelectric Power, Carrier Mobility, and Thermal Conductivity of Lead Telluride [in Russian], Izd. AN SSSR, Moscow (1953).
211. F. J. Morin and J. P. Maita, Phys. Rev., 96:28 (1954).
212. E. Hirahara and M. Murakami, J. Phys. Soc. Japan, 9:671 (1954).
213. S. J. Silverman and H. Levinstein, Phys. Rev., 94:871 (1954).
214. R. L. Petritz and W. W. Scanlon, Phys. Rev., 97:1620 (1955).
215. C. W. Ludwig and R. L. Watters, Phys. Rev., 101:1699 (1956).
216. N. V. Kolomoets, T. S. Stavitskaya, and L. S. Stil'bans, Zh. Tekh. Fiz., 27:73 (1957).
217. R. L. Petritz, W. W. Scanlon, and F. L. Lummis, Bull. Am. Phys. Soc., 2:32 (1957).
218. E. Z. Gershtein, T. S. Stavitskaya, and L. S. Stil'bans, Zh. Tekh. Fiz., 27:2472 (1957).
219. K. Shogenji and S. Uchiyama, J. Phys. Soc. Japan, 12:252, 431 (1957).
220. R. S. Allgaier and W. W. Scanlon, Phys. Rev., 111:1029 (1958).
221. M. N. Vinogradova, O. A. Golikova, B. A. Efimova, V. A. Kutasov, T. S. Stavitskaya, L. S. Stil'bans, and L. M. Sysoeva, Fiz. Tverd. Tela, 1:1333 (1959).
222. T. S. Stavitskaya and L. S. Stil'bans, Fiz. Tverd. Tela, 2:2082 (1960).
223. R. S. Allgaier and B. B. Houston, Jr., Proc. Sixth Intern. Conf. on Physics of Semiconductors, Exeter, England, 1962, publ. by the Institute of Physics, London (1962), p. 172.

LITERATURE CITED

224. T. S. Stavitskaya, V. A. Long, and B. A. Efimova, Fiz. Tverd. Tela, 7:2554 (1965).
225. I. V. Mochan and T. V. Smirnova, Fiz. Tverd. Tela, 3:2659 (1961).
226. D. H. Parkinson and J. E. Quarrington, Proc. Phys. Soc. (London), A67:569 (1954).
227. N. S. Baryshev, A. S. Makarov, K. Ya. Shtivel'man, and I. S. Aver'yanov, Fiz. Tekh. Poluprov., 2:1024 (1968).
228. L. S. Stil'bans, B. I. Bok, and É. L. Lifshits, Dokl. Akad. Nauk SSSR, 111:1011 (1956).
229. N. G. Lang, Fiz. Tverd. Tela, 3:2573 (1961).
230. I. A. Smirnov, B. Ya. Moizhes, and E. D. Nensberg, Fiz. Tverd. Tela, 2:1992 (1960).
231. R. S. Allgaier, J. Appl. Phys., 32, Suppl. to No. 10, 2185 (1961).
232. R. S. Allgaier and B. B. Houston, Jr., J. Appl. Phys., 37:302 (1966).
233. C. Herring and E. Vogt, Phys. Rev., 101:944 (1956).
234. L. G. Ferreira, Phys. Rev., 137:A1601 (1965).
235. H. B. Callen, Phys. Rev., 76:1394 (1949).
236. D. J. Howarth and E. H. Sondheimer, Proc. Roy. Soc. (London), A219:53 (1953).
237. Yu. I. Ravich, J. Phys., 29, Suppl. C4 to No. 11-12, 114 (1958).
238. H. Ehrenreich, J. Phys. Chem. Solids, 2:131 (1957); Ph. D. Thesis, Cornell University (1955).
239. D. M. Bercha and M. V. Tarnovskaya, Ukr. Fiz. Zh., 9:642 (1964).
240. Y. Kanai, R. Nii, and N. Watanabe, J. Appl. Phys., 32, Suppl. to No. 10, 2146 (1961).
241. A. Kobyashi, Y. Sato, and M. Fujimoto, Proc. Seventh Intern. Conf. on Physics of Semiconductors, Paris, 1964, Vol. 1, Physics of Semiconductors, (ed. by M. Hulin), publ. by Dunod, Paris; Academic Press, New York (1964), p. 1257.
242. Y. Kanai and K. Shohno, Japan. J. Appl. Phys., 2:6 (1963).
243. H. M. Day and A. C. Macpherson, Proc. IEEE, 51:1362 (1963).
244. R. Mansfield, Proc. Phys. Soc. (London), B69:79 (1956).
245. C. Erginsoy, Phys. Rev., 79:1083 (1950).
246. T. Shimizu, J. Phys. Soc. Japan, 18:1838 (1963).
247. A. Morita, J. Phys. Soc. Japan, 18:1437 (1963).
248. C. Herring, J. Appl. Phys., 31:1939 (1961).
249. B. Ya. Moizhes and Yu. I. Ravich, Fiz. Tekh. Poluprov., 1:188 (1967).
250. W. Zawadzki and J. Kolodziejczak, Phys. Status Solidi, 6:419 (1964).
251. T. P. McLean and E. G. S. Paige, J. Phys. Chem. Solids, 16:220 (1960).
252. R. N. Hall and J. H. Racette, J. Appl. Phys., 32, Suppl. to No. 10, 2078 (1961).
253. S. V. Airapetyants, B. A. Efimova, T. S. Stavitskaya, L. S. Stil'bans, and L. M. Sysoeva, Zh. Tekh. Fiz., 27:2167 (1957).
254. B. A. Efimova, T. S. Stavkitskaya, L. S. Stil'bans, and L. M. Sysoeva, Fiz. Tverd. Tela, 1:1325 (1959).
255. B. A. Efimova and L. A. Kolomoets, Fiz. Tverd. Tela, 7:424 (1965).
256. A. A. Machonis and I. B. Cadoff, Trans. Met. Soc. AIME, 230(2):333 (1964).
257. T. Irie, T. Takahama, and T. Ono, Japan. J. Appl. Phys., 2:72 (1963).
258. Y. Makino, J. Phys. Soc. Japan, 19:1755 (1964).
259. S. A. Semiletov and I. P. Voronina, Kristallografiya, 9:486 (1964).

LITERATURE CITED

260. H. E. Spencer and J. V. Morgan, J. Appl. Phys., 31:2024 (1960).
261. J. N. Zemel and J. D. Jensen, Bull. Am. Phys. Soc., 8:517 (1963).
262. K. Shogenji and S. Uchiyama, J. Phys. Soc. Japan, 12:1164 (1957).
263. K. Shogenji, J. Phys. Soc. Japan, 14:1360 (1959).
264. P. M. Starik, Fiz. Tverd. Tela, 7:2246 (1965).
265. S. V. Airapetyants, M. N. Vinogradova, I. N. Dubrovskaya, N. V. Kolomoets, and I. M. Rudnik, Fiz. Tverd. Tela, 8:1334 (1966).
266. N. V. Kolomoets, M. N. Vinogradova, E. Ya. Lev, and L. M. Sysoeva, Fiz. Tverd. Tela, 8:2925 (1966).
267. A. A. Andreev, Fiz. Tverd. Tela, 8:2818 (1966); A. A. Andreev and V. N. Radionov, Fiz. Tekh. Poluprov., 1:183 (1967).
268. G. T. Alekseeva, L. V. Prokof'eva, and T. S. Stavitskaya, Fiz. Tverd. Tela, 8:2819 (1966).
269. B. Lavelic, Physics Letters, 16:206 (1965).
270. A. I. Ansel'm, Introduction to the Theory of Semiconductors [in Russian], Fizmatgiz, Moscow (1962).
271. V. L. Bonch-Bruevich, in: Solid State Physics: Theory [in Russian], Izd. VINITI, Moscow (1965), p. 129.
272. L. V. Keldysh and G. P. Proshko, Fiz. Tverd. Tela, 5:3378 (1963).
273. L. W. Aukerman and R. K. Willardson, J. Appl. Phys., 31:939 (1960).
274. V. A. Saakyan, E. D. Devyatkova, and I. A. Smirnov, Fiz. Tverd. Tela, 7:3136 (1965).
275. G. L. Pearson and H. Suhl, Phys. Rev., 83:768 (1951).
276. M. Shibuya, Phys. Rev., 95:1388 (1954).
277. R. S. Allgaier, Phys. Rev., 112:828 (1958).
278. T. Irier, J. Phys. Soc. Japan, 11:840 (1956).
279. R. S. Allgaier, Phys. Rev., 119:554 (1960).
280. R. S. Allgaier, Proc. Fifth Intern. Conf. on Physics of Semiconductors, Prauge, 1960, publ. by Academic Press, New York (1961), p. 1037.
281. A. Sagar and R. C. Miller, J. Appl. Phys., 32, Suppl. to No. 10, 2073 (1961).
282. I. Cadoff and E. S. Pan, Bull. Am. Phys. Soc., 11:755 (1966).
283. A. A. Averkin, Yu. V. Ilisavskii (U. V. Ulisavsky), and A. R. Regel' (A. R. Regel), Proc. Sixth Intern. Conf. on Physics of Semiconductors, Exeter, England, 1962, publ. by The Institute of Physics, London (1962), p. 690.
284. W. Paul, M. Demeis, and L. X. Finegold, Proc. Sixth Intern. Conf. on Physics of Semiconductors, Exeter, England, 1962, publ. by The Institute of Physics, London (1962), p. 712.
285. Y. Sato, M. Fujimoto, and A. Kobyashi, J. Phys. Soc. Japan, 19:24 (1964).
286. A. A. Averkin, B. Ya. Moizhes, and I. A. Smirnov, Fiz. Tverd. Tela, 3:1859 (1961).
287. A. A. Averkin, I. G. Dombrovskaya, and B. Ya. Moizhes, Fiz. Tverd. Tela, 5:96 (1963).
288. A. A. Averkin, S. Kasimov, and E. D. Nensberg, Fiz. Tverd. Tela, 4:3667 (1962).
289. A. A. Averkin and P. G. Dermenzhi, Fiz. Tverd. Tela, 8:103 (1966).
290. A. A. Semerchan, N. N. Kuzin, A. N. Drozdova, and L. F. Vereshchagin, Dokl. Akad. Nauk SSSR, 152:1079 (1963).

291. P. W. Bridgman, Proc. Am. Acad. Arts Sci., 76:55 (1948).
292. G. E. Pikus and G. L. Bir, Fiz. Tverd. Tela, 4:2090 (1962).
293. L. E. Hollander and T. J. Diesel, J. Appl. Phys., 31:692 (1960).
294. Yu. V. Ilisavskii, Fiz. Tverd. Tela, 3:1898, 3555 (1961).
295. Yu. V. Ilisavskii, Fiz. Tverd. Tela, 4:918 (1962).
296. Yu. V. Ilisavskii and É. Z. Yakhkind, Fiz. Tverd. Tela, 4:1975 (1962).
297. G. L. Bir and G. E. Pikus, Fiz. Tverd. Tela, 4:2243 (1962).
298. J. R. Burke, Jr., Bull. Am. Phys. Soc., 8:310 (1963); Phys. Rev., 160:636 (1967).
299. R. Ito and K. Shogenji, J. Phys. Soc. Japan, 18:1343 (1963).
300. K. Shogenji and R. Ito, J. Phys. Soc. Japan, 20:172 (1965).
301. A. A. Chudinov, Fiz. Tverd. Tela, 4:755 (1962); 5:1458 (1963).
302. H. B. Huntington, Solid State Physics, 7:213 (1958).
303. S. Bhagavantam and T. S. Rao, Nature, 168:42 (1951).
304. P. W. Bridgman, Am. J. Sci., 10:483 (1925).
305. P. Aigrain, M. Balkanski, C. Benoil a la Guillaume, R. Coehlor, O. Garreta, H. Guennoc, C. Sébenne, and J. Tavernier, Selected Constants Relative to Semiconductors, Pergamon Press, Oxford (1961).
306. E. Burstein and P. H. Egli, Adv. Electronics Electron Phys., 7:1 (1955).
307. Yu. V. Ilisavskii, Semiconductor Strain Gauges [in Russian], LDNTP, Leningrad (1963).
308. A. N. Arsen'eva-Geil', Zh. Tekh. Fiz., 17:903 (1947).
309. A. N. Arsen'eva-Geil' and N. B. Dzhelenova, Fiz. Tverd. Tela, 4:2714 (1962).
310. A. N. Arsen'eva-Geil', in: Semiconductors in Science and Technology [in Russian], Vol. 2, Izd. AN SSSR, Moscow (1958), p. 299.
311. F. Braun, Ann. Phys. Chem., 153:556 (1874).
312. C. A. Hogarth, Proc. Phys. Soc. (London), B66:216 (1953).
313. W. B. Pennebaker, Solid State Electronics, 18:509 (1965).
314. A. F. Gibson, Proc. Phys. Soc. (London), B65:196, 214 (1952).
315. A. F. Gibson, Proc. Phys. Soc. (London), B65:378 (1952).
316. C. A. Hogarth, Proc. Phys. Soc. (London), B66:216 (1953).
317. R. F. Brebrick and W. W. Scanlon, Phys. Rev., 96:598 (1954).
318. T. S. Moss, Proc. Phys. Soc. (London), B68:697 (1955).
319. V. G. Bhide, J. N. Das, and P. V. Khandekar, Proc. Phys. Soc. (London), B69:245 (1956).
320. J. Bloem, Appl. Sci. Res., B6:92 (1956).
321. R. N. Hall, Proc. Fifth Intern. Conf. on Physics of Semiconductors, Prague, 1960, publ. by Academic Press, New York (1961), p. 193.
322. R. H. Rediker and A. R. Calawa, J. Appl. Phys., 32, Suppl. to No. 10, 2189 (1961).
323. A. G. Aronov and G. E. Pikus, Zh. Eksp. Teor. Fiz., 51:281 (1966).
324. J. F. Butler, J. Electrochem. Soc., 111:1150 (1964).
325. W. Shockley, Bell System Tech. J., 28:435 (1949).
326. C. T. Sah, R. N. Noyce, and W. Shockely, Proc. IRE, 45:1228 (1957).
327. L. Esaki, Phys. Rev., 109:603 (1958).
328. B. M. Vul, É. I. Zavaritskaya, and N. V. Zavaritskii, Fiz. Tverd. Tela, 8:888 (1966).

LITERATURE CITED

329. R. M. Williams and J. Shewchun, Phys. Rev. Letters, 14:824 (1965).
330. J. Stefan, Ann. Phys., 124:632 (1865).
331. I. M. Tsidil'kovskii, Fiz. Metallov Metalloved., 8:494 (1959).
332. Ya. S. Budzhak and H. Yu. Nemish, Urk. Fiz. Zh., 10:461 (1965).
333. Ya. S. Budzhak and K. D. Tovstyuk, Izv. Akad. Nauk SSSR, Ser. Fiz., 28:1318 (1964).
334. M. K. Zhitinskaya, V. I. Kaidanov, and I. A. Chernik, Fiz. Tverd. Tela, 8:295 (1966).
335. I. N. Dubrovskaya and Yu. I. Ravich, Fiz. Tverd. Tela, 8:1455 (1966).
336. I. N. Dubrovskaya, E. D. Nensberg, G. V. Nikitina, and Yu. I. Ravich, Fiz. Tverd. Tela, 8:2247 (1966).
337. T. S. Stavitskaya, L. V. Prokof'eva, Yu. I. Ravich, and B. A. Efimova, Fiz. Tekh. Poluprov., 1:1138 (1967).
338. D. M. Finlayson and D. Greig, Proc. Phys. Soc. (London), 73:49 (1959).
329. K. F. Cuff, M. R. Ellett, and C. D. Kuglin, Proc. Sixth Intern. Conf. on Physics of Semiconductors, Exeter, England, 1962, publ. by The Institute of Physics, London (1962), p. 316.
340. G. W. Johnson, J. Electronics Control, 16:497 (1964).
341. E. D. Devyatkova and I. A. Smirnov, Fiz. Tverd. Tela, 3:2310 (1961).
342. N. V. Kolomoets, M. N. Vinogradova, and L. M. Sysoeva, Fiz. Tekh. Poluprov., 1:1222 (1967).
343. E. D. Devyatkova and V. V. Tikhonov, Fiz. Tverd. Tela, 7:1770 (1965).
344. M. Rodot, in: Solid State Physics in Electronics and Telecommunications (Proc. IUPAP Conf., Brussels, 1958, ed. by M. Désirant and J. L. Michiels), Vol. 2, Semiconductors (Part 2), Academic Press, New York (1960), p. 680.
345. R. W. Keyes, J. Phys. Chem. Solids, 6:1 (1958).
346. J. M. Ziman, Electrons and Phonons, Clarendon Press, Oxford (1960).
347. G. E. Pikus, Zh. Tekh. Fiz., 26:49 (1956).
348. A. I. Akhiezer, J. Phys. USSR, 10:217 (1946); A. I. Akhiezer and L. A. Shishkin, Zh. Eksp. Teor. Fiz., 34:1267 (1958).
349. A. F. Ioffe, Physics of Semiconductors, Infosearch, London (1960).
350. P. Carruthers, Rev. Mod. Phys., 33:92 (1961).
351. K. Mendelssohn and H. M. Rosenberg, Solid State Physics, 12:223 (1961).
352. J. R. Drabble and H. J. Goldsmid, Thermal Conduction in Semiconductors, Pergamon Press, Oxford (1961).
353. P. G. Klemens, Solid State Physics, 7:1 (1958).
354. L. S. Stil'bans, in: Semiconductors in Science and Technology [in Russian], Vol. 1, Izd. AN SSSR, Moscow (1957), p. 86.
355. E. D. Debyatkova and I. A. Smirnov, Fiz. Tverd. Tela, 2:1984 (1960).
356. W. Zawadzki, Phys. Status Solidi, 2:385 (1962).
357. I. A. Smirnov and Yu. I. Ravich, Fiz. Tekh. Poluprov., 1:891 (1967).
358. V. M. Muzhdaba and S. S. Shalyt, Fiz. Tverd. Tela, 8:3727 (1966).
359. F. W. Sheard, Phil. Mag., 5:887 (1960).
360. Yu. I. Ravich, I. A. Smirnov, and V. V. Tikhonov, Fiz. Tekh. Poluprov., 1:206 (1967).
361. S. S. Shalyt, V. M. Muzhdaba, and A. D. Galetskaya, Fiz. Tverd. Tela, 10:1277 (1968).

362. B. I. Davydov and I. M. Shmushkevich, Usp. Fiz. Nauk, 24:21 (1940).
363. V. A. Kutasov, B. Ya. Moizhes, and I. A. Smirnov, Fiz. Tverd. Tela, 7:1065 (1965).
364. E. D. Devyatkova and V. A. Saakyan, Izv. Akad. Nauk Arm. SSR, Ser. Fiz., 2:14 (1967).
365. J. M. Honig and T. C. Harman, Quaterly Progr. Rep. Solid State Research, Lincoln Laboratory, Massachussetts Inst. of Technology (April, 1961), p. 8.
366. V. A. Saakyan and I. A. Smirnov, Fiz. Tverd. Tela, 8:3668 (1966).
367. I. A. Smirnov, M. N. Vinogradova, N. V. Kolomoets, and L. M. Sysoeva, Fiz. Tverd. Tela, 9:2638 (1967).
368. L. Genzel, Z. Physik, 135:177 (1953).
369. E. D. Devyatkova, B. Ya. Moizhes, and I. A. Smirnov, Fiz. Tverd. Tela, 1:613 (1959).
370. F. Simon, in: Handbuch der Physik (eds. H. Geiger and K. Schell), Vol. 10, Springer Verlag, Berlin (1926), p. 350.
371. M. Blackman, in: Handbuch der Physik (ed. S. Flugge), Vol. 7, Part 1, Springer Verlag, Berlin (1955), p. 325.
372. D. H. Parkinson and J. E. Quarrington, Proc. Phys. Soc. (London), A67:569 (1954).
373. E. D. Eastman and W. H. Rodebush, J. Am. Chem. Soc., 40:496 (1918).
374. C. T. Anderson, J. Am. Chem. Soc., 54:107 (1932).
375. G. Leibfried, in: Handbuch der Physik (ed. S. Flugge), Vol. 7, Part 1, Springer Verlag, Berlin (1955), p. 104.
376. J. Callaway, Phys. Rev., 113:1046 (1959).
377. M. G. Holland, Phys. Rev., 132:2461 (1963); M. G. Holland, in: Semiconductors and Semimetals (eds. R. K. Willardson and A. C. Beer), Vol. 2: Physics of III-V Compounds, Academic Press, New York (1966), p. 3.
378. H. B. G. Casimir, Physica, 5:595 (1930).
379. R. Berman, E. L. Foster, and J. M. Ziman, Proc. Roy. Soc. (London), A231:130 (1955).
380. P. G. Klemens, Proc. Phys. Soc. (London), A68:1113 (1955).
381. C. Herring, Phys. Rev., 95:954 (1954).
382. I. Pomeranchuk, Phys. Rev., 60:820 (1941).
383. J. M. Ziman, Phil. Mag., 1:191 (1956).
384. R. O. Pohl, Phys. Rev. Letters, 8:481 (1962); C. T. Walker and R. O. Pohl, Phys. Rev., 131:1433 (1963).
385. M. Wagner, Phys. Rev., 131:1443 (1963).
386. R. W. Keyes, Phys. Rev., 115:564 (1959).
387. J. S. Dugdale and D. K. MacDonald, Phys. Rev., 98:1751 (1955).
388. J. Tavernier, Proc. Fifth Intern. Conf. on Physics of Semiconductors, Prague, 1960, publ. by Academic Press, New York (1961), p. 638.
389. E. D. Devyatkova and I. A. Smirnov, Fiz. Tverd. Tela, 3:2298 (1961).
390. E. D. Devyatkova, A. V. Petrov, and I. A. Smirnov, Fiz. Tverd. Tela, 3:1338 (1961).
391. A. F. Ioffe and A. V. Ioffe, Dokl. Akad. Nauk SSSR, 98:757 (1954); Fiz. Tverd. Tela, 2:781 (1960).

LITERATURE CITED

392. E. D. Devyatkova, Zh. Tekh. Fiz., 27:461 (1957).
393. S. J. Angello, Elec. Eng., 79:353 (1960).
394. C. S. Duncan and C. W. Wilson, Elec. Eng., 79:372 (1960).
395. R. Nii, J. Phys. Soc. Japan, 13:769 (1958).
396. Y. Kanai and R. Nii, J. Phys. Chem. Solids, 8:338 (1959).
397. W. V. Huck, R. W. Fritts, and S. Karrer, Bull. Am. Phys. Soc., 2:185 (1957).
398. R. W. Fritts, Trans. AIEE, Part I, Comm. and Electronics, 78:817 (1960).
399. A. D. Stuckes and R. P. Chasmar, Rep. Meeting on Semiconductors, Ruby, England, 1956, publ. by Physical Society, London (1957), p. 119.
400. A. V. Ioffe and A. F. Ioffe, Izv. Akad. Nauk SSSR, Ser. Fiz., 20:65 (1956).
401. E. K. Iordanishvili, Semiconducting Thermoelectric Materials [in Russian], LDNTP, Leningrad (1963).
402. Yu. A. Dunaev, Zh. Tekh. Fiz., 16:1101 (1946).
403. E. D. Nensberg and A. V. Petrov, Dokl. Akad. Nauk SSSR, 160:1304 (1965).
404. B. K. Agrawal and G. S. Verma, Physica 28:599 (1962).
405. F. F. Steigmeier and I. Kudman, Phys. Rev., 141:767 (1966).
406. E. D. Devyatkova and I. A. Smirnov, Fiz. Tverd. Tela, 4:2507 (1962).
407. A. L. Éfros, Fiz. Tverd. Tela, 3:2065 (1961).
408. Ya. I. Frenkel', Introduction to the Theory of Metals [in Russian], Fizmatgiz, Moscow (1958).
409. V. P. Zhuze, Dokl. Akad. Nauk SSSR, 99:711 (1954).
410. T. A. Kontorova, Zh. Tekh. Fiz., 26:2021 (1956).
411. S. I. Novikova and N. Kh. Abrikosov, Fiz. Tverd. Tela, 5:1913 (1963).
412. J. Sharma, Proc. Indian Acad. Sci., A34:72 (1951).
413. E. Riano and J. L. Amoros Portolés, Bol. Real Soc. Españ. Hist. Nat., Secc. Geol., 56:345 (1958).
414. E. Grüneisen, Ann. Physik, 39:289 (1912).
415. E. Grüneisen, in: Handbuch der Physik (eds. H. Geiger and K. Schell), Vol. 10, Springer Verlag, Berlin (1926), p. 1.
416. A. F. Ioffe, Can. J. Phys., 34:1342 (1956).
417. V. Ambegaokar, Phys. Rev., 114:488 (1959).
418. P. G. Klemens, Phys. Rev., 119:507 (1960).
419. J. Callaway and H. C. von Baeyer, Phys. Rev., 120:1149 (1960).
420. B. Ya. Moizhes, R. V. Parfen'ev, F. A. Chudnovskii, and A. L. Éfros, Fiz. Tverd. Tela, 3:1933 (1961).
421. V. A. Saakyan, Fiz. Tverd. Tela, 9:2670 (1967).
422. W. Cochran, R. A. Cowley, G. Dolling, and M. M. Elcombe, Proc. Roy. Soc. (London), A293:433 (1966).
423. A. D. B. Woods, W. Cochran, and B. N. Brockhouse, Phys. Rev., 119:980 (1960); R. A. Cowley, W. Cochran, B. N. Brockhouse, and A. D. B. Woods, Phys. Rev., 131:1030 (1963).
424. W. Cochran, Proc. Roy. Soc. (London), A253:260 (1959).
425. G. Gilat and G. Dolling, Physics Letters, 8:304 (1964).
426. R. A. Cowley, Adv. Phys. 12:421 (1963).
427. E. R. Cowley and R. A. Cowley, Proc. Roy. Soc. (London), A287:259 (1965).
428. M. Matyáš, Czechoslovak J. Phys., 8:301 (1958).

429. M. Matyáš, Czechoslovak J. Phys. 8:309 (1958).
430. K. D. Tovstyuk, Ya. S. Budzhak, and M. V. Tarnovs'ka, Ukr. Fiz. Zh., 8:795 (1963).
431. P. J. Stiles, E. Burstein, and D. N. Langenberg, J. Appl. Phys., 32, Suppl. to No. 10, 2174 (1961).
432. I. Weinberg and J. Callaway, Nuovo Cimento, 24:190 (1962).
433. Y. Kanai, R. Nii, and N. Watanabe, J. Phys. Soc. Japan, 15:1717 (1960).
434. K. F. Cuff, M. R. Ellett, and C. D. Kuglin, J. Appl. Phys., 32, Suppl. to No. 10, 2179 (1961).
435. R. S. Allgaier, B. B. Houston, Jr., R. F. Bis, J. Babiskin, and P. G. Siebenmann, Proc. Seventh Intern. Conf. on Physics of Semiconductors, Paris, 1964, Vol. 1, Physics of Semiconductors (ed. by M. Hulin), publ. by Dunod, Paris; Academic Press, New York (1964), p. 659.
436. R. S. Allgaier, B. B. Houston, Jr., J. R. Burke, J. Babiskin, and P. G. Siebenmann, Bull. Am. Phys. Soc., 8:517 (1963).
437. C. D. Kuglin, M. R. Ellett, and K. F. Cuff, Phys. Rev. Letters, 6:177 (1961).
438. M. R. Ellett, K. F. Cuff, and C. D. Kuglin, Bull. Am. Phys. Soc., 8:246 (1963).
439. K. F. Cuff, M. R. Ellett, C. D. Kuglin, and L. R. Williams, Proc. Seventh Intern. Conf. on Physics of Semiconductors, Paris, 1964, Vol. 1, Physics of Semiconductors (ed. by M. Hulin), publ. by Dunod, Paris; Academic Press, New York (1964), p. 677.
440. E. Yamada and K. Shogenji, J. Phys. Soc. Japan, 16:1475 (1961).
441. K. Shogenji and T. Sato, J. Phys. Soc. Japan, 18:146 (1963).
442. S. S. Shalyt and A. L. Éfros, Fiz. Tverd. Tela, 4:1233 (1962).
443. M. J. Stephen and A. V. Lidiard, J. Phys. Chem. Solids, 9:43 (1958).
444. A. K. Walton, T. S. Moss, and B. Ellis, Proc. Phys. Soc. (London), 79:1065 (1962).
445. E. D. Palik, S. Teitler, B. W. Henvis, and R. F. Wallis, Proc. Sixth Intern. Conf. on Physics of Semiconductors, Exeter, England, 1962, publ. by The Institute of Physics, London (1962), p. 288.
446. D. L. Mitchell, E. D. Palik, and P. F. Wallis, Phys. Rev. Letters, 14:827 (1965).
447. E. A. Stern, Phys. Rev. Letters, 15:62 (1965).
448. D. L. Mitchell, E. D. Palik, J. D. Jensen, R. B. Schoolar, and J. N. Zemel, Physics Letters, 4:262 (1963).
449. D. L. Mitchell, E. D. Palik, J. D. Jensen, R. B. Schoolar, and J. N. Zemel, Bull. Am. Phys. Soc., 9:292 (1964).
450. D. L. Mitchell, E. D. Palik, and J. N. Zemel, Proc. Seventh Intern. Conf. on Physics of Semiconductors, Paris, 1964, Vol. 1, Physics of Semiconductors (ed. by M. Hulin), publ. by Dunod, Paris; Academic Press, New York (1964), p. 325.
451. R. J. Stiles, E. Burstein, and D. N. Langenberg, Phys. Rev. Letters, 9:257 (1962).
452. W. M. Walsh, Bull. Am. Phys. Soc., 7:202 (1962).
453. U. J. Hansen and K. F. Cuff, Bull. Am. Phys. Soc., 8:246 (1963).
454. R. Nii, J. Phys. Soc. Japan, 18:456 (1963); 19:58 (1964).
455. R. Nii and A. Kobayashi, Proc. Seventh Intern. Conf. on Physics of Semiconductors, Paris, 1964, Vol. 2, Plasma Effects in Solids (ed. by J. Bok), publ. by Dunod, Paris; Academic Press, New York (1965), p. 65.

LITERATURE CITED

456. M. Fujimoto, J. Phys. Soc. Japan, 21:1706 (1966).
457. E. Burstein, P. J. Stiles, D. N. Langenberg, and R. F. Wallis, Proc. Sixth Intern. Conf. on Physics of Semiconductors, Exeter, England, 1962, publ. by The Institute of Physics, London (1962), p. 345.
458. M. Ya. Azbel' and E. A. Kaner, Zh. Eksp. Teor. Fiz., 30:811 (1956).
459. H. Numata and Y. Uemura, Physics Letters, 9:227 (1964); J. Phys. Soc. Japan, 19:2140 (1964).
460. Y. Sawada, E. Burstein, D. L. Carter, and L. Testardi, Proc. Seventh Intern. Conf. on Physics of Semiconductors, Paris, 1964, Vol. 2, Plasma Effects in Solids (ed. by J. Bok), publ. by Dunod, Paris; Academic Press, New York (1965), p. 71.
461. O. V. Konstantinov and V. I. Perel', Zh. Eksp. Teor. Fiz., 38:161 (1960).
462. P. Aigrain, Proc. Fifth Intern. Conf. on Physics of Semiconductors, Prague, 1960, publ. by Academic Press, New York (1961), p. 224.
463. Y. Kanai, Japan. J. Appl. Phys., 2:137 (1963); Proc. Seventh Intern. Conf. on Physics of Semiconductors, Paris, 1964, Vol. 2, Plasma Effects in Solids (ed. by J. Bok), publ. by Dunod, Paris; Academic Press, New York (1965), p. 45.
464. Y. Shapira and B. Lax, Physics Letters, 7:133 (1963).
465. O. Beckman and Y. Shapira, Physics Letters, 21:282 (1966).
466. D. G. Bell, D. M. Hum, L. Pincherle, D. W. Sciama, and P. M. Woodward, Proc. Roy. Soc. (London), A217:71 (1953).
467. J. B. Conklin, Jr., L. E. Johnson, and G. W. Pratt, Phys. Rev., 137:A1282 (1965).
468. S. Rabii, Phys. Rev., 167:801 (1968).
469. L. Kleinman and P. J. Lin, Proc. Seventh Intern. Conf. on Physics of Semiconductors, Paris, 1964, Vol. 1, Physics of Semiconductors (ed. by M. Hulin), publ. by Dunod, Paris; Academic Press, New York (1964), p. 63.
470. P. J. Lin and L. Kleinman, Phys. Rev., 142:478 (1966).
471. D. L. Mitchell and R. F. Wallis, Phys. Rev., 151:581 (1966).
472. V. Heine, Group Theory in Quantum Mechanics: An Introduction to Present Usage, Pergamon Press, London (1960).
473. L. D. Landau and E. M. Lifshitz, Quantum Mechanics: Non-Relativistic Theory, 2nd ed., Pergamon Press, Oxford (1965), Ch. XII; L. D. Landau and E. M. Lifshitz, Statistical Physics, 2nd ed., Pergamon Press, Oxford (1968), Ch. XIII.
474. G. K. Koster, Solid State Physics, 5:174 (1957).
475. J. O. Dimmock and G. B. Wright, Phys. Rev., 135:A821 (1964); Proc. Seventh Intern. Conf. on Physics of Semiconductors, Paris, 1964, Vol. 1, Physics of Semiconductors (ed. by M. Hulin), publ. by Dunod, Paris; Academic Press, New York (1964), p. 77.
476. L. E. Johnson, J. B. Conklin, Jr., and G. W. Pratt, Jr., Phys. Rev. Letters, 11:538 (1963).
477. F. Herman, C. D. Kuglin, F. K. Cuff, and R. L. Kortum, Phys. Rev. Letters, 11:541 (1963).
478. J. M. Ziman, Principles of the Theory of Solids, Cambridge University Press (1964).
479. G. W. Pratt, Jr., and L. G. Ferreira, Proc. Seventh Intern. Conf. on Physics of Semiconductors, Paris, 1964, Vol. 1, Physics of Semiconductors (ed. by M. Hulin), publ. by Dunod, Paris; Academic Press, New York (1964), p. 69.

LITERATURE CITED

480. P. N. Nikiforov, in: Chemical Bonds in Semiconductors and Thermodynamics (ed. by N. N. Sirota), Consultants Bureau, New York (1968), p. 30. Abstracts of Papers presented at Seventh All-Union Conf. on Theory of Semiconductors [in Russian], Tartu (1966).
481. J. Kolodziejczak and S. Zukotynski, Phys. Status Solidi, 5:145 (1964).
482. W. Zawadzki, R. Kowalczyk, and J. Kolodziejczak, Phys. Status Solid, 10:513 (1965).
483. M. H. Cohen, Phys. Rev., 121:387 (1961).
484. P. S. Allgaier, Phys. Rev., 152:808 (1967).
485. E. G. Strel'chenko, Fiz. Tverd. Tela, 8:965 (1966).
486. M. H. Cohen and E. I. Blount, Phil. Mag., 5:115 (1960); Y. Yafet, Solid State Physics, 14:2 (1963).
487. Yu. I. Ravich, Fiz. Tverd. Tela, 7:1821 (1965).
488. A. F. Ioffe, Semiconductor Thermoelements and Thermoelectric Cooling, Infosearch, London (1957).
489. A. F. Ioffe, S. V. Airapetyants, A. V. Ioffe, N. V. Kolomoets, and L. S. Stil'bans, Dokl. Akad. Nauk SSSR, 106:981 (1956).
490. L. S. Stil'bans, Author's, Abstract of Dissertation for Candidate's Degree [in Russian], Physics Institute, Academy of Sciences of the USSR, Moscow (1961).
491. See [488].
492. R. W. Fritts, in: Direct Conversion of Heat to Electricity (eds. J. Kaye and J. A. Welsh), J. Wiley, New York (1960), Ch. XVII, p. 1.
493. D. A. Wright, in: Direct Generation of Electricity (ed. K. H. Spring), Academic Press, London (1965), Ch. V, p. 285.
494. R. W. Fritts, in: Thermoelectric Materials and Devices, Reinhold, New York (1960), p. 143.
495. F. D. Rosi, E. F. Hockings, and N. E. Lindenblad, RCA Rev., 22:82 (1961).
496. M. Fleischman, H. Luy, and J. Rupprecht, Z. Naturforsch., 18a:646 (1963).
497. G. A. Alatyrtsev, and Yu. N. Malevskii, Teploénergetika, 3:61 (1961).
498. Patent No. 7281, class 100 DI (Japan).
499. Patent No. 3165615, class 219-9.5 (USA).
500. M. Weinstein and A. I. Mlavsky, Rev. Sci. Instr., 33:1119 (1962).
501. E. W. Bollmeier, in: Direct Conversion of Heat to Electricity (eds. J. Kaye and J. A. Welsh), J. Wiley, New York (1960), Ch XVI, p. 1.
502. E. W. Bollmeier, Elec. Eng., 78:995 (1959).
503. E. K. Iordanishvili, Thermoelectric Electrical Power Sources [in Russian], "Sov. radio," Moscow (1968).
504. R. T. Carpenter and D. G. Harvey, Astronautics, 7(5):30, 31, 58, 60, 61 (1962).
505. R. T. Carpenter, Astronaut. Aerospace Eng., 1(4):68 (1963).
506. J. R. Wetch, Nucleonics, 24(6):33 (1966).
507. K. E. Buck, Nucleonics, 24(6):38 (1966).
508. R. A. Smith, F. E. Jones, and R. P. Chasmar, Detection and Measurement of Infrared Radiation, Clarendon Press, Oxford (1957).
509. M. S. Sominskii, in: Semiconductors in Science and Technology [in Russian], Vol. 1, Izd. AN SSSR (1957), p. 338.
510. P. W. Kruse, L. D. McGlaughlin, and R. B. McQuistan, Elements of Infrared Technology, J. Wiley, New York (1962).

511. Yu. A. Ivanov and B. V. Tyapkin, Infrared Technology in Military Science [in Russian], "Sov. radio," Moscow (1963).
512. G. A. Morton, RCA Rev., 26:3 (1965).
513. J. N. Humphrey, Applied Optics, 4:665 (1965).
514. D. G. Coates, W. D. Lawson, and A. C. Prior, J. Electrochem. Soc., 108:1038 (1961).
515. B. T. Kolomiets, Élektrichestvo, No. 3, 57 (1949), No. 11, 44 (1951); Zh. Tekh. Fiz., 21:3 (1951).
516. T. H. Johnson, H. T. Cozine, and B. N. McLean, Applied Optics, 4:693 (1965).
517. R. P. Day, R. A. Wallner, and E. A. Lodi, Infrared Phys., 1:212 (1961).
518. C. Hilsum and O. Simpson, Proc. Inst. Elec. Engrs., 106, Part B, Suppl. 15, 398 (1959).
519. V. V. Orlov, Fiziolog. Zh. SSSR, 44:258 (1958).
520. A. J. Strauss, Trans. Metall. Soc. AIME, 239:794 (1967).
521. J. F. Miller, J. W. Moody, and R. C. Himes, Trans. Metall. Soc. AIME, 239:342 (1967).
522. P. M. Starik, Ukr. Fiz. Zh., 11:265 (1966).
523. N. I. Akimenko, Ukr. Fiz. Zh., 13:1009 (1968).
524. H. Gobrecht, S. Seeck, and M. Hofmann, J. Phys. Chem. Solids, 29:627 (1968).
525. C. A. Nanney, Proc. Ninth Intern. Conf. on Physics of Semiconductors, Moscow, 1968, Vol. 2, publ. by Izd. "Nauka," Leningrad (1968), p. 910.
526. O. S. Gryaznov and Yu. I. Ravich, Fiz. Tekh. Poluprov., in press.
527. A. N. Mariano and K. L. Chopra, Appl. Phys. Letters, 10:282 (1967).
528. S. S. Kabalkina, N. R. Serebryanaya, and L. F. Vereshchagin, Fiz. Tverd. Tela, 10:733 (1968).
529. B. Houston, R. E. Strakna, and H. S. Belson, J. Appl. Phys., 39:3913 (1968).
530. I. N. Dubrovskaya, B. A. Efimova, and E. D. Nensberg, Fiz. Tekh. Poluprov., 2:530 (1968).
531. I. A. Chernik, V. I. Kaidanov, M. N. Vinogradova, and N. V. Kolomoets, Fiz. Tekh. Poluprov., 2:773 (1968).
532. I. A. Chernik, V. I. Kaidnov, and E. P. Ishutinov, Fiz. Tekh. Poluprov., 2:995 (1968).
533. I. A. Chernik, M. N. Vinogradova, N. V. Kolomoets, and D. P. Sementsina, Fiz. Tekh. Poluprov., 2:1173 (1968).
534. I. A. Chernik, V. I. Kaidanov, E. D. Nensberg, and V. V. Polyakov, Fiz. Tekh. Poluprov., 2:142 (1968).
535. S. A. Aliev, I. N. Dubrovskaya, L. L. Korenblit, B. Ya. Moizhes, V. M. Muzhdaba, Yu. I. Ravich, I. A. Smirnov, V. I. Tamarchenko, and S. S. Shalyt, Proc. Ninth Intern. Conf. on Physics of Semiconductors, Moscow, 1968, Vol. 2, publ. by Izd. "Nauka," Leningrad (1968), p. 659.
536. M. M. Elcombe, Proc. Roy. Soc. (London), 300A, 100 (1967).
537. G. Coste, Phys. Status Solidi, 20:361 (1967).
538. K. J. Button, B. Lax, M. H. Weiler, and M. Reine, Phys. Rev. Letters, 17:1005 (1966).
539. C. Nanney, Phys. Rev., 138:A1484 (1965).
540. W. Schilz, Solid State Commun., 5:503 (1967); Phys. Status Solidi, 29:559 (1968).

541. J. N. Walpole and A. L. McWhorter, Phys. Rev., 158:708 (1967).
542. D. E. Aspnes and M. Cardona, Phys. Rev., 173:714 (1968).
543. C. Keffer, T. M. Hayes, and A. Bienenstock, Phys. Rev. Letters, 21:1676 (1968).
544. I. Melngailis and A. R. Calawa, Appl. Phys. Letters, 9:304 (1966).
545. M. L. Shultz and R. Dalven, Bull. Am. Phys. Soc., 12:139 (1967).
546. J. F. Butler, A. R. Calawa, and T. C. Harman, Appl. Phys. Letters, 9:427 (1966).
547. J. F. Butler, A. R. Calawa, I. Melngailis, T. C. Harman, and J. O. Dimmock, Bull. Am. Phys. Soc., 12:384 (1967).
548. A. J. Strauss, Phys. Rev., 157:608 (1967).
549. G. Martinez, J. M. Besson, M. Balkanski, and M. Moulin, Proc. Ninth Intern. Conf. on Physics of Semiconductors, Moscow, 1968, Vol. 2, publ. by Izd. "Nauka," Leningrad (1968), p. 1190.
550. T. S. Stavitskaya, Author's Abstract of Dissertation [in Russian], Leningrad (1968).
551. Aviat. Week and Space Technol., 86(10):161 (March 6, 1967).
552. G. Lapidus, Electronic News, 12:4 (1967).
553. R. W. Fritts, IEEE Trans. Ind. and Gen. Appl., 3(5):458 (1967).
554. N. Fuschillo and R. Gibson, Adv. Energy Conversion, 7:43 (1967).
555. R. X. Flaherty and G. E. King, Semicond. Prod. and Sol. State Techn., 8(6):18 (1965).

SUBJECT INDEX

Unless otherwise stated, entries apply to all lead chalcogenides

A

absorption of light and e.m. waves 43-48
    fundamental edge 51-56
    long wavelengths 56-63
    magnetoabsorption 244-249
    short wavelengths 63-66
absorption of ultrasound in magnetic field 261-262
anisotropy coefficient of effective masses 128-129, 226, 233, 248, 257
applications
    photoelectric effects 334-343
    piezoresistance 345-346
    stimulated recombination radiation 343-345
    thermoelectric properties 323-334

B

band nonparabolicity 233-235, 299-306
    effective width of interaction forbidden band 305-306
    formulas 347-350
barrier capacitance 145-147
Bridgman–Stockbarger crystal growth method 21-22

C

carrier density
    doped crystals 35
    intrinsic carriers, temperature dependence 123

carrier density (continued)
    pure crystals 34
carrier lifetime
    dependence on carrier density 75
    measurement methods 67-69, 71
    values 71-72
carrier mobility 85-110
    at low temperatures 100-104
    at moderate and high temperatures 90-100
    effect of carrier scattering 85-89, 106-108
    films 110
    solid solutions 108-110
    theory 85-89
carrier recombination
    laser effect 79-82, 249
    mechanisms 72-76
    radiation spectra 77-79
    rate 67
    surface velocity 71
carrier scattering
    by carriers 106-108
    by dislocations 104
    by ionized impurities 100-103
    by lattice vibrations 94-100
    comparison with experiment 319-322
    determination from thermoelectric and thermomagnetic effects 171-175
    effect on mobility 85-89, 106-108
chemical binding 14-15
cleavage planes 14

Cohen model of band structure 304
constant-energy ellipsoids 55, 226, 233, 302
critical points in energy spectra 64-65
crystal growth methods *see* preparation methods
crystal structure 13-14
current-voltage characteristics 140-145
cyclotron resonance 250-257
Czochralski crystal growth method 24

D

de Haas-van Alphen effect 225-226
Debye temperature 196, 209
deformation potential constants 135-136, 138, 311
diffusion of atoms 36-41
   in doped single crystals 40
   in polycrystalline samples 40-41
   in undoped single crystals 39-40
   table of coefficients 38
dispersion, optical *see* optical dispersion
dissociation energies 16
dopants, properties in Pb chalcogenides 34-35
doping methods 34-36

E

effective mass determination
   from cyclotron resonance 252-257
   from electric susceptibility 58
   from infrared reflection 59
   from longitudinal Nernst-Ettingshausen effect 166, 168
   from magnetoabsorption 248
   from Shubnikov-de Haas effect 233
   from thermoelectric power 159-162
   from transverse Nernst-Ettingshausen effect 169
effective masses
   anisotropy coefficient 128-129, 226, 233, 248, 257
   conductivity mass 58
   cyclotron mass 221

effective masses (continued)
   density-of-states mass 54
   determination *see* effective mass determination
   electric susceptibility mass 58, 60
   longitudinal mass 87
   pressure dependence 132, 162, 311-312
   table of values 298
   temperature dependence 314-316
   theory and calculations 292-299
   transverse mass 87
elastoresistance
   constants 136-137, 139
   tensor 135
electric susceptibility 57
electrical conductivity
   effect of magnetic field 124-129, 227-237
   effect of pressure 131-139
electrical properties 85-147
electromagnetic waves
   absorption *see* absorption of light
   reflection *see* reflection of light
energy band structure *see also* effective mass determination, effective masses, forbidden band width
   band edges 292-309
   calculated results 280-289
   calculation methods 267-279
   Cohen model 304-306
   composition dependence of parameters 55
   constant-energy ellipsoids 55, 226, 233, 302
   critical points 64-65
   distorted bands 117
   Hall effect data 115-116
   interpretation of experimental data 263-267
   Kane model 301-303, 305, 306, 347-350
   Landau bands and quantum numbers 222, 236-237, 245-246
   nonparabolicity 233-235, 299-306, 347-350
   PbTe, p-type, valence band 118-122, 189-192

SUBJECT INDEX

energy band structure (continued)
   pressure dependence of parameters 53-54, 131-132, 138, 162, 310-312
   relativistic corrections 275
   spin-orbit interaction 274-276, 278
   temperature dependence of parameters 53, 123, 312-317
   theory—experiment comparison 289-292
   thermoelectric power data 162-169
epitaxial crystal growth method 26

F
Faraday effect 241-244
films, carrier mobility 110
forbidden band width
   composition dependence 55
   determination from optical absorption 52
   determination from Shubnikov—de Hass effect 236
   determination from thermal and electrical properties 185-188
   interaction band, effective width 305-306
   pressure dependence 53-54, 131-132, 138, 310-312
   solid solutions 317-319
   temperature dependence 53, 123, 312-317
fundamental absorption edge 51-56

G
g factors 236-237
   calculations 306-309
   table of values 298
Grüneisen constant 208

H
Hall effect 110-118, 122-124
   effect of pressure 131-133
   planar effect in PbTe 129

Hall effect (continued)
   p-type PbTe 118-122
helicon waves 259-261
historical review 5-11

I
infrared detectors 334-343
interaction forbidden band 305-306
interaction of light with matter 48-51

K
Kane model of band structure 301-303
   formulas 347-350
Knight shift 226-227

L
Landau bands 222
   spin splitting 236-237, 245
Landau quantum numbers 222, 246
laser effect 79-82
   in magnetic field 249
lasers 343-345
lead telluride
   phonon spectrum 216-219
   planar Hall effect 129
   p-type
     Hall effect and valence band structure 118-122
     thermal conductivity and valence band structure 189-192
   thermoelectric generators 330-334
Lorenz number 174, 177-184

M
magnetic properties 221-227
   de Haas—van Alphen effect 225-226
   Knight shift 226-227
   magnetic susceptibility 223-226
   spin paramagnetism 226
magnetic susceptibility 223-226
   de Haas—van Alphen effect 225-226
magnetoabsorption 244-249

magneto-optical effects 240-249, 257-259
  Faraday effect 241-244
  laser effect in magnetic field 249
  magnetoabsorption 244-249
  magnetoreflection
    infrared 244
    microwave 257-259
magnetoreflection
  infrared 244
  microwave 257-259
magnetoresistance 124-129
  Shubnikov—de Haas effect 227-237
measurement methods
  carrier lifetime 67-69, 71
  effective mass 58, 59, 159-162, 166, 168, 169, 233, 252-257
  energy band structure 115-116, 162-169, 267-289
  forbidden band width 52, 185-188, 236
  optical properties 45-48
  permittivity 145-149, 259
  phonon energy 142-144
  scattering parameters 171-175
  table of methods 351
melt cooling crystal growth method 22

N
N (normal) processes 197-201
Nernst—Ettingshausen effect
  carrier-density dependence 156-157
  longitudinal, use in band structure determination 163-169
  definition 152
  transverse, use in effective mass determination 169-171
  transverse, use in scattering mechanism determination 173-175
nonstoichiometric compositions 30

O
optical dispersion
  integral relationships 46
  long wavelengths 56-63
  short wavelengths 63-66

optical properties 43-66
  definitions of constants 43-45
  interaction of light with matter 48-51
  measurement methods 45-48

P
Peltier coefficient 151-152
permittivity
  definition 44
  determination from barrier capacitance 145-149
  determination from microwave magnetoreflection 259
  dispersion equation 46
phase diagrams 26-31
phonons
  energy determination from tunnel effect 142-144
  N(normal) collision processes 197-201
  spectrum of PbTe 216-219
  U(umklapp) collision processes 198-201
photoconductivity 66-69
  applications 334-341
photodiodes 341-343
photoelectric effects 66-84
  applications 334-343
  spectra 76-77
photoelectric emission 82-84
photomagnetic effect 70-72
photoresistors 334-341
photovoltaic effect 69-70
  applications 341-343
physicochemical properties 13-41
piezoresistance 133-139
  applications 345-346
plasma frequency 57
p—n junctions 139-147
polycrystalline materials, preparation 16-21
preparation methods 15-26
  polycrystalline materials 16-21
  single crystals 21-26

R
radiative carrier recombination 73-75
  spectra 77-79

SUBJECT INDEX

reflection of light and e.m. waves 43-48
    long wavelengths 58, 59, 62
    magnetoreflection 244
    short wavelengths 63-65
refractive index
    definition 43
    determination 46
    long wavelengths 58, 62

S

self-diffusion of atoms 36-41
Shubnikov–de Haas effect 227-237
single crystals, preparation 21-26
    Bridgman–Stockbarger method 21-22
    Czochralski method 24
    epitaxial method 26
    slow cooling of melt 22
    vapor phase method 24-25
    zone leveling method 23-24
solid solutions
    carrier mobility 108-110
    forbidden band width 317-319
    thermal conductivity 214-216
    thermoelectric power 159
specific heat 195-196
spin paramagnetism 226
stimulated recombination radiation 79-82, 249
    applications 343-345
stoichiometry control 31-34
strain gauges 345-346

T

table of main properties 352-354
thermal conductivity 175-177
    ambipolar diffusion 184-189
    effect of impurities 209-214
    electromagnetic (photon) component 193-194

thermal conductivity (continued)
    electronic component 177-184
    lattice component 194-206
    PbTe, p-type, rel. to valence band structure 189-192
    solid solutions 214-216
thermal expansion 207-209
thermal properties 175-219
thermoelectric figure of merit 323-330
thermoelectric generators 330-334
thermoelectric power
    pressure dependence 131, 162
    temperature and carrier-density dependences 153-159
    theory 149-153
    use in band structure determination 162-169
    use in effective mass determination 159-162
    use in scattering mechanism determination 171-175
transmission of light 45
tunnel effect 142-144
    in magnetic field 237-240

U

U(umklapp) processes 198-201
ultrasound *see* absorption of ultrasound

V

vapor phase crystal growth method 24-25
vapor pressures, saturation values 16

Z

zone leveling crystal growth method 23-24

NS
11-6-70